Synthesis Series in Biomedical Engineering

Volume 1

Sensory Organ Replacement and Repair
Gerald E. Miller
2006

Artificial Organs
Gerald E. Miller
2006

Signal Processing of Random Physiological Signals
Charles S. Lessard
2006

Image and Signal Processing for Networked E-Health Applications
Ilias G. Maglogiannis, Kostas Karpouzis, Manolis Wallace
2006

Volume 2

Fundamentals of Respiratory Sounds and Analysis
Zahra Moussavi
2006

Basic Probability Theory for Biomedical Engineers
John D. Enderle, David C. Farden, Daniel J. Krause
2006

Intermediate Probability Theory for Biomedical Engineers
John D. Enderle, David C. Farden, Daniel J. Krause
2006

Advanced Probability Theory for Biomedical Engineers
John D. Enderle, David C. Farden, Daniel J. Krause
2006

Volume 3

Capstone Design Courses: Producing Industry-Ready Biomedical Engineers
Jay R. Goldberg
2007

BioNanotechnology
Elisabeth S. Papazoglou, Aravind Parthasarathy
2007

Bioinstrumentation
John D. Enderle
2006

Volume 4

Multimodal Imaging in Neurology: Special Focus on MRI Applications and MEG
Hans-Peter Müller, Jan Kassubek
2007

Estimation of Cortical Connectivity in Humans: Advanced Signal Processing Techniques
Laura Astolfi, Fabio Babiloni
2007

Brain–Machine Interface Engineering
Justin C. Sanchez , José C. Principe
2007

Introduction to Statistics for Biomedical Engineers
Kristina M. Ropella
2007

Volume 5

Introduction to Continuum Biomechanics
Kyriacos A. Athanasiou, Roman M. Natoli
2008

The Effects of Hypergravity and Microgravity on Biomedical Experiments
Thais Russomano, Gustavo Dalmarco, Felipe Prehn Falcão
2008

A Biosystems Approach to Industrial Patient Monitoring and Diagnostic Devices
Gail Baura
2008

Volume 6
Neural Interfacing: Forging the Human-Machine Connection
Thomas D. Coates Jr.
2008
Quantitative Neurophysiology
Joseph V. Tranquillo
2008
Tremor: From Pathogenesis to Treatment
Giuliana Grimaldi, Mario Manto
2008

Volume 7
Introductory Medical Imaging
A. A. Bharath
2008
Lung Sounds: An Advanced Signal Processing Perspective
Leontios J. Hadjileontiadis
2008
An Outline of Informational Genetics
Gérard Battail
2008

Volume 8
Basic Feedback Controls in Biomedicine
Charles S. Lessard
2009
Understanding Atrial Fibrillation: The Signal Processing Contribution
Luca Mainardi , Leif Sörnmo , Sergio Cerutti
2008

Volume 9
Landmarking and Segmentation of 3D CT Images
Graham Boag
2009
Phonocardiography Signal Processing
Rasha Bassam
2009
Introduction to Biomedical Engineering: Biomechanics and Bioelectricity Vol 1
Christensen
2009
Introduction to Biomedical Engineering: Biomechanics and Bioelectricity Vol 2
Christensen
2009

Synthesis Series in Biomedical Engineering 6

Neural Interfacing: Forging the Human-Machine Connection
Thomas D. Coates Jr.

Quantitative Neurophysiology
Joseph V. Tranquillo

Tremor: From Pathogenesis to Treatment
Giuliana Grimaldi, Mario Manto

ISBN: 9781608451890

www.morganclaypool.com

APublicationintheMorgan&ClaypoolPublishers

SYNTHESISSERIESINBIOMEDICALENGINEERING

SeriesEditors:JohnD.Enderle

Synthesis Series in Biomedical Engineering 6

This volume includes:

Neural Interfacing: Forging the Human-Machine Connection

Thomas D. Coates Jr.

Quantitative Neurophysiology

Joseph V. Tranquillo

Tremor: From Pathogenesis to Treatment

Giuliana Grimaldi, Mario Manto

Neural Interfacing: Forging the Human-Machine Connection

Thomas D. Coates Jr., Ph.D.

MORGAN & CLAYPOOL PUBLISHERS

ABSTRACT

In the past 50 years there has been an explosion of interest in the development of technologies whose end goal is to connect the human brain and/or nervous system directly to computers. Once the subject of science fiction, the technologies necessary to accomplish this goal are rapidly becoming reality. In laboratories around the globe, research is being undertaken to restore function to the physically disabled, to replace areas of the brain damaged by disease or trauma and to augment human abilities. Building neural interfaces and neuro-prosthetics relies on a diverse array of disciplines such as neuroscience, engineering, medicine and microfabrication just to name a few. This book presents a short history of neural interfacing (N.I.) research and introduces the reader to some of the current efforts to develop neural prostheses. The book is intended as an introduction for the college freshman or others wishing to learn more about the field. A resource guide is included for students along with a list of laboratories conducting N.I. research and universities with N.I. related tracks of study.

KEYWORDS

Neural Interface, Neuro-prosthetics, Neural Prosthetics, Neural Engineering

Contents

For all my mentors: past, present, and future.

FOREWORD

Art, Science, and Metaphysics

Understanding the human body is certainly one of the most daunting challenges facing science today. Although we tend to think of this undertaking in modern terms, it is not a new endeavor. Since the beginning of recorded history, we have been trying to understand ourselves from both a physiological and metaphysical standpoint. That is, what the different organs of the body do, what causes disease, what is intellect, where is the seat of "consciousness," and is reality "real." There have been many people both ancient and modern who have contributed to the body of knowledge: Plato, Hypocrites, Galen, Descartes, and Ramón y Cajal are a few that immediately come to mind. However, the advancement of human thought and knowledge has not been a linear process. There have been many false starts, periods of enlightenment, and periods of darkness where the quest for truth (scientific or otherwise) was nearly snuffed out. For millennia we have been asking the question "who and what are we?", but it is only in the last few centuries that we have begun to get some "real" answers.

Our ability to understand the world around us is bound by the limitations of our brain and five senses. While we know the limitations of our five senses, we cannot say with any great certainty what the limits of the brain are, especially with respect to cognition. Thus, it is difficult to ascertain the ultimate limits of our ability to understand. The advances in human knowledge that have occurred in the last several hundred years have happened for a number of reasons. Chief among these is the de-mystification (or perhaps the re-demystification following the middle ages) of the natural world. That is, moving away from the belief that nature, natural processes, and the universe are mysterious and cannot be understood by man. This was a vitally important step because our perceptions of the natural world and our will to explore it can be greatly influenced by our beliefs.

The ability to extend the range of our senses with tools such as the telescope, microscope, X-ray machines, and others has played an important part in the discovery and elucidation of natural processes. It is through our ability to create tools that we overcome the inherent limitations of our bodies. Although neural interfaces and neuro-prosthetics may initially be used for rehabilitation and restoration of function to the disabled, it is probably inevitable that they will become the means by which we will further extend our senses and augment our perceptual abilities. How might the human ability to understand the world around us be changed should it one day be possible to cybernetically augment our brains and radically alter our perceptual experience? How we view the universe and our relationship to it is important because it is through this lens that we interpret what we experience. Our knowledge, cultural precepts, religious beliefs, and personal philosophies both influence and are influenced by our day-to-day interactions with our environment and with each other.

As an example of how one's personal philosophy may influence how one interprets the world around them, consider this passage from the Tao Te Ching[1]:

[1]Tao Te Ching by Lao Tzu, p. 11, Translation by Gia Fu Sheng and Jane English.

Thirty spokes share the wheels hub;
It is the center hole that makes it useful.
Shape clay into a vessel;
It is the space within that makes it useful.
Cut doors and windows for a room;
It is the holes which make it useful.
Therefore profit comes from what is there;
Usefulness from what is not there.

Thus, where one person might see a room filled with furniture as a space defined by the objects it contains, a Taoist might see the same room full of objects as being defined by the spaces between the furniture. This may seem like a trivial distinction, however, it is germane to our discussion for it represents an alternate point of view applicable to a number of problems. For instance, in the following figure do you see a series of vertical lines occurring in a given space of time or a series of spaces of various size that are delineated by vertical lines?

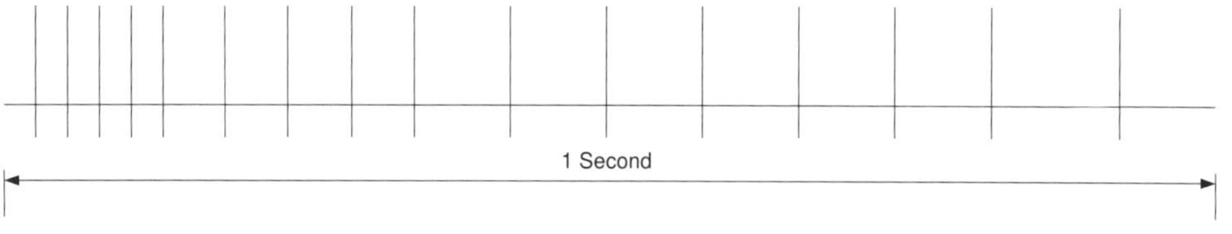

In this case, these two different points of view represent two different ways of looking at the issue of neural coding. That is, understanding the code used by brain cells to communicate with one another. The first point of view, the number of vertical lines (or spikes) occurring in a given amount of time, represents what is known as a rate code. The second point of view represents temporal coding which looks at the change in the intervals between the spikes. Thus, two different ways of looking at the world (spaces *vs.* objects) can translate into two different but equally relevant interpretations of the same data.

Science may seem the antithesis of art and other creative activities due to its heavy focus on logic and reasoning. This emphasis on logic and reason does not mean that there is no room for creativity in the sciences; in fact, creativity is essential. Regardless of your field of study, creative thinking is necessary for the generation of novel ideas and new points of view. In science, producing new techniques, new ways of looking at old problems, and new avenues of discovery may require that the person free themselves from conventional ways of thinking. That's not to say that one should ignore an existing body of work, but one should not be unnecessarily bound by it either. Creativity and thinking "out-of-the-box" is as crucial in the sciences as it is in any other discipline, although not as widely recognized as being so. Scientists often get stereotyped in pop-culture as the absent-minded professor, the arrogant genius, or the anti-social geek. These stereotypes are caricatures that do not reflect the wide range of individual variation. In my time in science I have met many diverse, talented people who do not fit this popular mold. Really, the *only* thing you can say of people in the sciences as a group is that they tend to be creative, intelligent, multi-talented individuals with a passion for discovery and invention.

In both modern and ancient culture science sometimes finds itself at odds with religion. In a politically charged environment where issues become rapidly polarized and the middle ground falls away, civil discourse is of paramount importance. Civilized society depends upon, at the bare minimum, tolerance for points of view different from ones own. Ideally, understanding and respect are preferable to mere tolerance. The intersection of science and religion is often a contentious one. On issues such as stem cell research, there are many opposing points of view and those who are very passionate in their assertions of what is morally and ethically correct. In such a milieu it is very difficult to achieve compromise. The goal of building a culturally diverse, civilized global society is not possible without the ability to find common ground, but to find something one must have the desire to look for it first. Compromise is not possible without a willingness to listen to and respect other points of view.

My interests tend to gravitate toward the center of the Art-Science-Metaphysics triad in an effort to achieve a balance in my own life. I believe that in the larger sense these three disciplines represent three different ways of exploring the same question: "What is the nature of existence?" Unfortunately, a harmony between these three disciplines at a societal level has often been problematic due to their vastly different ideologies. Science is primarily interested in the logically quantifiable aspects of characterizing natural processes through observation, formulation of testable hypotheses and experimentation. Art, a more difficult to define endeavor, generally focuses on communication of the internalized emotional and/or spiritual aspects of the artist's world. Metaphysics, with its two major subdivisions ontology and theology, ponders deeper and possibly un-answerable questions such as "Is there reality separate from mind?" and "What is the purpose of the universe and what is man's place in it?" To sum this up in a sound-bite:

Art, Inspires

Science, Explains

Metaphysics, Cogitates

Regardless of your religion, philosophy of life, or where your personal journey leads you, we all have something to contribute. It is as a collective that humanity will move forward and advance our understanding of both ourselves and the world in which we live. The specifics of how each of us proceeds in our own personal search for meaning in our lives may be very different. However, the underlying reason of *why* we do it is not. We are all on the same quest to try to understand our place in the universe. In that respect, we are not so different after all.

Thom Coates, August 8, 2008

CHAPTER 1

Introduction

"All our knowledge begins with the senses, proceeds then to the understanding and ends with reason. There is nothing higher than reason."

- Immanuel Kant -

From the lost civilization of the Krel in Cyril Hume's 1956 movie *Forbidden Planet*, to the cybernetically enslaved humans of Andy and Larry Wachowski's *The Matrix*, writers of science fiction have long imagined the interfacing of humans and machines. While machines (most notably computers) are often used as analogies for how the brain functions, no two systems could be more different. Machines, assembled from hard-wired component parts, rely on software to endow them with the flexibility to accomplish different processing tasks. Humans, on the other hand, reach their completed, adult state through a process of growth and development. Beginning in the womb and continuing until a few minutes after we draw our last breath, the nervous system is always on and always changing in response to a continual stream of sensory information. Thus, as we experience and learn, subtle changes in the brain incorporate this new information into the existing structure. A connection might be slightly altered here, a shift in the amount of a particular neurochemical there, minute changes all to preserve, for example, the sight, sounds, smells, and emotional response of seeing an elephant for the first time. While computers are lousy at writing a good poem about that elephant, they are very good at calculating the elephant's internal volume or analyzing elephant population statistics. The differences between biological neural systems and digital computers give rise to strengths and weaknesses unique to each. The phrase "brain-machine interface" (BMI) brings to mind the futuristic idea of seamlessly integrating computers and the brain for the purpose of augmenting human abilities. Presumably, such a human would be able to capitalize on the strengths of both systems. While this may someday be possible, there is a vitally important and immediate application of this technology.

Traumatic injury and diseases such as Parkinson's, Huntington's, and Alzheimer's disable millions of people every year, resulting in the loss of physical and/or mental function. The primary goal of neural interfacing and neuro-prosthetics research is to develop implantable systems to restore function to the mentally and physically disabled. Although still in its early stages, this research holds great promise for helping the blind to see again, paraplegics to walk, and quadriplegics to interact with the world using computers in ways not possible a few years ago [1]. One of the keys to achieving the lofty goal of restoration/augmentation of function using artificial means is the interface; the human-to-machine connection.

This first chapter is intended to introduce some of the terminology, neural physiology, and other background material germane to the discussion of neural interfaces. If you desire a more detailed discussion of the topics touched upon in this chapter, please refer to the numbered references in the bibliography. (If you would like to delve deeper into neuroscience or neurophysiology related topics, I suggest beginning with the latest edition of "From Neuron to Brain" by Nicholls et al. [2].) Chapters 2 and 3 explore interfacing methods and discuss some current neuroprosthetic applications. Chapter 4 speculates on some of the more far-reaching applications of neural interfacing (NI) and touches on ethical issues. Lastly, Chapter 5 provides information for students interested in pursuing a career in neural engineering.

1.1 THE ROAD AHEAD

One of the greatest challenges facing neuroscientists today is understanding in great detail how the brain works. In other words, how the structure, physiology and biochemistry of the brain relates to its high level function. Remarkable progress has been made in this area but there is still much that we do not understand. For example, is memory encoded as structural features, patterns of gene activation, as patterns of activation across the neural network, in the pattern of the electrical field itself or all (or none) of the above? We have identified many possible mechanisms that play a role in memory and other functions. But how these mechanisms fit into the bigger picture of how a specific memory is stored still eludes us. Fortunately, a highly detailed understanding may not be necessary to, for example, connect a fully functional prosthetic arm because there are areas of the brain where the neural activity correlates well with limb movement. Thus, by placing sensors in these areas and performing suitable processing on the acquired data, the user can direct the motions of a prosthetic limb or the cursor on a computer screen [1]. Unfortunately, there are few such correlates for memory and even fewer for cognition. If we eventually hope to build replacement parts for damaged brains, augment intelligence or integrate vast computer databases with human memory we must have a far greater understanding of the brain than we have at present. Advancing our understanding of the brain is pivotal to the continued advancement of neural interfaces and neuro-prosthetics.

Historically, the key to increasing our understanding has been the development of new and better tools with which to study the brain. The desire to comprehend our surroundings either on very small (e.g., the sub-atomic) or very large (e.g., the universe) scales and our lack of a natural means to do so, underscores the importance of tools and technology. While it is often the case that the development of new tools precedes major discoveries, it is also true that those same discoveries may further the development of the next generation of tools. In science there is not a clear division between tool-building and tool-using activities since scientists have typically been both the builders and users. This is a role that has arisen out of necessity because the needed tools often don't exist (try finding a particle accelerator at your local hardware store) and it's up to the researchers to create them to facilitate their work. Simply stated, tools are the means by which we bring into our conscious perception the processes occurring at normally imperceptible temporal and/or spacial scales; without them, there is little hope of elucidating mysteries of the brain. Perhaps I am biased since I spend much of my time designing and building tools for others to use, but I believe that there is something primal, something uniquely human (or at least uniquely primate) surrounding the activity of making and using tools. After all, it *is* one of the key things that sets us apart from nearly all other animals on this planet.

How information processing and storage might be accomplished in the brain is a topic to which whole books have been dedicated. Research continues to uncover new information which sometimes forces us to re-evaluate our theories on brain function. Unlike computers, in the brain there is no distinction between software and hardware. The physical, chemical, molecular, and perhaps even quantum [3, 4] structure of the brain alters itself from moment to moment to store and/or process information. It is not known how this process is directed. What are the underlying rules (if any) that dictate the proper way to modify structure to accomplish the given processing task? To use the computer analogy, what we are looking for is the brain's operating system. Realistically, what we are finding is something very different. In fact, it is almost certain that no two brains process the same information in exactly the same way.

There *are* commonalities between all normal human brains such as cytoarchitecture, regional organization or the organization of fiber tracts (see below), but the commonality apparent at the gross anatomical level may not be relevant to determining how information processing and storage is accomplished in an individual brain. For example, vision processing is localized to specific regions of all brains; however, the specific way a given visual experience is perceived, understood, and integrated into memory by a given individual is unique. It's as if each brain develops its own unique operational schema (conceptual framework) within the bounds common to all human brains. So, while it is possible to place an electrode in the brain that will receive a signal whenever you move your right index finger, it is not possible to do the same for recalling the experience of listening to your grandmother tell you a bedtime story. This is because episodic memory (memory of places, events, emotions, etc.) results in a pattern of activation across many brain regions. Using tools which allow visualization of whole-brain activity, it may be possible to teach a system to recognize the unique pattern of activation that corresponds to a mental state or (in the future) an already stored experience [5]. The inverse operation of predicting the unique pattern of brain activation that will result from a given experience or being able to look at brain activity to tell what someone is thinking, may not be possible for a very long time to come.

Computers are often used as a convenient analogy for describing how the brain operates; in actuality, digital computers and the brain have little in common. Whether you are a fly or a human your brain makes use of massively paralleled architecture, electro-chemical communication, genetic regulation, and a host of other complex biological processes. The human brain and nervous system make decisions about a continual stream of information, coordinate responses to that information, manage all biological functions, and are responsible for the epiphenomena we refer to as self awareness. Computers, on the other hand, operate by performing a set of pre-determined, sequential operations on an input to produce an output. While modern computers permit certain types of parallel processing they are not truly parallel. They simply break a larger problem into smaller chunks and operate on the chunks semi-independently (and sequentially) often waiting for other parts of the program to finish before the next series of parallel processing operations can continue. Only certain types of problems can be solved in this manner, so the parallel processing strategies of the present are not even generalizable to all types of problems that can be solved on a computer.

In terms of raw number crunching ability, a handheld calculator can far outmatch most humans in a contest of speed at performing complex mathematical operations. However, the human contestant with their paper and pencil is actually performing a far more complicated feat. The brain is processing various sensory input streams, calculating how hard to press the pencil to the paper, controlling the muscles of the arm and hand, maintaining posture, using language to create the characters that are written on the paper, and controlling a host of other necessary functions that are all completely transparent to the human contestant scribbling away on the paper. While our brain was not specifically "designed" for number crunching, such computation is one of a great number of tasks that it can perform.

The chess playing computer Deep Blue that outmatched world champion Gary Kasparov in 1997 was heralded by some in the artificial intelligence community as a sign that machines which rival the human brain in ability were not far away. However, Kasparov was defeated not by a "thinking" synthetic opponent but by custom hardware designed *only* to play chess and with enough speed that it could evaluate over 200 million positions per second. In short, Kasparov was defeated by brute force rather than intelligence. I would also like to point out that following the match, Gary Kasparov with his "slower" human brain was able to stand, walk from the building, and continue with his life. Deep Blue, on the other hand, was powered down, disassembled, and is now on display in a museum. Why? Because despite its parallel processors, custom chips, and clever software it was useful for no other purpose than defeating the chess master. Thus, Deep Blue was more a testament to its designers engineering ability and determination to build the "ultimate" chess-playing computer than a milestone on the road to adaptable, "intelligent" machines.

1.2 THE BRAIN AND NERVOUS SYSTEM, IN A NUTSHELL

The brain is the organ of thought just as the heart is the organ of blood circulation. This statement may seem obvious but the idea is of relatively recent origin. Beginning with the ancient Egyptians and well into the Middle Ages, it was the heart that was believed to be the organ of thought. The brain was a minor organ that filled up ones skull or became infested with daemons that had to be released through trepaning (drilling holes in ones head). Although the ancient Egyptians ascribed only minor importance to the brain, they were the first to begin describing its anatomy[1]. Later, in 450 B.C., the Greek physician Alcmaeon of Croton [6], based on his dissections of animals, was the first to suggest that it is the brain, not the heart that is the seat of intelligence. This notion was accepted by some such as Plato but rejected by Aristotle who contended that the brain is a radiator which cools the blood. By 300 B.C., the prevailing opinion had shifted to the ventricles of the brain being the seat of intelligence. While still incorrect, at least it was in the right general area. The Middle Ages saw a complete halt to brain research because of the church's ban on the study of anatomy. It wasn't until the 1600's that the French philosopher Reneé Descartes[2] properly identified the brain as the seat of intellect and 18 years later that Thomas Willis, M.D. at Oxford wrote the first neuroanatomy text[3]. The Willis text is considered the beginning of the modern descriptions of brain anatomy.

There are different levels to the structure/organization of the brain, including: gross anatomical (Fig. 1.1), cytoarchitectural, and functional. "Gross anatomical" refers to the unique features of the brain and nervous system that are observable with the naked eye. This might also include visible connections between different brain regions or between the brain and the parts of the body to which it connects. This was the first type of description we have of the brain since early anatomists did not have tools like the microscope and staining techniques to study it in greater detail. It wasn't until the time of Camillo Golgi and Santiago Ramón y Cajal (mid to late 1800's) that microscopes had become sufficiently powerful and stains were developed which permitted the visualization of the fine structure of the brain. Gross anatomy tells us little about what the various structures do, but it does provide landmarks to aid us in finding our way around the landscape of the brain. The classic, gross anatomical reference for the brain and body, *Anatomy Descriptive and Surgical*, was first published by Henry Gray in 1858.

Histology is a branch of anatomy which studies the minute structure of animal and plant tissues. Histology, as a field of scientific inquiry, began with the invention and refinement of the light microscope. Cytoarchitecture refers to the use of histological features to differentiate brain regions from one another based on cell composition and arrangement. At the beginning of the 20th century, Korbinian Broadmann used cytoarchitectonic studies performed by himself and others to divide the brain into 52 regions referred to as Broadmann areas [2, p. 364]. Some Broadmann areas (BA) were later found to correspond to regions that were identified during functional mapping of the brain. For example, the somatic sensory map (Fig. 1.2) corresponds to areas 3, 1, and 2, and the primary motor map (Fig. 1.2) is located in BA 4.

Initial descriptions of the function of the various regions were based on studies of brains damaged by injury and disease which led to, for example, aphasia or loss of the ability to use and/or comprehend words [8, pp. 391-393]. Also, electrical probing of the cortex (the surface of the brain) by researchers in the 1800's revealed areas which seemed to play a role in specific functions like movement and vision. These methods, though useful for early descriptions of functional localization, were fraught with uncertainties. In the case of brain injury, for example, was the damaged area the locus of the lost functionality or did the injury simply damage connections passing through that area on the way to another area of the brain?

[1] The Edwin Smith Surgical Papyrus.
[2] Les passions de l'ame, 1649.
[3] Cerebri Anatome, 1664.

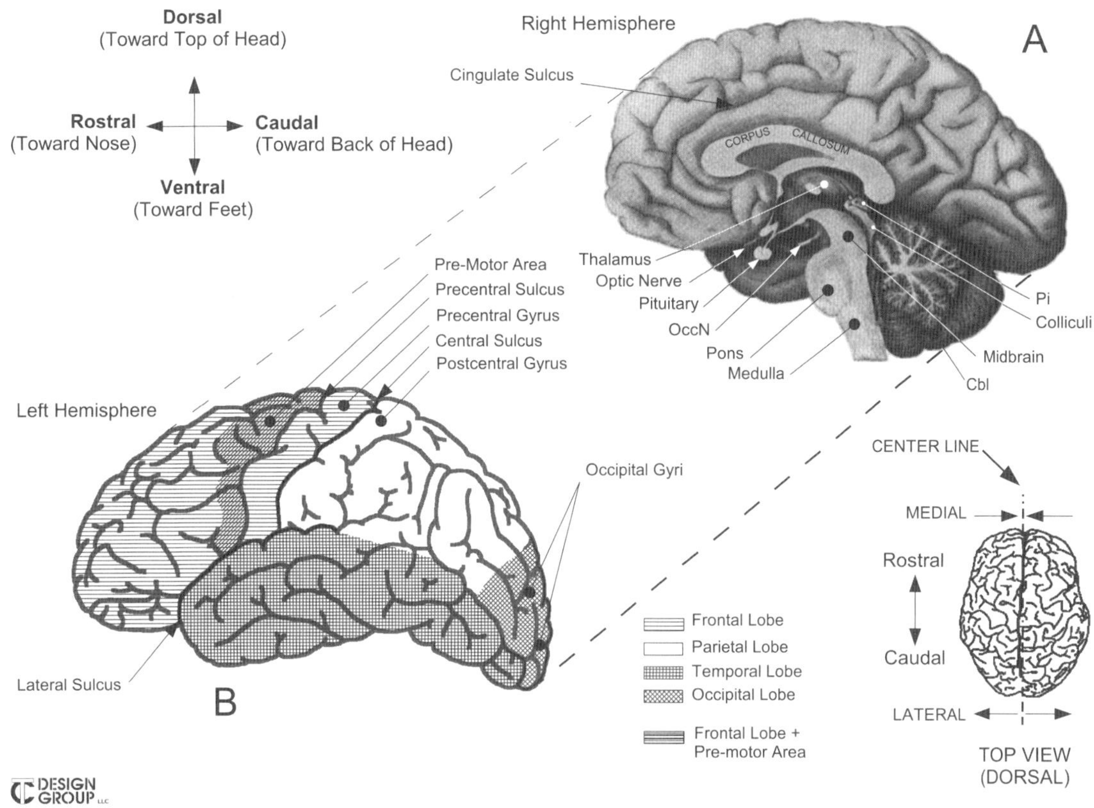

Figure 1.1: Illustration showing some of the gross anatomy of (A) the medial aspect (sectioned along the center line) of the right cerebral hemisphere, midbrain, and brain stem. OccN: Occulomotor Nerve; Cbl: Cerebellum; Pi: Pineal Body (adapted from [7, Fig. 715]). (B) Drawing of the left cerebral hemisphere of the human brain showing the approximate locations of the lobes (based on [7, Figs. 728 and 744]).

More recent methods using electrical/chemical stimulation and recording via microscopic electrode arrays have helped to form a better picture of the brain's functional localization and connectivity.

One of the best known examples of a functional map is the sensory-motor representation located on either side of the central sulcus on both hemispheres of the brain (Fig. 1.2). The neurons of corresponding areas of the motor map (on the pre-central gyrus, Fig. 1.1) are active during volitional movement of the given part of the body, while the neurons of the sensory map (on the post central gyrus, Fig. 1.1) are active when sensory input (pressure, hot/cold, etc.) is received from the corresponding area of the body. Most of the areas represented on the sensory-motor map in the left cerebral hemisphere are connected to the right side of the body and vice versa. Referring again to Figure 1.1-B, we see another area of particular interest in neuro-prosthetic limb research, the pre-motor area. This area is involved in movement planning and encodes information about joint angles, limb trajectory and other information useful for command and control of a prosthetic limb.

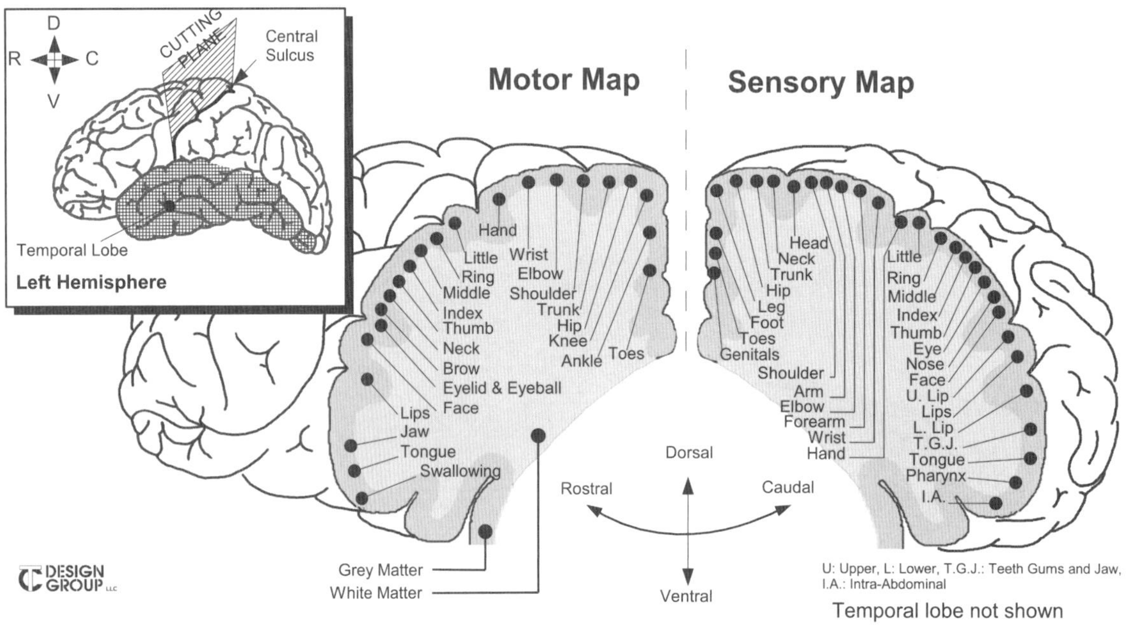

Figure 1.2: Diagram of the motor and sensory representations in the left hemisphere of the brain. Note: Eye, Nose, and Tongue on the sensory map refer to sensations of heat, pressure, etc. Vision, olfaction, and taste processing are located in other sense specific brain areas.

The human brain receives sensory input from all parts of the body and exerts control over all bodily systems both voluntary and autonomic. The bulk of the connections to and from the brain are made by nerve fibers that traverse the spinal cord. Injury to the spinal cord interrupts these connections causing varying degrees of functional loss below the site of the damage. The spinal cord is more than just a conduit for these fibers. Like the brain, it has white matter composed of fibers and gray matter containing the neuron cell bodies and their local interconnections. Unlike the brain, the spinal cord's white matter is located at the periphery while the gray matter is centrally located (compare Figs. 1.2 and 1.3). The cord contains neurons involved in, for example, automatic reflexes responsible for coordinated walking/running, flexor withdrawal (withdrawing your hand from a hot surface), bladder control, and sexual function. Efferent fibers traveling from the cord to the periphery have their cell bodies in the gray matter of the cord. Afferent fibers from the somatic sensory system have their cell bodies in the dorsal root ganglia (see Fig. 1.3). In addition to their connections with the brain, both efferent, and afferent fibers make synapses with other spinal neurons to enable the previously mentioned reflex responses. The projections from spinal neurons enter/exit the spinal column via the dorsal/ventral horns located on either side of each segment of the spinal cord. The spinal nerves are formed by the merging of fibers from the dorsal root ganglia and ventral horn (Fig. 1.3). Spinal nerves branch and merge many times along their path in the body as groups of fibers split off to connect many of the body's systems to the brain. Some of these fibers, such as those connecting the legs to the spinal cord, can be over a meter in length. Like the brain, the spinal cord is highly organized and this organization can be seen at both the gross and microscopic level (Fig. 1.3).

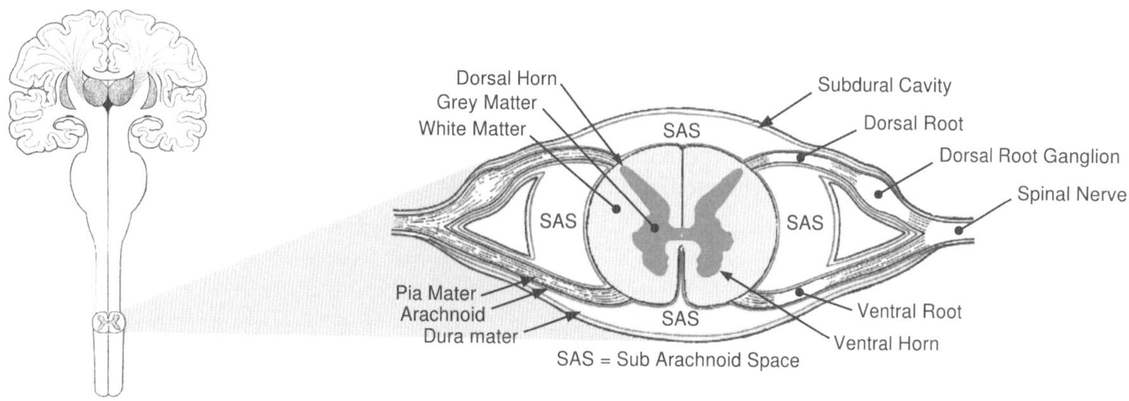

Figure 1.3: Cross-section of the spinal cord. Adapted from [7, Figs. 759 and 770].

The entire nervous system functions as a unified whole, blissfully unaware of any divisions that we might assign. In essence, this is much like the lines drawn on a map to delineate the borders of countries, states, and cities; the divisions are largely for our convenience in discussing the brain, spinal cord, and all of the attached systems.

1.2.1 NEURONS

The brain and nervous system are composed of trillions of cells and there are many ways to classify the different type of cells which comprise this total. For the purpose of our discussion, it is sufficient to say that there are many different types of cells and each cell type plays a unique role (either directly or indirectly) in information processing. The major cell type that is directly involved in information processing is the neuron and there are approximately 100 billion of these in the adult brain. Neurons communicate with each other through their axons and dendrites (Fig. 1.4-A). In general, dendrites conduct impulses toward the cell body (soma) while axons conduct impulses away from the soma (Fig. 1.4-A). It has been estimated that in the adult brain there are approximately 4–6 quadrillion interconnections between neurons alone. The actual number of interconnected cells is far greater when you consider the neuroglia which also have a role in information processing in the brain.

A *synapse* is the point of contact and information exchange between neurons. The brain "processes" approximately 1 quadrillion (1×10^{15}) synaptic events per second. These events are not confined to the simple relaying of messages from one side of the synapse to the other; rather, complicated chemical, cellular, molecular, and even genetic mechanisms can modify the message at the synapse in ways only the brain understands. Any comparisons between the "processing" speed of the brain and that of computers is purely a guess since (1) the two systems are so different and (2) it is presently not possible to know all the processing-relevant cellular/molecular events in the brain that are taking place in a given interval of time. Even considering only the number of synaptic events per second, the brain far exceeds even the fastest modern computer.

A synapse is composed of a presynaptic membrane (on an axon), the synaptic space, and the post-synaptic membrane (on a dendrite). Dendrites may be irregular or "bumpy" in appearance due to the presence of synaptic "spine." The spine is formed due to varicosities along the length of the axon where they synapse with dendrites. The dendrite in turn develops a warp in the post-synaptic membrane to "clasp" the axonic varicosity. This gives the appearance of a bump on the dendrite since the smaller portions of the axon are not as easily visualized as the larger varicosity. The term "neurite" is used to refer to either an axon or a dendrite while the plural "neurites" may refer to both simultaneously. When an electrical impulse (termed an action potential or AP) arrives at the pre-synaptic membrane of a chemical synapse it triggers the release of small molecules (neurotransmitters) which cross the synaptic space, bind to receptors on the post-synaptic membrane (Fig. 1.4-C), and either increase or decrease the probability that an action potential will be generated at the axon hillock (Fig. 1.4-A) of the post-synaptic neuron.

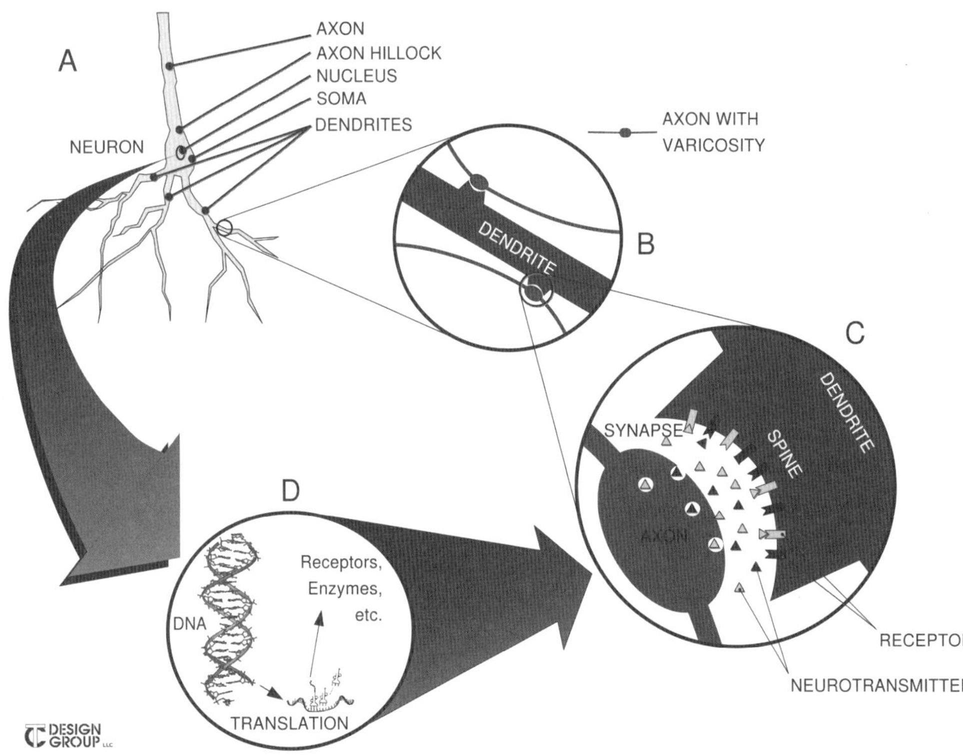

Figure 1.4: Diagrammatic representation of complexity at the level of the neuron: (A) The neuron shown above has a dendritic tree formed by three main branches projecting from the bottom of the cell and sends out a single axon from the apex. The axon hillock is the junction of the soma and axon. (B) Dendrites are studded with synaptic spines and receives information from other neurons. At each synapse, (C) neurotransmitters are released from the pre-synaptic membrane and bind receptors on the post synaptic membrane. (D) Within the somas of pre- and post-synaptic cells, genes that code for receptors and neurotransmitters may be up or down regulated in response to activity or other events.

Neuroglia, while long thought to only provide metabolic support to neurons, are now known to be involved in synaptic transmission and information processing. Astrocytes respond to the level of synaptic activity and exhibit reciprocal communication with neurons via calcium signaling [9]. It has long been known that both neurons and neuroglia can produce neuromodulatory substances which effect the probability that activity at proximally located synapses will result in the generation of action potentials. The effect neuromodulation has on information processing and the exact mechanisms whereby it is accomplished are still the subject of investigation.

Every living cell has a charge difference across its plasma membrane due to differences in the concentration of ions between the inside and outside of the cell. Neurons contain specific cellular mechanisms to manipulate this charge differential to accomplish signaling between cells. In a neuron in its resting state (not conducting a signal) the inside of the cell is about -70 millivolts with respect to the outside. This potential difference, measured across the plasma membrane, is created largely due to a higher concentration of cations (positively charges ions) in the extracellular fluid and a higher concentration of anions (negatively charges ions) in the intracellular fluid (cytoplasm). On the outside, the principle ions are Sodium (Na^+) and Chloride (Cl^-) with Potassium (K^+) present in low concentration. Inside the cell the situation is reversed: (K^+) is the principle cation while (Na^+) and Chloride (Cl^-) are present in low concentration. The potential difference is created by the presence of anions in the form of proteins, amino acids, phosphate, sulfates, etc., in the cytoplasm [10].

Cells are leaky; sodium and potassium constantly diffuse across the plasma membrane following the concentration gradient. That is, without a mechanism to constantly maintain the ion (charge) gradient the neuron would not have a strong potential difference across its plasma membrane. Without this potential neurons would be unable to propagate action potentials and thus unable to convey information. To maintain the ion gradient Na^+/K^+ pumps are interspersed along with other cellular apparatus throughout the plasma membrane. These pumps are constantly moving Na^+ ion from the inside of the cell to the extracellular space and moving K^+ the opposite direction. In addition to diffusion the membrane is also selectively permeable to certain ions. This selective permeability and its ability to be modulated by both internal and external influences is the basis for information processing and signaling in neurons.

Action potentials (a.k.a. spikes) are generated at the axon hillock in response to the summated electrical activity occurring at all of the neuron's synapses (Fig. 1.4-A). When this summated activity results in the membrane potential at the axon hillock shifting positive from -70 mV (resting) to about -50 mV (threshold potential), voltage gated ion channels open allowing a mass influx of sodium ion into the cell. This decreases the potential difference between the inside and outside and is termed depolarization. An above threshold depolarization in one area of the axon causes adjacent channels to open which depolarizes the next area and so on in a cascade effect which results in action potential propagation down the axon. When the depolarized area has reached about +35mV, the Na^+ channels have closed and the K^+ channels have opened allowing the flow of positive charge out of the cell to temporarily restore the local resting potential. Since the K^+ channels close slowly this results in undershoot and slight hyper-polarization (Fig. 1.6). After a few milliseconds the Na^+/K^+ pumps have restored the ionic concentrations to their resting values. During the re-polarization phase and undershoot the neuron is not capable of generating an AP in response to stimulus. This time interval is referred to as the refractory period. The refractory period places a fundamental limit on the maximum rate of firing of a given neuron.

Trains of action potentials manifest as moving waves of focal depolarization down the axon. The spikes you see on the oscilloscope result from the flow of electric current created by the movement of ions as these waves pass the stationary electrode. Multiple spikes are referred to as a *spike train* and when a neuron is producing spikes we say that it is "firing." Depending on the location of the electrode you will see the depolarization as a positive- or negative-going event (Fig. 1.5). That is, if your electrode is on the inside of the cell you will see the resting potential at -70 mV and the peak of depolarization at approximately +35mV just as discussed above.

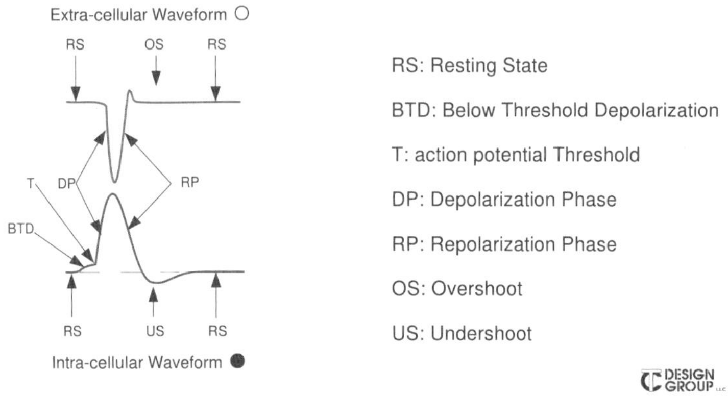

Figure 1.5: Comparison of extracellular and intracellular waveforms.

The values of potentials, shapes of waveforms, etc., are based on the study of neurons using intracellular recording techniques. All of the neural interfacing applications that I am aware of at the time of this writing, use extracellular electrodes which are located in the fluid filled space outside the cell. The waveform recorded by such electrodes will be inverted and its shape/amplitude will be altered as the distance from the source increases. The extracellular and intracellular waveforms are not mirror images of one another. The extracellular waveform is the result of the bulk current flow created by the movement of ions in the vicinity of the electrode. As such, details like the below threshold depolarization and undershoot which are visible using intracellular recording techniques are not visible in the activity recorded by an extracellular electrode (Fig. 1.5).

Action potentials are generated in an "all-or-none" fashion. This means that once the stimulus is sufficient to trigger an AP, a greater intensity of stimulus does not equate to a larger amplitude spike. Rather, the intensity of the stimulus is encoded by the relative timing characteristics of the individual spikes comprising a spike train. It is a popular mis-conception to think that neurons communicate like components in computers, using "digital" signals. This probably arose from a mis-understanding of the all-or-none characteristics of AP propagation and the fact that at first glance spike trains somewhat resemble digital pulse trains. However, if you are searching for an analogy, analog frequency modulation (FM) is probably a better (though not entirely accurate) one for the way neurons communicate with one another.

Communication between neurons occurs through a combination of electrical and chemical signal transduction. The reason for this is relatively straightforward, the all-or-none method of electrical transmission is well suited for rapid signaling over distance but lacks a robust method for modification of the message. Chemical signaling permits robust information processing at the level of the synapse through alteration of the message by amplification, attenuation or modulation of its content. For example, when an action potential arrives at the pre-synaptic membrane a quantity of neurotransmitter (NT) is released. The amount of NT released could be dictated by genetic factors (i.e., up- or down-regulation of the gene for that particular NT thus altering the amount available for release), the number of action potentials arriving at the pre-synaptic membrane during a given time, (i.e., high-frequency activity might cause more release than low-frequency activity), the speed with which NT is made available for release, etc. Once the NT has been released it will diffuse across the synaptic space and bind the receptors on the post

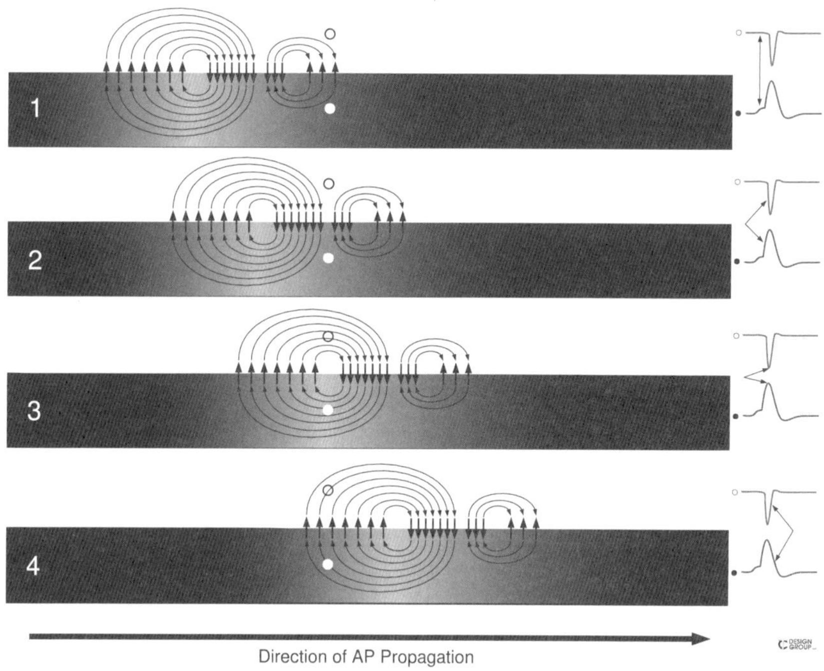

Figure 1.6: Diagrammatic representation of action potential (AP) propagation along a small section of an axon and its relationship to the recorded waveform. The "wave" of depolarization is depicted as a lighter area moving along the axon. Current flow is shown as circular arrows. Note that depending on whether the recording electrode is located inside (•) or outside (○) the cell the detected waveform will be inverted. (1) An AP arrives from the left and triggers the opening of voltage gated ion channels (VGIC's) which permit the influx of positive ions (Na^+) into the lumen. This depolarizes the membrane forward of the site of opened VGIC's (smaller circular arrows). When the depolarization reaches threshold the VGIC's open, the process begins again and the AP continues to propagate down the axon. Due to the the refractory period, the AP only propagates in one direction (to the right) as seen in 2–4.

synaptic membrane. These receptors open ion channels which can either hyperpolarize the cell (making it less likely to fire) or depolarize the cell sending the axon hillock closer to the -50 mV threshold needed for the generation of an action potential. Thus, the number and type of receptors (as well as other factors) influence the probability that the post-synaptic neuron will generate an action potential in response to the pre-synaptic activity.

In Figure 1.7 we see a simple example of how a neural message might be modified by modulation at chemical synapses. *Attenuation* (Fig. 1.7, left) results when the number of spikes produced at the post-synaptic neuron is less than the corresponding activity at its pre-synaptic inputs. In other words, the strength of the elicited response is less than that of the stimuli. *Amplification* (Fig. 1.7, right) refers to the opposite situation where the post-synaptic response is greater than the pre-synaptic stimuli. Furthermore, the post-synaptic response can be temporally altered, that is, the inter-spike intervals (ISI) in the response do not necessarily have to be a fixed ratio of the ISIs of the stimulus.

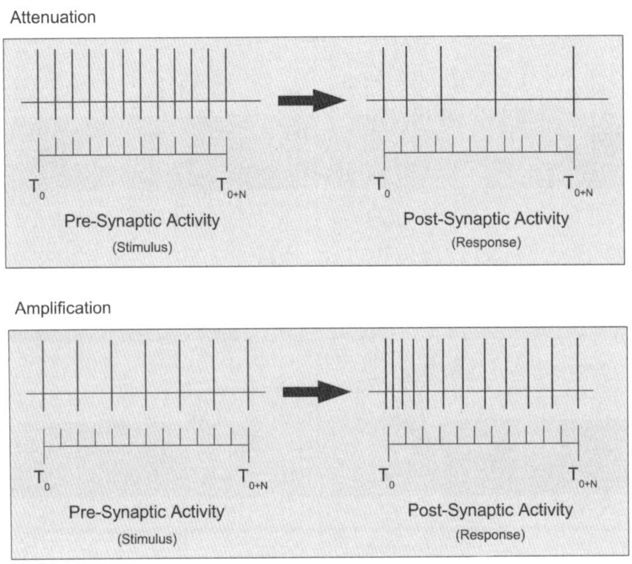

Figure 1.7: Example of attenuation and amplification of a message.

There are over 40 known types of neuro-active substances and even more types of receptors. Some of these neurotransmitters are listed in Table 1.1 along with their receptors and where they are found in the nervous system. Although the basis for chemical transmission was proposed by John Langley nearly 100 years ago, we are still discovering new substances involved in neurotransmission. For example, nitric oxide whose function as a neurotransmitter was still debated when I was in graduate school 8 years ago, has been found in many brain areas and is believed to be involved in learning and memory in the hippocampus[4]. Other molecules such as Oxytocin have dual functionality as both a hormone and a neurotransmitter.

It is likely that there are many more neuroactive substances awaiting discovery. Moreover, continued research is necessary to gain a clear picture of role that each of the *known* substances plays in brain function. The key to understanding the molecular and chemical basis of neural processing is better tools. One example of such tools is the present work being conducted to build Micro-ElectroMechanical Systems (MEMS) capable of delivering drugs, neurotransmitters, or other substances to highly focused areas within the nervous system. MEMS push the envelope of our ability to fabricate very complex systems on very small scales. Continued development and refinement of MEMS is linked to the continued advance of microfabrication technology. A system capable of delivering controlled amounts of different substances *in vivo*, with high spatial accuracy and the ability to monitor the electrical/chemical activity of the surrounding neurons would be an invaluable tool for discerning the role of those substances in neural processing.

[4]An area in the brain involved in memory.

Table 1.1: Examples of some currently known eurochemicals and their receptors

Name	Receptor	Location*
Acetylcholine	Muscarinic acetylcholine, Nicotinic acetylcholine	C,P
Aspartate	NMDA	C
Dopamine	Dopamine	B
Epinephrine	α_1, β-adrenergic, β_2	P
Gamma-aminobutyric acid	GABA A,B and C	C,Retina
Glutamate	glutamate (metabotropic), NMDA, Kainate, AMPA	C,P
Glycine	Glycine, NMDA	C,P
N-Acetylaspartylglutamate	mGluR3, NMDA	C,P
Neuropeptide Y	Neuropeptide Y (5 subtypes)	B
Norepinephrine	α_1 and α_2 adrenergic, et al.	C,P
Serotonin (5-HT)	5-HT	C,P
Nitric oxide	?	B

*C=Central Nervous System (B=Brain), P=Peripheral Nervous System

1.2.2 NEURAL CODING

The term *information*, in the context of our discussion, refers to the intelligence, (i.e., knowledge about something) that is being exchanged, represented, signaled, or relayed, within or between neurons; information is what is conveyed by a spike train. The coding of stimuli into trains of action potentials is complex and better understood in some systems than in others. The following discussion covers neural coding related to the topics covered in the following chapters; it only scratches the surface of neural coding theory and research. The term *Neural code* refers to a scheme or heuristic that is used by neurons to convert their inputs into trains of action potentials. In 1961, Horace Barlow postulated in his "Efficient Coding Hypothesis" that neurons use a code which minimizes the number of spikes needed to transmit a given signal [11]. Since then it has been demonstrated that single neurons can convey information with exceptionally high efficiency that approaches the theoretical limits on information transmission [12, p. 17] and it is theorized that significant information can be relayed by as few as one or two spikes. Furthermore, Barlow suggested that this code was particularly well suited for the representation of visual and auditory stimulus such as that encountered in the natural environment. This is not surprising considering that nervous systems have been optimized through millions of years of evolution in the "great outdoors" to make maximal use of their biological components.

There are a number of theories regarding how neurons or ensembles of neurons represent their information. The earliest work on neural coding was conducted by Adrian et al. [13] in 1926 who described what is known as the "rate code model." This type of coding was observed in muscle stretch receptors whereby the rate of firing of the receptors increased as the amount of weight hung from an excised muscle was increased. Thus, the frequency of the generation of action potentials was directly proportional to the stretch of the muscle. The rate code disregards the fluctuations that occur in the spacing (inter-spike interval) of spikes in a spike train. Rate codes are usually calculated over relatively long time periods which minimizes the impact of inter-spike variations.

In contrast to the rate code, the temporal coding model posits that information about the stimulus is encoded in the precise inter-spike timing of spike trains. For example, in Figure 1.8 we see a train consisting of six spikes (numbered 1–6) which occur over a 1 sec interval. If we were only to consider the rate we would say this is a 6 Hz spike train. However, we immediately notice that the intervals between the spikes are not consistent. Furthermore, there are three identical intervals of 166.7 ms and two 250 ms intervals. Using this information we could then construct a histogram of the inter-spike intervals shown at the right of Figure 1.8. The theory that information about the applied stimulus is conveyed in recurring timing motifs present in the inter-spike intervals of the spike train is referred to as correlation coding. As can be seen in the example, both rate and temporal codes can be incorporated into the same

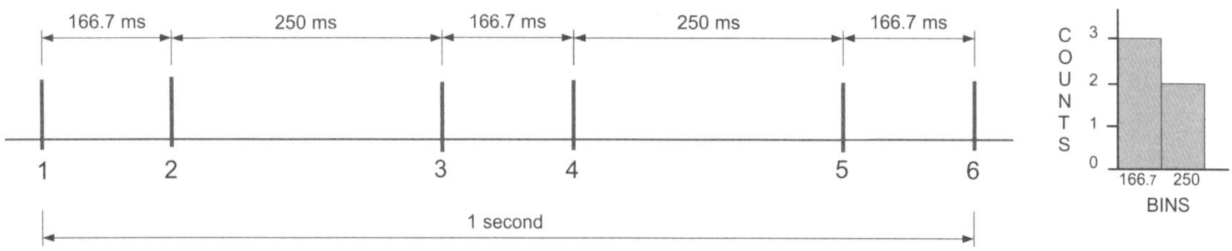

Figure 1.8: Rate and temporal coding example.

spike train since variations in the inter-spike timing can occur without changing the mean rate taken over a given time interval. This type of dual coding has been observed in the human cochlea (the part of the ear that encodes sound), the retina, and other sensory systems of the human body. Coding of auditory signals by the cochlea is of particular importance to cochlea prosthesis research (discussed in Ch. 3) which attempts to restore a type of hearing to the profoundly deaf. It should be noted that the example shown in Figure 1.8 paints an ideal picture of spike timing. In practice, there is considerable “jitter” in the inter-spike timing such that the histogram would contain bins which span a range of time values instead of the single values shown in the example.

The representation of movement direction is important for a number of systems including the vestibular[5] and motor[6] systems. The full range of all possible movement directions is represented in the activity of populations of neurons, where each individual neuron in the population responds best to movement in a specific direction. In Figure 1.9-A, we see a single neuron with a receptive field[7] whose center is 270°. The tuning curve below it indicates that while its peak response (and preferred direction) is 270°, it also responds to movement ± 15° from its center but with decreasing strength. This neuron is part of a larger population of neurons (Fig. 1.9-B), each having their own preferred direction. If we think of each neuron's preferred direction as a vector whose magnitude (length) is equal of the strength of the response, we can represent the population response as shown in Figure 1.9-C. Summation of all the vectors will produce a net population vector whose direction matches the direction of movement of the limb. Figure 1.9-D illustrates how the activity of the population and the net population vector might look for different directions of movement. The population response of neurons in the primary motor,

[5]Involved in balance and determining head position.
[6]Directs the voluntary movement of limbs.
[7]The region of physical space in which a stimulus will produce a response in a given neuron.

pre-motor, and supplementary motor cortices is under investigation by a number of laboratories as the basis for direct neural control of a prosthesis and other brain-computer interactions.

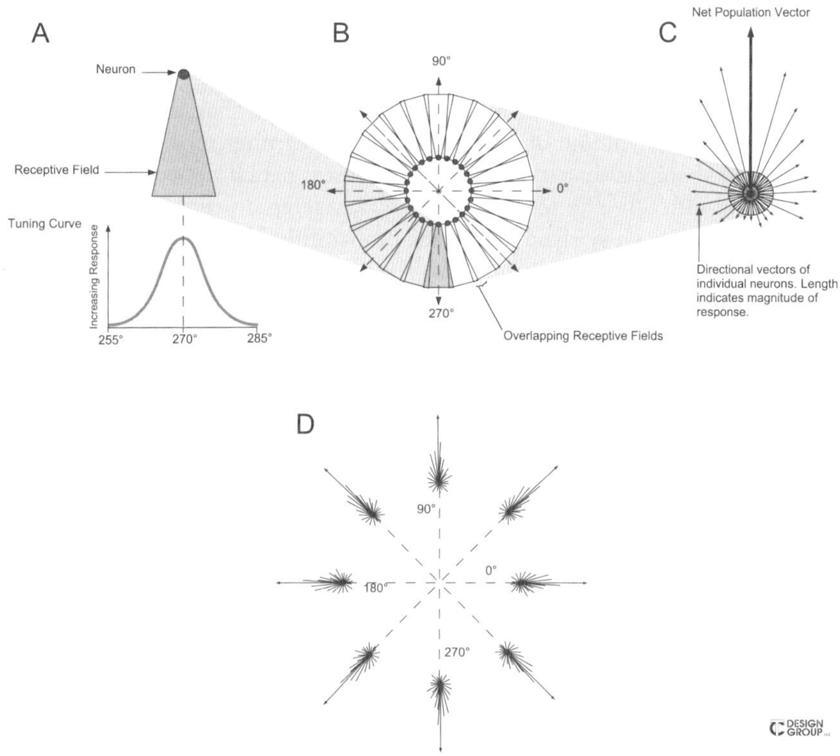

Figure 1.9: Motor cortex population coding example.

Neural coding is usually discussed from the standpoint of electrical activity vis-á-vis the spiking activity of neurons. It is important to note that the electrical activity which is so obvious and relatively easily to record, is the end product of a myriad of underlying chemical, molecular, and genetic processes which have only been discovered in the past 80 years or so. It is only in the last 40 years that we have been able to study neurochemical processes directly in living tissue and tools for the *in vivo* study of molecular/genetic events are still in the early stages of development. However, it is these underlying processes that endow the brain with its unique abilities and when things go wrong, lead to neurological deficits. To elucidate the rules which govern neural coding, we will have to dig deeper than just the electrical activity. Understanding the neural code is only the beginning, our true interest lies not with the neural code *per se* but the information that is contained within. That information, like encrypted e-mail, can't be viewed until the code is cracked. Neural coding is an important piece of the "brain puzzle" that must be deciphered in order for us to gain a better understanding of the language neural interfaces need to use when "talking" to living neurons[8].

[8]For further reading on neural coding consider, "Spikes: Exploring the Neural Code"[12].

1.2.3 LEARNING AND PLASTICITY

In terms of learning, two questions of importance to neural interfacing are:

1. What are the cellular mechanisms of learning? and,

2. How is memory represented in the brain?

The first question relates to determining the physical change that occurs in the neural substrate when we learn something. The volume of literature on how learning may occur in the brain is large. In general, the prevailing theory is that learning in a biological neural network manifests as changes in synaptic connectivity and efficacy. The term *synaptic efficacy* refers to the relative amount of influence that pre-synaptic activity at a given synapse has on the probability that the post-synaptic neuron will fire in response. As previously discussed, chemical synapses have the ability to modify the neural message through various mechanisms. Neurons also have the ability to add or delete (prune) synapses from their arbors. It is believed through these two mechanisms—changes in synaptic connectivity and modification of synaptic efficacy—formation of memory in the brain is accomplished. The more difficult part of the answer to the first question is not *what* changes but *why*.

The Canadian psychologist Donald Hebb postulated that:

> *"When an axon of cell A is near enough to excite cell B and repeatedly or persistently takes part in firing it, some growth process or metabolic change takes place in one or both cells such that A's efficiency, as one of the cells firing B, is increased"* [14, p. 62].

This idea is often expressed by the shorter phrase, "Neurons that fire together, wire together," and is termed *Hebbian Learning*. Hebbian learning pre-supposes the existence of a mechanism (or mechanisms) whereby the postsynaptic cell could back-propagate a signal to the active synapses which would indicate that their efficacy should increase (Fig. 1.10). In our earlier discussion, I indicated that axons conduct action potentials with no mention regarding the same for dendrites. The axon-only theory of AP conduction was the prevailing view until 1994 when research led by Bert Sakmann [15] discovered that certain neurons are able to back-propagate action potentials through their dendritic arbors. This discovery provided the basis for which Hebbian learning might occur in biological neural networks. That is, the coincidence (or near coincidence) of arrival of an AP at both the pre- and post-synaptic membranes could signal changes in the synapse whereby its efficacy would be increased. Alternately, one could speculate that lack of coincidence might, over time, result of the weakening of synaptic efficacy. Three years later, both Hebbian learning and back-propagation were demonstrated in mammalian neural networks by Markham et al. an Stuart et al. [16, 17]. However, only certain types of neurons have been shown to propagate dendritic action potentials. Since other neuron types would also have to be capable of modifying their synaptic efficacy, learning through back-propagation of action potentials is probably not the only type of learning that occurs in biological neural networks.

The second question, "How is memory represented in the brain?", may prove much harder to answer because it relates to how the synaptic modifications discussed above are used to encode either the external sensory stimulus or internal thoughts/ideas that are to be stored for later recall. Unlike the synaptic learning that occurs at the neuronal level, memory representation may not have a easy, direct physical correlate. There is a great deal of speculation regarding memory representation and cognition but little hard fact indicating how this is accomplished. Using scalp electrodes to record the EEG[9] we can see that various parts of the brain are active during different cognitive tasks. However, the spatial

[9]EEG: Electroencephalography. A technique which noninvasively records the brain's electric field using electrodes placed on the scalp.

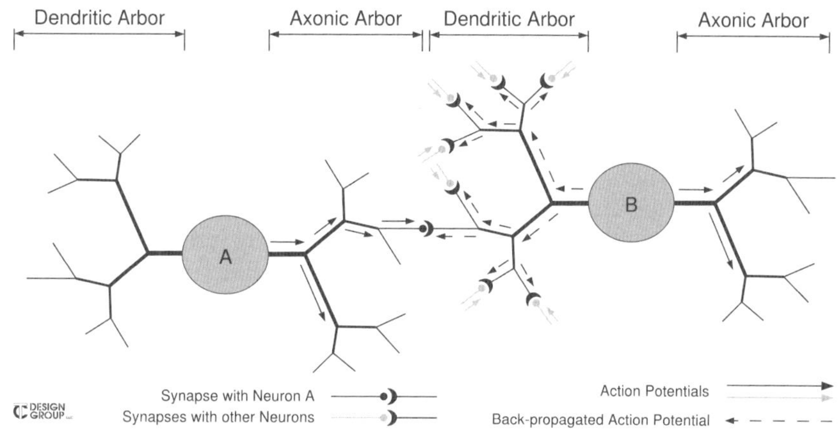

Figure 1.10: Back-propagation and Learning. An axon from neuron A synapses with a dendrite from neuron B. The sum of all the activity in the dendritic arbor of cell B, including that of other neurons that synapse with B, results in an action potential being generated a the axon hillock of B. An action potential propagates through the axonic arbor of B and back-propagates up the dendritic arbor. When the AP arrives at the terminal ends of the dendritic arbor it triggers a change in efficacy at the synapses which were active around the time of the events which triggered the AP.

resolution (pixels per square cm) of EEG is not sufficient to determine what is happening in those brain areas in great detail. Conversely, functional Magnetic resonance imaging (fMRI) has better spatial resolution but poor temporal resolution (pixels per second). Thus, smaller regions can be imaged but the image is a summation of all the activity occurring in a given time window. The current tradeoff between these and other available techniques for studying whole brain activity force the researcher to choose between either spatial or temporal resolution. What is needed is a technique that can provide both and at significantly better resolution than what is available at present. In part, the answer to the memory encoding/representation question is one which is awaiting better technology. Theories relating to memory representation will be discussed in the section on cognition.

"Plasticity" refers to the brain's ability to change and/or adapt. Synaptic plasticity discussed above, is necessary for learning while *neuro-plasticity* refers to the ability of the brain's regions to alter their function or organization. This can happen following brain damage as uninjured areas attempt to assume the damaged area's function and is usually more pronounced in younger patients. Neuroplasticity is also observed in children that are born without a particular sensory modality. For example, in children that are blind from birth there seems to be a re-tasking of some of the parts of the brain normally involved in visual processing to sub-serve other functions. However, individuals who were born sighted and blinded later in life do not exhibit as great a degree of functional re-assignment. While the brain of a child exhibits greater plasticity than that of an adult, current research [18] suggests that the adult brain may be far more plastic than was originally thought.

Both synaptic- and neuro-plasticity are important to neuroprosthesis research because the brain's ability to adapt can assist with the integration of, and communication with, artificial components. It is well known that there are critical periods during development where sensory input is required to assure the proper development of the brain. For example, children who are born with congenital cataracts must have them removed before 5 years of age or their visual perception skills will not develop normally. There is also a critical period for language development which is why cochlear implants have the highest rate of success if implanted prior to 2 years of age. In the future, children who are born blind may have the best outcome if they are fitted with a vision prosthesis early in life. This would permit the brain to adapt to the new input and integrate it as a normal part of the patient's sensorium. It is likely that this will be true for all persons who were born without a given sensory modality; sensory replacement at an early age produces the best outcome. Perhaps future biotechnology will produce a way to induce a development-like plastic state in the adult brain for the purpose of neuroprosthetic integration; until then we must rely on the brain's natural abilities. The story is different for those who were born with a modality and later lost it because their brains have already learned to process the sensory information. In this case, the sensory input must be restored before functional re-assignment begins to occur.

1.2.4 COGNITION

The term "Cognition" is an abstract concept which collectively refers to the mental processes of learning, memory, perception, judgment, and reasoning. In psychology and the neurosciences, it is usually used to refer to human (or human-like) processing and application of knowledge. Cognition is related to the idea of "mind" in that both terms refer to the conscious stream of thought. Both terms have also been applied to metaphysical constructs which posit the existence of a "soul." For the purposes of our discussion, the term "mind" will be used to refer to aspects (both known and unknown) of cognitive function which act in synergy to produce the self-aware, human mental experience.

Augmenting human cognition using artificial means is not as cut-and-dried as is often portrayed in popular literature. There are many competing theories on how the sensory inputs, biological processes, thoughts, and emotions work together to create what we refer to as our cognitive experience. Restoration/augmentation of the senses or connecting end-effectors to the motor centers of the brain may prove to be far easier than augmenting cognition. In the near future, we may discover drugs which improve memory, heighten alertness, or sharpen mental skills. Cognitive supplements are sought after by a number of government agencies and the military to improve safety for personnel who must endure long shifts with little or no sleep. Such conditions are encountered during natural disaster relief operations and on the battlefield. The more ambitious goal of building replacement parts for the brain or correcting defects of cognition such as hyper-violent behavior and schizophrenia rely on a firm understanding of neural coding and neurophysiology. The far-reaching, science fiction-like applications such as computer database integration with memory, seamless integration of digital and mental computation, or the connecting of minds together to form a powerful meta-consciousness capable of understanding far more than a single individual, will not occur without an understanding of how the physiology of the brain relates to the emergent property of mind. Ultimately, there may not be a easy way to separate physiology from cognition; the two may be inseparably intertwined. One thing is certain though, until we understand cognition it will be difficult to determine how to augment it in any practical fashion or cure cognitive defects such as schizophrenia.

Scientists often create models (computer simulations) to test their theories; most of the models of cognition discussed below involve deconstructionist methodologies (the whole is equal to the sum of its parts). Under this approach a complete description of the brain and cognitive function, relies on proper description of all of the units of which it is comprised. This begs the question, "How can we be sure that we know what all the parts should be?" Ostensibly the answer would be, "When the model produces results that match observed behavior." However, we should consider the possibility that a model might produce results that corroborate the observed behavior without accurately modeling the underlying processes that generated the behavior. That is, just because we can make a robot behave like a bug doesn't necessarily mean that we understand how or why a real bug's brain creates those behaviors. There are *many* theories regarding how the brain's physiology relates to mental function and cognition; four of them (of which I am most familiar) are discussed below.

1.2.4.1 Connectionisim

Connectionists posit that the brain's cognitive phenomena is an emergent property attributable to its network of interconnected simple units. Transformation of information occurs through the associative connections between the units comprising the network instead of symbolic manipulation. Connectionism is in opposition to the classical theory of neural processing which states that information processing in the brain occurs through the manipulation of symbols like in a digital computer. Connectionist models employ artificial neural networks (discussed in Sec. 1.3.1) to model their biological cousins. The artificial neurons can be as simple as an equation which governs how their output is related to their input. Alternately, sophisticated modeling software such as NEURON[10] or MCell[11] permits the inclusion of details such as ion channels, neurotransmitters, axonic/dendritic arborizations, and more.

Neural networks based on Connectionist models have many uses including identifying trends in data for financial forecasting, interpreting data from chemical sensing arrays for toxin detection, and other pattern recognition tasks. In terms of cognition, there have been many notable successes of these networks for simulating certain aspects of human behavior. One of first is NETtalk[12] [19] which simulates language acquisition in a small child. When the simulation begins it mostly sounds like phonemic nonsense reminiscent of the babbling sounds made by babies. The system rapidly progresses to correctly assembling the phonemes into words then connecting the words into sentences. Finally, the system is able to read text with suprising accuracy and sounds eerily like a 7-year old child, although the voice pitch, and timbre is the intentional choice of the designers.

Connectionist models have been used to model simple behaviors and have found application in a number of other areas. The successes of artificial neural networks based on Connectionists theories might lead one to conclude that this is the correct explanation of how information processing is accomplished in the brain. However, as is frequently the case in science, things are not so straightforward because other competing, and/or complementary theories have also produced positive results.

1.2.4.2 Modularity of Mind

Modularity of mind theory (MoMT) attempts to define a cognitive architecture based on the hypothesis that the brain is composed of innate, functionally distinct modules that interact to produce complex behavior. According to the theory, this architecture arose as a result of evolutionary pressures as natural selection acted to produce many of the modules [6]. MoMT proponents fall into two camps: those who believe *all* cognitive processes are modular, termed *massive modularity* and those who believe that only low level cognitive processes are modular. An example of one such cognitive module is language acquisition. All normal humans are born with the innate ability to acquire language. Damage to the language areas of the brain results in problems with linguistic expression and/or comprehension of spoken language. According to the theory, we are genetically "hardwired" with pre-defined circuitry for various modules (e.g., the language acquisition module) because of evolutionary processes.

[10] `http://neuron.duke.edu/`
[11] `http://www.mcell.cnl.salk.edu/`
[12] `http://www.cnl.salk.edu/ParallelNetsPronounce/index.php`

1.2.4.3 Subsumption Architecture

Subsumption architecture (SA) is similar to MoMT in that it too posits that complicated behavior is composed of simpler, modular elements. The difference lies in the way the modules are organized and interact to produce the observed behavior. Whereas MoMT suggests an additive and perhaps nonlinear process, SA is based on the idea that behaviors compete for dominance from moment to moment. The dominant behavior is the one that is selected by internal goals and the external environment. For example, a "search" behavior might rely on a number of simpler behaviors such as those that control movement and obstacle avoidance. When an obstacle is encountered the move forward behavior is "subsumed" by the obstacle-avoidance behavior. Once clear of the obstacle, the move forward behavior then re-assumes dominance.

Subsumption architecture has been implemented in a number of robots produced by Rodney Brook's lab at MIT. In 1990, Dr. Brooks and others from MIT founded iRobot Corporation. The company produces a number of commercially available robots including the Roomba™vacuuming appliance. The software which guides Roomba™is an implementation of the Subsumption architecture that mimics insect behavior. For those who like to tinker, iRobot sells developer kits, add-ons, and other accessories which permit users to hack the software and modify the hardware[13].

1.2.4.4 Holonomic Theory

The Holonomic Theory of Cognition [3, 4] is probably the most difficult to understand of all the theories discussed here. In addition to neuroscience, it borrows ideas from laser holography, gestalt psychology, and quantum mechanics. The theory was first proposed by Karl Pribram in the early 1960's based upon his studies of cortical [dendritic] receptive fields whose properties he described using Gabor[14] functions.

We have all seen holograms before, from the small squares making our credit cards more secure to novelty portraits whose eyes seem to follow us around the room, they have become a standard feature of life in the 21st century. These holograms, which are viewable in polychromatic[15] light, are called "sun-lit holograms." However, the first holograms were made with lasers[16] and lasers were needed to view them. To make this type of hologram the beam from a laser was split in two. One of the beams (termed the reference) illuminated a photosensitized glass plate directly. The other beam illuminated the object being photographed. The laser light reflected from the object is focused on the plate where it mixes with the reference beam to create an interference pattern which encodes both the scene and the viewing angles. After the plate is developed it can be re-illuminated using only the reference beam to reveal the 3D image of the object. One of the curious properties of the holographic plate is the distributed nature of the information it contains. That is, the interference pattern does not represent the corresponding points in space of the visual scene like a photograph. If the plate is broken and one of the pieces is illuminated, it does not contain a fragment of the image. Rather, it contains the whole image but viewed from an angle that corresponds to the piece's place in the original unbroken plate.

Gestalt psychology attempts to define basic principles or laws of cognition which determine the way in which we perceive objects. Gestalt theory maintains that the brain is analog, self organizing, and must be studied as a unit rather than as the sum of its parts. A gestalt is, "... a configuration, pattern, or organized field having specific properties that cannot be derived from its component parts[17]." Thus, gestalt psychology treats the mind as a unified whole comprised of gestalts which may function separately or

[13] `http://irobot.com/sp.cfm?pageid=373`

[14] Dennis Gabor (1900-1979): Nobel laureate, mathematician, and the father of laser holography.

[15] Light composed of many different wavelengths/colors (e.g., sunlight).

[16] A coherent (one phase), monochromatic (one wavelength) light source.

[17] *Websters Unabridged Dictionary*, 2nd ed., Random House Inc., New York, 2001.

together. Gestalt systems exhibit one or more of the following four key properties: emergence, reification, multi-stability, and invariance.

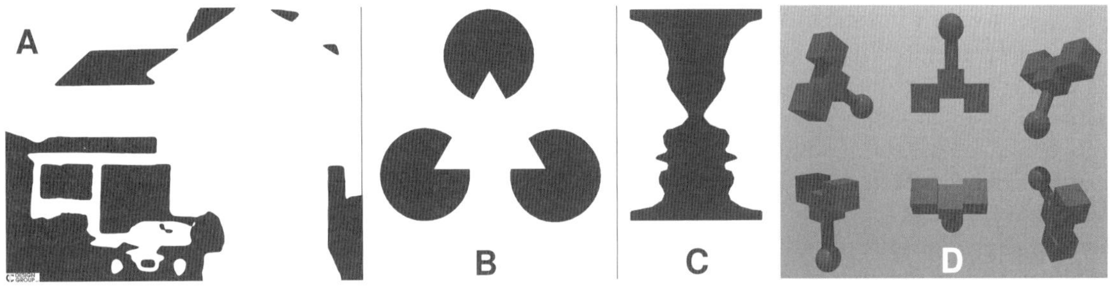

Figure 1.11: Examples of the Four primary Gestalt properties. (A) Emergence, (B) Reification, (C) Multi-stability, and (D) Invariance.

Emergence is a property that describes how our brains extract meaningful patterns when presented with incomplete information. The thresholded picture of a house with a car parked in front (Fig. 1.11-A) lacks most of the detail present in the original photograph. Yet we are easily able to identify what is depicted. This is because we have pre-existing abstract concepts (or gestalts) of "house" and "car" which our brains are able to apply to extract meaning from the incomplete/low detail patterns. *Reification*, or the property of illusory contours, can be seen in Figure 1.11-B. Although the drawing consists only of three circles with missing wedges, our mind fills in the missing contours allowing us to perceive a triangle where none actually exists. *Multi-stability* relates to perceptually unstable objects which seem to "flip" between one interpretation and another. For example, in Figure 1.11-C do you see two faces or a vase? Lastly, *invariance* describes our ability to recognize objects regardless of their rotation or scale (Fig. 1.11-D).

Simply stated, the Holonomic Theory of cognition describes a theoretical system where information is represented in gestalt form as holograph-like interference patterns. The major points of the theory are:

1. Holonomic processes are created by the electrical activity occurring in the smallest fibers of axonic and dendritic arbors in certain areas of the brain.

2. The receptive fields of these arbors are mathematically described by Gabor functions.

3. The interference patterns do not exist as "waves," instead they represent the intersection of waves and take the form of Fourier[18] coefficients.

4. All memory is not holonomic. Memory has layers, the deepest of which is the "deep holonomic store." The holonomic store is separate from the surface pattern created by the neurological storage circuitry. The neural circuitry of the brain contains the "indexing information" needed to access the deep store. For example, if you had a cube filled with thousands of different images all jumbled together, and each of the images was stored using a unique process whereby it could be retrieved using the identical process that stored it, then a great number of images might be stored in the same medium so long as each was stored using a unique encoding process.

[18]Fourier Analysis: the represention of a wave function as the sum of sine and cosine functions.

If the purpose of the neural substrate is to both support holonomic processes and permit access to the information stored within them, we must study the relationship between physiology and holonomy. Karl Pribram stated in a recent article on the holonomic theory that, "...holographic and holonomic processes are truly holistic in that they spread patterns everywhere and everywhen to entangle the parts with one another. In this domain, space, and time no longer exist and therefore neither does causality"[19]. If Dr. Pribram is correct, especially from the standpoint of the lack of causality, it could make direct, artificial access to long-term memory extremely complicated. In general, the way we learn about brain function is to observe the response elicited by a given stimulus; that is, the effect that is produced by a cause. If cause and effect break down at the level of holonomic processes then a new paradigm for their *in vivo* study must be found. It would not be enough to implant an array of electrodes in a key area of the brain, produce random stimuli and learn from the responses. In order to access the information we would have to understand how it is encoded, stored, retrieved, and decoded, then determine the exact set of neural signals needed to manipulate the underlying neural substrate to accomplish reading/writing of the deep store. Without such an understanding it would be like trying to write a computer program by randomly pressing the keys on the keyboard.

1.2.4.5 Which Theory is Correct?

I selected these four examples to give you an idea of the diversity of opinion that exists regarding cognition and the neurophysiological basis for cognitive processes. We must also consider the possibility that two or more of the above theories (or others that I did not discuss) could be correct but for different areas of the brain or for different "levels" of processing. For example, MoMT may describe phenomena like our language acquiring ability while Subsumption might be the right model for the organization of simpler behaviors, especially in the primitive brain areas and the spinal cord. Connectionism may describe the way neural circuitry organizes to process sensory data, produce modules, or create surface patterns which access the deep holonomic store, etc., etc. Based on the history of our understanding of cognition, it is likely that these theories are neither entirely correct nor entirely incorrect. Rather, like the works of Plato, Aristotle, and Descartes, they represent milestones along the road toward a better understanding of the mind and brain. No doubt future scientists will look back at our work an marvel at what we managed to get right even with our primitive techniques and have a good laugh at how wrong we were in other regards; c'est la science!

1.2.4.6 The Brain as a Mirror for itself

Ultimately, it may not be a question of whether or not we *will* understand the brain rather *can* we gain a complete understanding of brain function. Hence, the question:

Is there a fundamental limit to human-kind's ability to understand our own brains?

There is an enormous amount of complexity in both the way the brain is organized at the macro/micro-scopic level and what we think we know of its function. To understand the issue of the decidability of brain function you must first understand the problem space. One way of characterizing the complexity of a system is to estimate the degrees of freedom, that is, the number of ways things can change at any given instant in time. We can only make a guess at this since our understanding of the brain is far from complete and it is almost certain that we do not know all the ways things can change. Given what we currently know, the number of parameters which are freely available to change at any given moment are immense.

One way to visualize the present state of our knowledge of the brain is as a continuum that ranges from what we know with 99.99% certainty to what has yet to be discovered (Fig. 1.12-A). Among the

[19]Karl Pribram: `http://www.scholarpedia.org/article/Holonomic_brain_theory`

things we know are facts such as these: the brain contains neurons, neurons have axons and dendrites and pre-synaptic membranes secrete neurotransmitters in response to action potentials. In Figure 1.12-A, as we move to the right our knowledge is less certain. When we discuss theories that are based on this area of knowledge we tend to increase our use of qualifiers such as "possibly", "may", or "... evidence suggests that... ." As we move further to the right the probability of error in our knowledge base increases and we tend to be more abstract in our discussions. So the real question is, "How far along the continuum will we be able to progress while maintaining a high degree of certainty in our knowledge?" Unfortunately this is not a question that is easily decided, if at all.

The mathematician Kurt Gödel demonstrated via his two "Incompleteness Theorems"[20] the inherent limitations of mathematical proof. Simply put, this means that there are certain mathematical statements that cannot be decided. An example of one of these undecidable problems used by Dr. Stanislaw Ulam [21] when discussing his thoughts on the decidability of brain function uses the phenomena of twin prime numbers. Prime numbers are those integers that are divisible only by themselves and one. Some prime numbers like 3,5,7, ... , 41,43, ... differ only by two. These prime number pairs are referred to as twins. The question is: are there a finite or infinite number of twins? The fact that there are an infinite number of primes has been known since Euclid's time, however, the question of the number of twins is not known and may be undecidable since the problem space is potentially infinite. In fact, the question of decidability may itself be undecidable. This is philosophically relevent to our understanding the human brain and mind because the same question occurs: is its scope finite or infinite.

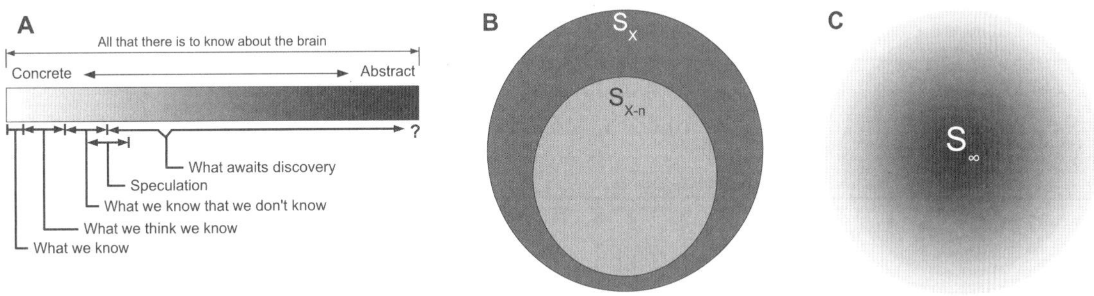

Figure 1.12: What can the brain know about itself? (A) Knowledge as a continuum from the known to the unknown, (B) The brain/mind as a finite (bounded) system, (C) The brain/mind as an unbounded, infinite system.

For example, in the case of a bounded, finite system "S" with complexity "X" shown in Figure 1.12-B the system is unable to fully describe itself because its capacity is being utilized by the very cognitive processes it seeks to describe. Thus, while the system might be self aware and able to ponder it's own existence, a full description of itself is not within its scope. However, a description of a less complex system (S_{X-n}) is within it's scope. On the other hand, if cognitive processes are not bounded (Fig. 1.12-C) as suggested by holonomic and other holistic theories of cognition, then we may be able to more closely approach a full [practical] description of the brain and all its processes. Still another possibility is that this is not the correct way to think of cognitive processes and a complete description may or may not be possible for completely different reasons than those discussed above. The problem with the "problem" is that apart from divine intervention we have no way of determining the depth of our ignorance or the height of our knowledge relative to the totality of what there is to be known about the brain/mind. At this point in time, the question of the limits to our understanding is largely a philosophical matter. I'm reasonably certain that we are not near these limits and the coming decades (perhaps centuries) will, no doubt, see remarkable progress in the area of neural interfacing and neuroengineering.

1.2.5 MACHINE INTELLIGENCE AND MODELLING

Artificial Intelligence (AI) or machine intelligence is not a field which studies human brain function per-se but is often interested in modeling cognitive processes. Such cognitive models may utilize computer simulations that are derived from, or inspired by, biological neural systems. As such, AI has a close association with neural modeling in which brain function is studied by creating computer simulations of biological neural systems. The simulations embody various properties of the system being modeled and allow researchers to study its function in detail, in real time. As computer power increases over time more complex models will become possible, however, it is doubtful that computers, as we know them today, will ever be able to simulate the entire human brain with all of its electrical, chemical, genetic, and other processes. That is, using computer models to try to understand everything about the brain is a lot like trying to use a magnifying glass to understand everything about a bacterium. The tool is simply not adequate for the job.

Dr. Hans Moravec predicted in his book, *Robot* [22, p. 63], during a discussion on the next generation of computer chips that, "As production techniques for those tiny components are perfected... The 100 million MIPS to match human brain power will then arrive in home computers by 2030." This statement is predicated on the belief that we understand or can quantify the processing capacity/bandwidth of the brain as we can with computers; this belief is quite possibly in error. While the power of computers *has* grown rapidly over the past 60 years the predictions of machines imbued with human intelligence made by various AI researchers since the 1950's have not come to pass. Don't get me wrong, some really remarkable innovation has and will continue to come from AI research. Despite 60 years of innovation, we are still not close to having an intelligent computer like the HAL-9000 of *2001: A Space Odyssey*. Not that this is a bad thing since that particular fictitious computer went insane and killed most of it's human shipmates.

All jokes aside, the reason for the languid progress of implementing human intelligence in a machine is relatively straightforward—those early predictions were made without an adequate understanding of the robustness/complexity of neural computation, in general, and of the human brain in specific. The fact that the available power of computer hardware is ever-increasing does not solve the problem of implementing human-like intelligence in software. Until we understand brain function far better than we do at present it will be very difficult if not impossible to implement that in a machine. What we are more likely to end up with are machines with a type of limited and/or highly specialized intelligence that makes them far easier to use than today's home computers. In addition to the implications for the advancement of AI, such machines will be extremely useful as the processing intermediary for neural-prostheses. Neural data processing strategies used today rely on software that interprets the detected neural activity into control directives. Smaller, faster computers, and the ability to design "intelligence" into the software and/or hardware should prove extremely useful to neuro-prosthetics.

1.3 ACQUIRING AND USING NEURAL DATA

Real-world systems are dynamic and often nonlinear. That is, they are always changing and do not respond in exactly the same way every time when given the same inputs. A single neuron, when stimulated, will not produce exactly the same response even if the stimulus is identical from one trial to the next. This seeming unreliability in the spike timings is referred to as "jitter." It is possible that what we call noise (or jitter) in spike timings is a purposeful feature of neural processing that we do not as yet fully understand [23]. It has been suggested that noise in the neural response prevents the system from falling into local minima during various types of operations. This is analogous to rolling a marble down a bumpy incline where the valleys between the bumps are the local minima. The goal is to get the marble all the way down the incline without getting stuck along the way. To prevent the marble from stopping short of its goal we add "noise" to the system by shaking the incline. The noise keeps the marble bouncing around and permits gravity to move it to its final goal.

This jitter in the spike timing means that we cannot rely on strict uniformity in the neural response. So a paradigm of, "IF you see a spike train with this timing THEN do this," will not work reliably enough to be useful. Instead, we have to utilize methods of analysis which do not depend on strict repeatability. Such methods might include those based on statistics, nonlinear dynamics, and information theory.

There are many ways of visualizing patterns in data derived from nonlinear systems, two examples are scatterplots and attractors. A scatterplot is comprised of measurements of two parameters taken repeatedly over time. If some of the instantaneous values of one parameter have associations/dependencies on the other, the data points would cluster as can be seen in Figure 1.13-A. If the values are not correlated they would scatter, forming a random pattern of dots (Fig. 1.13-B).

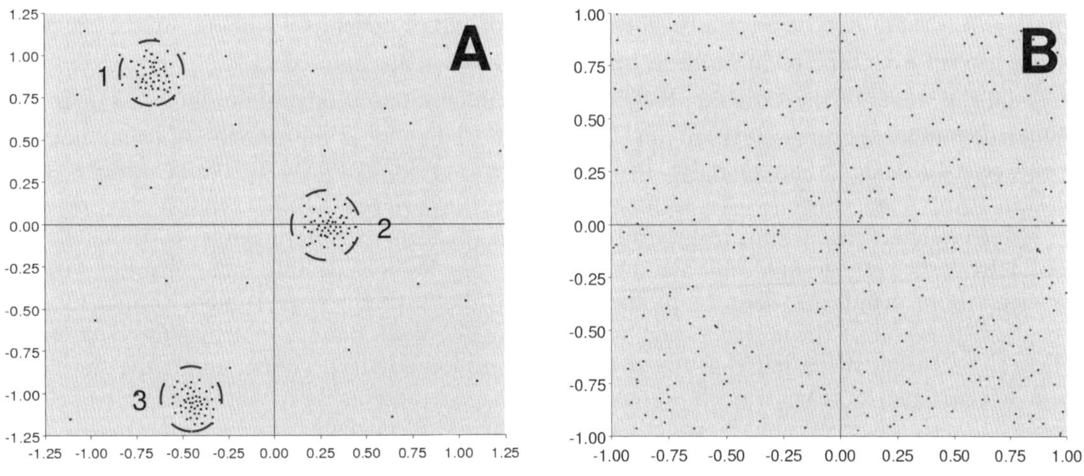

Figure 1.13: Example showing a scatter plot (A) of data which exhibits a high degree of association and (B) a plot of random (nonassociated) data. In (A), many of the recorded measurements cluster in three distinct regions of the plot. While no two measurements are exactly the same we can see both a pattern to their distribution and their bounds represented by the circles. Scatter plots are useful for showing nonlinear associations between different variables of a dynamic system.

Like scatterplots, attractors are also useful for visualizing relationships between variables in a nonlinear system. An attractor is the set of values to which the system evolves over time. Data from dynamic systems like the brain form *strange attractors*. Strange attractors are attractors which have chaotic dynamics, that is, when visualized the sets often look like noisy orbits which seem to move about one or more centers (Fig. 1.14) or "basins of attraction." The coordinate space in which the attractor exists is termed its state space and represents all the possible values that the points which comprise the attractor may take on. Each degree of freedom (a variable) is assigned an axis in the multidimensional space.

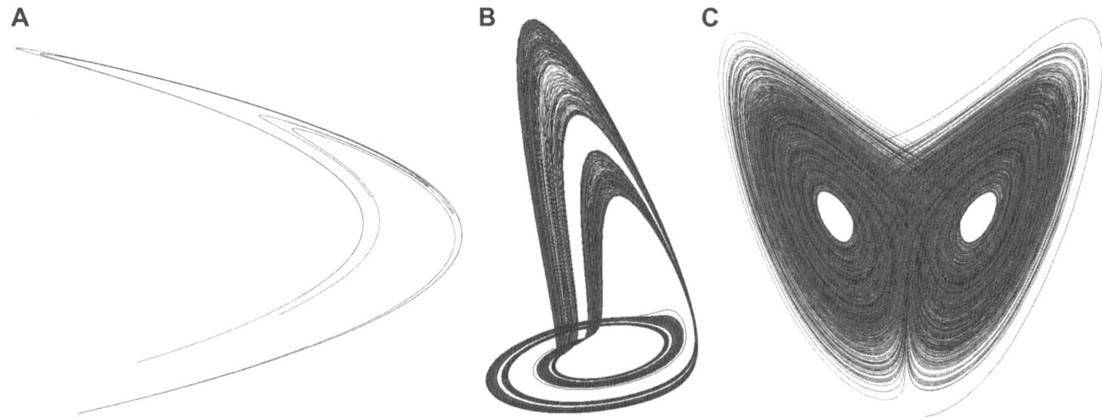

Figure 1.14: Examples of strange attractors: (A) Henon, (B) Rossler, (C) Lorenz. As the plot of the attractor evolves over time (accumulating many points), the shape formed by the points indicates the bounds on the values of the variables. If the system is in continuous operation then the values of the variables are measured over some finite time interval. Thus, the plot of the attractor is a snapshot of the state space of the dynamic system during that interval. *Generated using FractInt.*

Measurements taken from real-world systems might form attractors similar to those shown in Figure 1.14 with easily recognizable features, or they may be considerably more chaotic. Attractors constructed from neural data look more like noisy (but interestingly shaped) blobs similar to a plate of spaghetti or a roll of string that's been pulled apart. The main point is not that the data form an aesthetically pleasing shape but that the shape represent something pertinent about the state of the system. For example, let's say the the bowl of spaghetti shape represents the activity of a group of neurons during an awake, attentive state. Now, let's suppose that the messy ball of string shape represents the same group of neurons during epileptiform (seizure) activity. The state space of the attractor could be used as the input to an artificial neural network which could learn to classify the shapes into two outputs: normal and seizure. Thus, the attractor is able to represent the unique neural activity corresponding to two different states of the system.

Computers are used to acquire neural data, extract meaningful features, make decisions, and produce output for either analytical or control purposes. The initial steps in this process are illustrated in the block diagram shown in Figure 1.15. There we see a micro-electrode inserted into the brain (represented as a cylindrical volume). The electrode is connected to a headstage located a very short distance away which amplifies the detected signals. Since the signals are very small in amplitude, the headstage must be located as close to the micro-electrode as possible to minimize noise. Without some initial amplification the signals would not be detectable after traversing the longer distance to the amplifier. The amplifier boosts the strength of the signal to a range compatible with the data acquisition card installed in the computer.

The output of the amplifier is an analog waveform, that is, a signal which is continuous with respect to both time and amplitude. The DAQ card samples its inputs, periodically taking "snapshots" of the value of the function. Thus, the digitized signal is a discrete representation of a continuous function. *Aliasing* occurs when frequency components of the waveform exceed 1/2 the sampling rate. This results in an incomplete, distorted representation of the original continuous function. Most DAQ cards have built in lowpass[20] or "anti-aliasing" filters to minimize this problem.

Often amplifiers contain notch or "band-reject" filters[21] to compensate for common sources of interference such as the 50/60 cycle noise present in AC powered devices. Most amplifiers also contain at least one lowpass filter per channel. The filtering on-board the amplifier is usually intended to function as a coarse, first pass at cleaning up the signal. While the elements of the system being measured (say, two connected neurons) may exchange information with very high fidelity, noise is often introduced into the measurement, not by the cells but by our measurement technique and/or amplifiers. Once the signal has been digitized (Fig. 1.15-2), further signal processing using a computer is usually necessary to remove noise or other unwanted components (Fig. 1.15-3) and produce data that is ready to be passed to the next step.

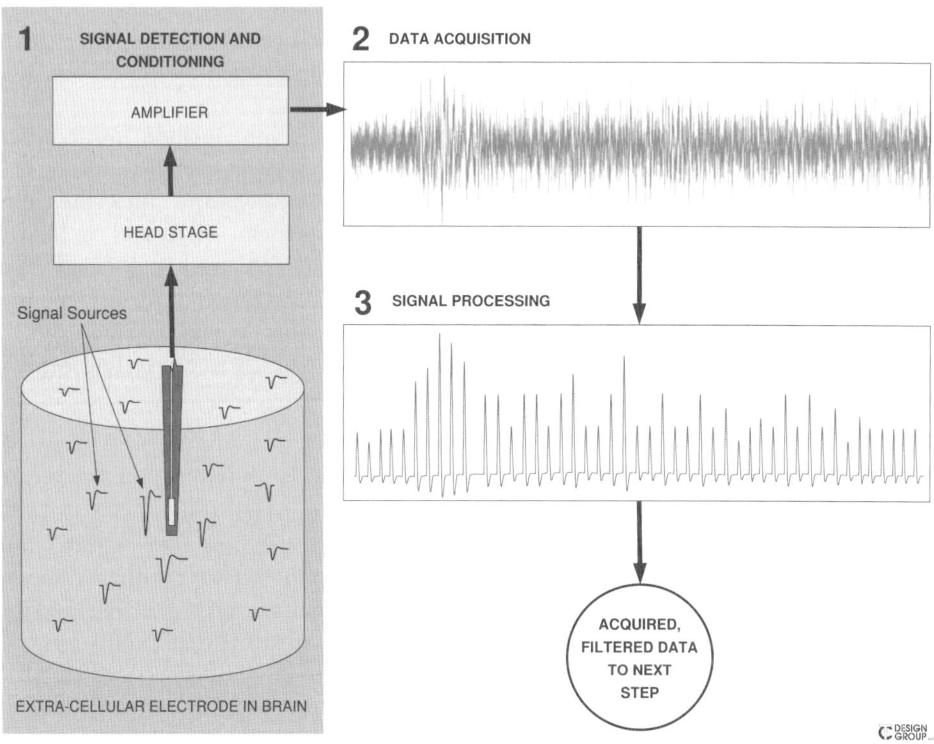

Figure 1.15: An extra-cellular electrode in the brain (shown here as a cylindrical volume) detects signals from multiple sources. The head stage is usually located very near the electrode or may even be part of the electrode. The primary function of the head stage is to amplify the waveform to a level well above the noise that will be introduced as the signal traverses the long connection to the amplifier.

[20]A filter which eliminates all frequency components above a given frequency (termed the cutoff frequency).
[21]A filter which eliminates a particular band of frequencies components.

The signal detected by the extra-cellular electrode in Figure 1.15-1 consists of spike trains from a number of sources (neurons) summed together into one composite waveform. In order to use this data to control a neuro-prosthesis with any degree of accuracy we must recover information on the activity of the individual neurons (or small ensembles) that went into the admixture. This does not necessarily mean that we need to reconstruct the exact waveform of all the sources in the mixture, although this is sometimes the goal. We do, at least, need to quantify some meaningful feature of the activity that will allow us make decisions about what to do when that particular neuron or group of neurons is active.

The general term for separating the signals from an admixture without having a-priori knowledge of the mixing process and the number of signals, is referred to as "blind source separation." The classic example of blind source separation is the "Cocktail Party Problem." In this problem there is a room filled with conversing people and microphones placed around the periphery. The goal is to separate the individual conversations from the audio recorded by each microphone. Although each microphone picks up the din of many people all speaking at once, some people are closer to some microphones than to others. Thus, the sound picked up by a given microphone would contain a different mixture of people's voices than that of a another microphone. Since the distance of a given speaker from each of the microphones varies, they are subject to delays whose phase varies as a function of distance. These differences in amplitude and phase are important features which are used to separate the individual conversations. The same is true for neural activity recorded by multiple electrodes.

Popular methods for accomplishing blind source separation include *independent component analysis* (ICA), *artificial neural networks* (ANNs), and *template matching*. ICA is a statistical technique which can be used to decompose a complex waveform into the independent sources of which it is comprised. This highly useful technique can be found in add on toolboxes for popular numerical processing software such as MatLab[22] and LabView[23]. An explanation of ICA can be found at `http://www.cis.hut.fi/projects/ica/icademo` and an on-line, interactive demonstration of its use in solving the cocktail party problem can be found at `http://www.cis.hut.fi/projects/ica/cocktail/cocktail_en.cgi`. The remaining two methods: ANNs and template matching are discussed below.

1.3.1 SPIKE IDENTIFICATION USING ANNS

Among the many applications for artificial neural networks are pattern recognition (Fig. 1.16) and blind source separation. The learning algorithm (rule) of the ANN dictates how the weights (connections) between elements (neurons) are modified to accomplish learning in the network. Other parameters of the ANN include the type of network (backpropagation, radial basis function, self organizing) and the number of inputs, neurons, layers, and outputs. The learning rule along with the other parameters are selected depending on the task the ANN must accomplish. *Self organization* is a property of certain dynamic systems whereby the complexity of the system may increase over time without external influence. When applied to neural networks it is used to indicate a neural network where some or all of the aforementioned parameters are determined by the network itself as it operates. Self-organizing ANNs are guided by unsupervised learning rules. One such learning rule was conceived by Linsker et al. [24] further developed by Bell et al. [25] and is based on principles taken from the field of information theory.

To begin to understand how blind source separation is accomplished using ANNs, you need a basic familiarity with *information theory*. Information theory began as a branch of mathematics used in the communications industry for studying the limits on data compression. Today, it is a broad discipline

[22]MatLab is a trademark of The Mathworks.

[23]LabView is a trademark of National Instruments.

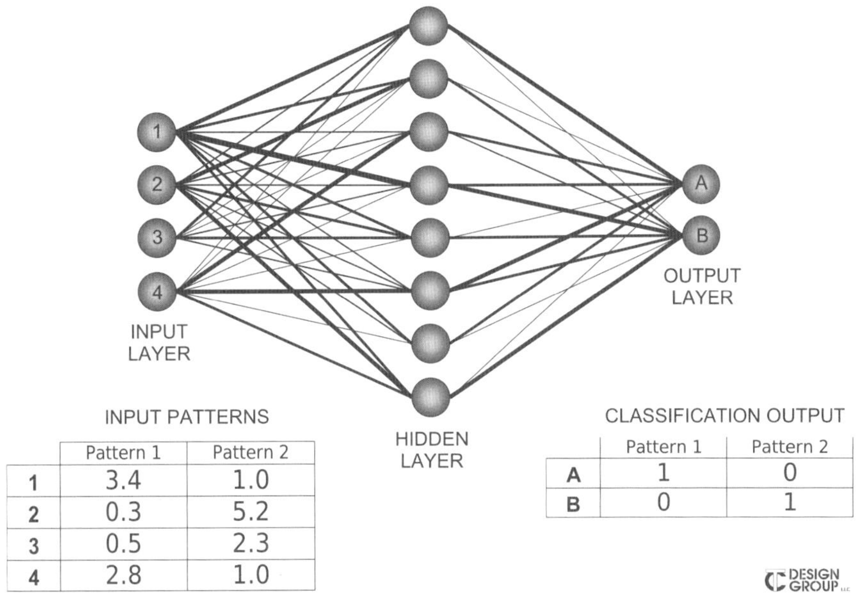

INPUT PATTERNS

	Pattern 1	Pattern 2
1	3.4	1.0
2	0.3	5.2
3	0.5	2.3
4	2.8	1.0

CLASSIFICATION OUTPUT

	Pattern 1	Pattern 2
A	1	0
B	0	1

Figure 1.16: Illustration showing the parts of a three-layer artificial neural network or "Multi-layer Perceptron" and its use for pattern recognition. In this example the network has four input neurons each of which connect to eight neurons in the hidden layer then two neurons in the output layer. The thickness of the lines connecting the layers indicates the relative importance (weight) that the receiving neuron places on the input from the originating neuron. The weights are determined during training of the neural network. The task of the network in this example is to recognize and classify a variety of patterns (two are shown) as belonging to either group "A" or group "B". The network's "decision" is indicated by a "1" at one of the output neurons.

that has found use in many other fields including neuroscience where it has been applied to the study of neural communication. One of the key information theoretic measures is the *Shannon entropy* [26] which is a measure of the complexity of a given signal. The greater the complexity of a signal the more bits that are needed to represent it across a communications channel. For example, a coin toss has an entropy of 1 bit since a single bit is all that is needed to represent the two possible outcomes (0=heads or 1=tails). If noise is added to the coin, say the coin lands on a bumpy surface where some outcomes are not clearly heads or tails, then more bits are required to accurately represent all the possible states of the coin. In other words, noise increases signal complexity thus increasing entropy.

In terms of our discussion on neural activity, the communication channel is the electrode and the signal is the waveform consisting of the summated activity of an unknown number of sources. When this detected activity consists mostly of noise the distribution of the signal is broad and has a high entropy (Fig. 1.17-A). Conversely, when the signal consists mostly of high amplitude spikes its distribution is narrow and has a lower value for the entropy (Fig. 1.17-B). The distribution of a signal that contains little noise and only a single source would be even narrower than that shown in Figure 1.17-B. However, noise is not the only source of complexity, in the brain where there are many active sources contributing to the summated waveform, the activity itself can be very complex. In this case, much of the low amplitude activity (activity that is more distant from the electrode) cannot be distinguished from noise and therefore cannot be separated.

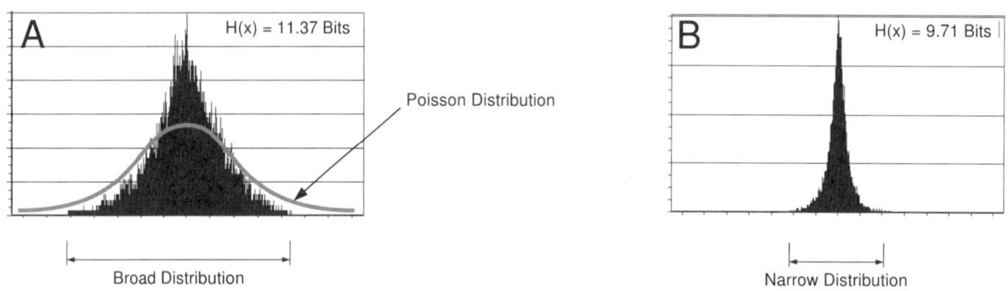

Figure 1.17: Comparison of the entropy H(x) and distribution of two signals. The distribution at the left is of a poor quality signal that contains mostly noise. As the neural data becomes more noisy its distribution approximates the shape of the Poisson distribution. The signal on the right is of a "clean" recording containing little noise. The signals were obtained from recordings of the activity in the lateral eye optic nerve of the Horseshoe crab *Limulus polyphemus*. *Adapted from [27].*

As in the cocktail party problem discussed earlier, electrodes placed in different locations around a group of neurons in the brain will "see" the summated electrical activity originating from all of the nearby sources. However, since some sources are closer to certain electrodes than others, the mixture will not be identical for each electrode. When there are two or more electrodes recording in proximity to one another, the entropies of pairs of recorded signals can be used to compute the Mutual Information or *I(X;Y)* which provides a measure of the similarity between the distributions of two signals. In this case, the signals are different "views" of the same signal-scape as seen from the points of view of different electrodes. Using Figure 1.17 as an example, the I(X;Y) between distributions "A" and "B" would be small, while the I(X;Y) between the distribution of a noiseless signal that contains only a single source and that of "A" would be even smaller. The mutual information provides us with a measure of the redundancy between two signals. InfoMax incorporates information theoretic measures into the unsupervised learning algorithms of a self-organizing ANN to extract the independent components of the waveforms using redundancy reduction. The inputs to the network are waveform data collected from multiple electrodes. The ANN then iterates with this data and learns the proper network configuration to extract the nonredundant components. The outputs of the ANN are waveforms which contain little or no redundant information. Depending on the complexity of the initial waveforms (number of summated sources + noise) the number of recoverable source waveforms will vary. At least one of the output waveforms will contain the residual components of signals which were inseparable.

One of the first of these ANN-based, blind source separation techniques was "InfoMax" which was successfully demonstrated by Bell and Sejnowski at the Salk Institute in 1995. Subsequent efforts by a number of researchers [28, 29, 30, 31] have demonstrated blind source separation using other types of neural networks and learning algorithms. The principle of redundancy reduction used in INFOMAX is reminiscent of that proposed by Barlow [11] with respect to sensory systems, in general, and by Atick et al. to retinal processing in specific. As such, this method of source identification/separation may be more "brain-like" than other methods. Since the network must re-organize (re-train) with each new data

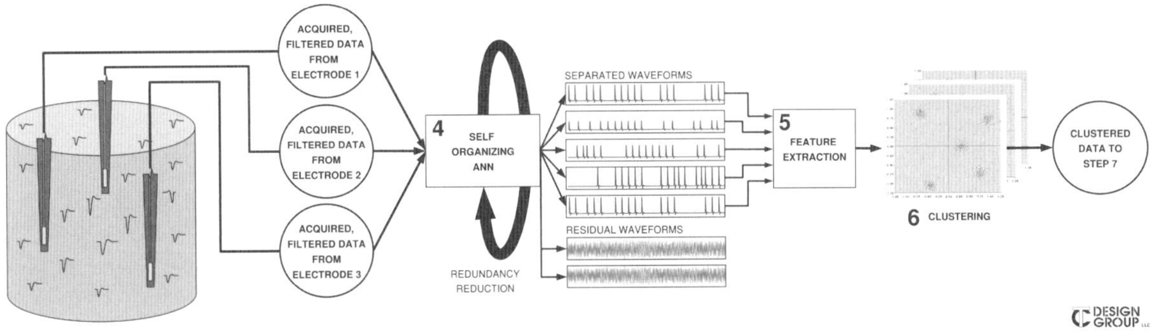

Figure 1.18: Block diagram of the technique of blind source separation by redundancy reduction using a self organizing network. The waveform data is passed to the initialized network which then interates with the data producing separated and residual waveforms. The separated waveforms are passed to step 5 while the residuals are discarded.

set, it is too slow to be of use in an overall scheme for real-time control of a prosthesis. I mention it here because it is the most robust of the three techniques and the speed issues are largely technological in nature and will likely be solved given further advances in computer hardware. The most widely used strategy at present for accomplishing spike sorting in real time is template matching.

1.3.2 SPIKE SORTING USING TEMPLATE MATCHING

In the previous section we discussed the process of de-constructing the recorded waveform using artificial neural networks. This process assumed that we knew nothing about the nature of the sources. Template matching, on the other hand, takes advantage of the fact that we *do* know something of the source; we know it is a spike train. As such, we know the general shape of the spikes which comprise the train and how their shape may be distorted by the C-R[24] properties of the extracellular medium. Additionally, signals originating from sources close to the electrode will be greater in amplitude than those from more distant sources. With this knowledge we can derive a large library of templates which allow the signal shape to be used as the basis for assigning detected activity to individual neurons.

Spikes are detected using amplitude thresholding, that is any signal above a pre-determined voltage level is considered a spike, anything below is noise. Then, using the template library the algorithm searches through the identified spikes trying to match features of the spike to the templates. When a match is found that section of the waveform is marked and the process continues. The marked segments are then overlaid as shown in Figure 1.19-4. The operator then uses a box drawing tool to indicate which group of spikes seems to have originated from the same (or similar) sources. The marked segments are then re-assembled into spike trains which contain spikes from only a single source. The individual spike trains are then processed further (Fig. 1.19-5) to extract relevant features such as timing characteristics, amplitude, etc. Those features might then used to plot the data on a Cartesian coordinate system where, hopefully, the data will cluster (Fig. 1.19-6).

[24]C-R: Capacitive-Resistive. In essence, the extracellular medium acts as a lowpass filter whose cutoff frequency is inversely related to the distance from the detector.

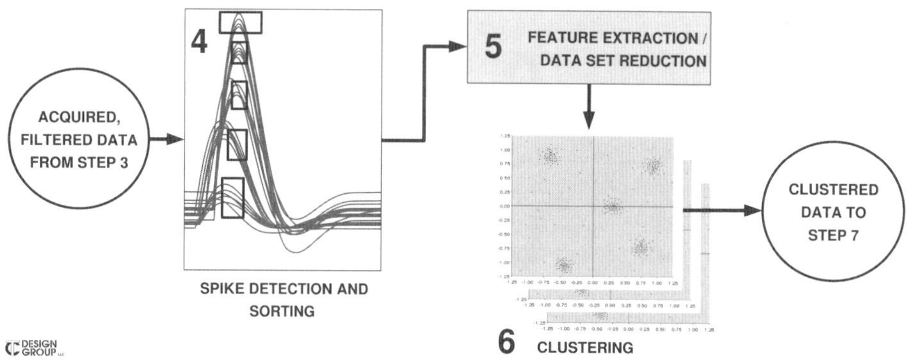

Figure 1.19: Diagram of the technique of source separation using spike sorting. Steps 1-3 are shown in Figure 1.15.

1.3.3 USING NEURAL DATA

The output of step 6 (Fig. 1.18-6 or 1.19-6) in this example are arrays of (x,y) values. Each array (or dataset) corresponds to relevent features of the detected neural activity during the time period of interest. These values form well defined clusters on scatterplots of each dataset and represent the unique state of the system (the brain region being monitored) when, for example, the subject was asked to perform a certain task. Using these datasets we can train an artificial neural network to recognize the pattern of clustering that is unique to each state. The ANN will map these states to binary output values which correspond to commands to operate a computer mouse. Thus, when one pattern of activity is recognized the move-up output might be active, another pattern of activity might correspond to move-down and so forth. The output of the neural net might then be translated into standard Human Interface Device (HID) commands and converted to the Universal Serial Bus (USB) interface protocol which would allow it to be connected to a computer and substitute for the mouse.

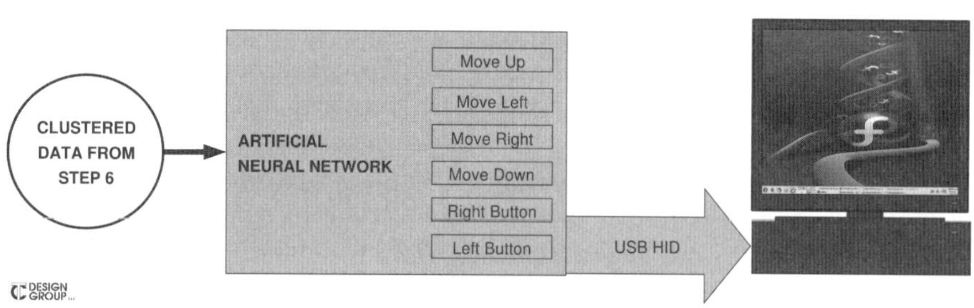

Figure 1.20: In the final step of the example the clustered data is used to train an artificial neural network (ANN). The network learns to classify the pattern of activity in the clustered data into six outputs which correspond to mouse controls and movement. The system running the ANN then outputs these commands over a USB interface to a PC to allow the user to move the cursor and interact with the desktop.

In this example, the neural activity is used to control a cursor on a computer screen. However, the neural net might have been trained to produce output suitable to control a prosthetic arm, the controls of a wheelchair or some other application. While ANNs are not the only method for performing pattern recognition they do have the unparalleled advantage of adaptability.

CHAPTER 2

Neural Interfaces: Past and Present

Through evolutionary processes biological organisms have come to make maximal use of the building blocks of their biological substrate. At the genetic level, for example, homeobox sequences that guide body patterning in early development are highly conserved (relatively unchanged) across seemingly unrelated species. In the nervous system, certain amino acids, the building blocks of protein found throughout the body, also function as neurotransmitters. In essence, biological organisms are the living embodiment of "Kaizen"[32], the principle of incremental positive change toward a goal of optimization of a system or process. Multi-use molecules are but one feature of this evolutionary optimization.

The lessons learned from the study of biological systems have diverse application in many areas including neural interfacing. The field which studies and applies these lessons is *cybernetics*. The inception of cybernetics as a discipline occurred in the early 1940's with the work of Norbert Weiner [33]. It includes areas of communication and control theory concerned with the comparative study of biological and man-made automatic control systems. The term is now also used to refer to a field which seeks to apply the knowledge gained by this comparative study to build better machines or to build equipment which in some way mimics human function. The term "Cyborg" (CYBernetic ORGanism) was coined by Clynes and Kline in their 1960 article in *Astronautics* to refer to a human connected to a machine for some adaptive purpose such as functioning in outer space [34]. In pop-culture the term Cyborg has been applied to not only the use of technology to restore function in the physically disabled but to make those restored individuals, "better, faster, and stronger" than before.

The term *Neural Interface* refers specifically to a category of technologies which are concerned with connecting excitable tissue (nerves, brain cells, muscles, etc.) to machines and is part of the broader field of Neural Engineering which is, "... an emerging interdisciplinary research area that brings to bear neuroscience and engineering methods to analyze neurological function as well as to design solutions to problems associated with neurological limitations and dysfunction"[35]. The field includes (but is not limited to) disciplines which study neural systems such as cybernetics and neuroscience, and aspects of other disciplines such as biomedical engineering, mathematics, prosthetics, and modeling as they apply to the definition of the field given above. The highly interdisciplinary nature of neural- interfacing/engineering stems from the need to pull together a diverse array of technologies from sometimes seemingly unrelated disciplines to accomplish its goals.

Human augmentation is not just on the minds of science fiction writers. The Defense Advanced Research Project Administration (DARPA) in the U.S., known for funding the more "far-reaching" research efforts, issued a solicitation on "Augmented Cognition" in 2001 which advocated research using a number of different technologies for the explicit purpose of tapping into and augmenting human sensory and cognitive function. In 2002, Lt. Cmdr. Dylan Schmorrow, Ph.D. (DARPA) said in a speech at DarpaTech that, "We are entering an era of unprecedented human advancement in which Darwinian principles of evolution may begin to show signs of artificial self-acceleration"[36] referring to the potential application of technological, biological, and psychological research to precipitate the emergence of the next stage in human development. While we are many years away from even remotely achieving Dr. Schmorrow's (and others) vision of artificially accelerating the evolution of the species, it is an intriguing idea which feeds the sometimes heated debate on the ethics of neural engineering. The present Augmented Cognition program capitalizes on the billions of dollars that have been spent by various government agencies (including DARPA) over the past 20 years, funding various segments of this work.

The National Institutes of Health (NIH), National Science Foundation (NSF), and other agencies in the U.S. have a myriad of research projects which either directly or indirectly advance neural interfacing research. Outside the U.S., a number of universities have their own independent projects underway and there are also a number of fledgling commercial efforts both in the U.S. and abroad. The primary focus of much of the work is rehabilitation and recovery of function following injury/disease or from birth defects. However, an interface developed for a prosthetic arm or a cochlear (hearing) implant might also be used as a means of directly interacting with a computer for expansion of function rather than just restoration.

Researchers have been electrically probing and recording from the brain for over 100 years, hoping to unlock its secrets. The first brain-to-machine interfaces were developed in the early 1960's and implanted in the first human volunteer in 1968 [37, 38]. However, it wasn't until the advent of microfabrication technology in the 1970's that it became possible to put large numbers of very small transducers into brain tissue. Since that time such devices have been decreasing in size, increasing in complexity and have enabled the study of networks of neurons *in vivo*. Some of the first individuals to experiment with brain-to-machine interfaces and neuro-prosthetics were Dr. José M.R. Delgado, Dr. Giles S. Brindley, and Dr. Walpole S. Lewin.

In the 1930's, Walter Hess[1] demonstrated that motor and emotional responses could be produced by electrical stimulation of the brain of cats. In the 1950's, building on this earlier work, Dr. José Delgado refined the procedure for the implantation of very small diameter wire electrodes into brain tissue using a stereotaxic (three-dimensional positioning) instrument and aseptic (sterile) procedures. He then used these intra-cerebral electrode assemblies, sometimes containing upwards of 100 wires, for the study of the electrical activity in various deep-brain regions. Coupled to wireless recording and stimulation devices he invented (termed stimocievers), Delgado was able to study the cerebral mechanisms of behavior in awake, freely moving animals, and humans.

The ability to both record and stimulate offered a number opportunities for experimentation including the possibility of automated learning[2] using a reciprocal brain-to-computer-to-brain connection. Demonstration of this concept came in the late 1960's when Delgado, working with other researchers, connected a female chimpanzee named Paddy (Fig. 2.1) to an analog computer. Paddy had electrodes implanted in her left and right amygdaloid nuclei[3] and her Reticular formation[4]. Using data telemetered via a stimociever the computer was able to detect the occurrence of a specific type of neural waveform (a spindle) in her amygdala then stimulate her reticular formation in response. After two hours on this paradigm the occurrence of spindles in her amygdala was reduced by half. After 6 days with treatments lasting 2 hours each day the occurrence of spindle waveforms was reduced to less than 1 percent of the untreated value. Paddy was quieter, less attentive, and less motivated during behavioral testing but still performed her given tasks without error. When the treatments were discontinued, she returned to normal after about 2 weeks. These experiments provided strong evidence that behavioral modification through automated learning was indeed possible. He predicted in his 1969 book, *Physical Control of the Mind* [39, p. 91], that this new technology could "...provide the essential link from man to computer to man, with a reciprocal feedback between neurons and instruments..."

[1]Nobel Prize in Medicine, 1949.

[2]Learning which occurs artificially, without the subjects participation.

[3]A part of the brain involved in the processing and memory of emotional responses to stimuli.

[4]A part of the brain involved in alertness, fatigue, and motivation.

Figure 2.1: Paddy (left) and Carlos (right) outfitted with early versions of stimocievers. Later, animals were implanted with versions of this technology which was miniaturized and contained completely under the skin. These miniature units would later be used in humans [39]. *Photo courtesy of J.M.R. Delgado.*

In addition to automated learning which had potential application in the detection and prevention of epileptic seizures, another of the goals was to identify areas of the brain involved in aggressive or anti-social behavior. Dr. Delgado and his collaborators performed extensive experiments in animals and humans throughout the 1960's to determine if electrical stimulation could be used to modify behavior. His promising results were reported in a number of papers and in the previously mentioned book. At a time when lobotomies were commonly used for treating severe forms of mental illness, Delgado's work held the promise of providing a safer and more effective alternative.

In 1966, at a ranch outside of Cordoba Spain Dr. Delgado and his team implanted wireless, brain stimulators (Fig. 2.2-A) into several adult bulls. The stimulation electrodes were implanted in the region of the caudate nucleus which is one of the structures found in the mid-brain. In a dramatic demonstration of the potential of technology to modify behavior Delgado halted a Brave[5] bull in mid-charge using nothing more than electrical stimulation. After several stimulations there was a lasting inhibition of aggressive behavior (Fig. 2.2-B) [39].

Dr. Delgado's work stirred up a furor of controversy in the 60's and was largely misrepresented by the media of the day. His work was the subject of articles and editorials in a number of publications including the New York Times magazine. Conspiracy theorists believed his work was evidence of government plots to control its population with covertly implanted devices. Conservative ethicists railed against the potential use of such technology to forcibly modify peoples behavior. The truth was (and still is) that our knowledge of the brain is far too primitive to allow this type of manipulation. However, deep brain stimulation has helped many sufferers of Parkinson's disease, epilepsy, chronic pain, and other illnesses.

[5]"Brave" refers to the breed of bulls used for bull fighting. Brave bulls are dangerous animals that have been bred for their aggressive tendencies over numerous generations.

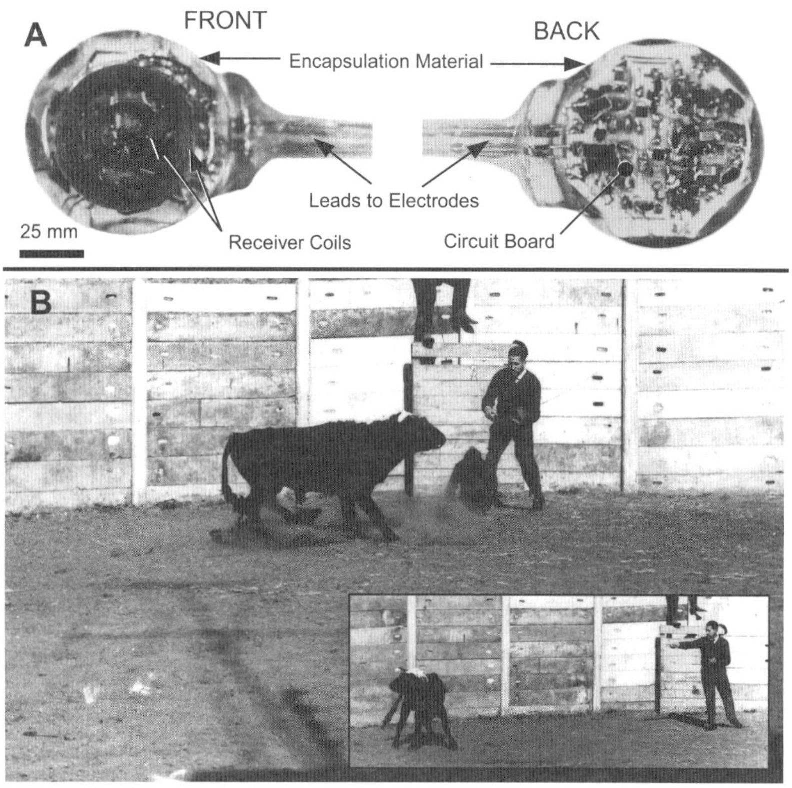

Figure 2.2: (A) Example of a four-channel wireless, deep-brain stimulator developed by Dr. Delgado. (B) Bull halted in mid-charge by inhibition of aggressive behavior using brain stimulation. Dr. Delgado can be seen holding the remote control which actuated the implanted stimulator. (B Inset) The bull not only stops but turns away from the object of his rage (the investigator) [39]. *Photos courtesy of J.M.R. Delgado.*

Delgado, now 93 and living in the U.S., says that the purpose of his work was always, "...to understand how the brain works and understand which areas of the brain were involved in aggressive behavior"[40]. He has devoted a large part of his career to understanding the psychophysiological roots of anti-social behavior. He originated a number of techniques and devices including deep brain stimulation electrodes, radio-controlled stimulators, brain pacemakers, and brain-to-computer-to-brain communication.

Around the same time as Dr. Delgado's work in the U.S., Drs. Giles Brindley and Walpole Lewin were experimenting with artificial vision in the U.K. Their work to create a vision prosthesis for implantation on the surface of the brain [37] built upon the earlier work of Foerster [41] and Krause et al. [42], both of whom experimented with electrically evoked visual sensations by direct brain stimulation. The device built by Brindley and Lewin (Fig. 2.3) consisted of an array of 81 small coils with platinum electrodes. The entire array, with the exception of the working surfaces of the electrodes, was encapsulated in a sheet of silicone rubber that allowed the array to flex and conform to the surface of the brain. The entire device was contained within the skull and had no external connections. The coils could be individually activated by placing a transmitter coil over the desired location and emitting a radio frequency pulse. The receiver converted this pulse into an electric current which stimulated the brain region directly under the electrode.

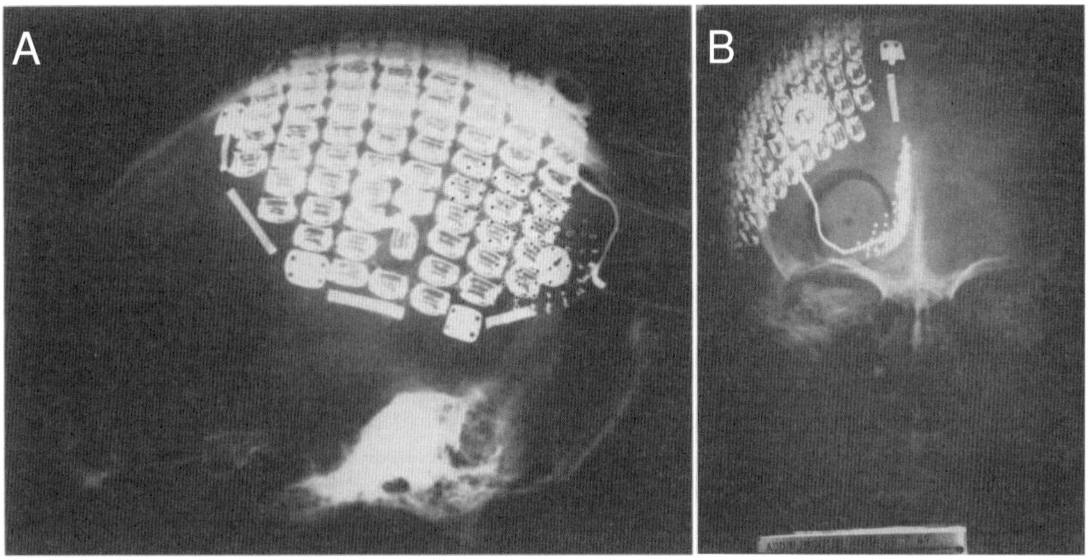

Figure 2.3: X-Ray of implanted electrode sheet. (A) Left side view. (B) Back view [37]. *Photos used with permission of Blackwell Publishing.*

When a coil in the array was activated the patient experienced the sensation of light. The sensation took the form of a small spot or spots of light (termed "phosphenes") in a constant position in her left visual field. The location of the phosphenes in the visual field corresponded to cortical visual maps already established by Holmes in 1945 [43]. The phosphenes track with the eyes during voluntary eye movement but not during reflex eye movements. Brindley and Lewin demonstrated that through activation of multiple coils simultaneously the patient could be made to see simple shapes and patterns. Later work by Brindley focused on characterizing the cortical response to stimulation [44] and its sensory effects [45]. Dr. Brindley obtained a patent for the device in 1972 [46] but his work on this vision prostheses did not continue much past 1973. However, work continued on the development of cortical vision prosthetics by others [47, 48, 49] and with the advent of microchip fabrication technologies, researchers developed high-density electrode arrays [50] for implantation in the brain. In January 2000, researchers at the Dobelle Institute in Lisbon, Portugal announced an implanted vision system which is able to restore rudimentary vision to totally blind individuals [49].

William H. Dobelle, MD devoted much of his career to cortical prosthesis research. He founded the Dobelle Institute in Lisbon, Portugal where he developed a vision prosthesis that was implanted in about six patients with good results. The system permits the user to read two inch tall letters at a distance of 5 feet through a narrow visual tunnel with acuity roughly equivalent to 20/400 vision. One of those patients, a Canadian man named Jens who had been totally blind for nearly 20 years, was able to drive a car around a parking lot using the prosthesis. With the passing of Dr. Dobelle in 2004, 30 years of work developing the "Dobelle Artificial Vision System" came to a halt. There are a few groups who continue to experiment with cortical vision implants [51] but no system like Dobelle's is presently available.

Over the intervening 40 or so years since the pioneering work of Drs. Delgado, Brindley, Lewin, and others, research has produced many technologies which are likely candidates for directly linking biological organisms to computers. The creation of a bi-directional neural interface relies not on a single technology but devices and expertise from a diverse mix of disciplines. Research efforts are necessarily multifaceted consisting of groups conducting bio-materials, interfacing, neural coding, and other research. While some groups are developing specific applications (e.g., cochlear implants to restore hearing) others are accomplishing the basic research upon which future applications will be built. Interfacing technologies can be grouped into two broad categories: those which aim to connect the Central Nervous System (CNS) and those that connect the Peripheral Nervous System (PNS) to a computer or other hardware. However, some technologies have applications in both areas.

2.1 INVASIVE INTERFACES

The term "invasive interfaces" is used here to refer to any technology which requires breaking the skin to accomplish implantation. The ability to place a sensor directly into the tissue you wish to monitor allows far more accuracy than can be obtained with noninvasive techniques. The disadvantage is the risk of sequelae such as infection, tissue reactions, etc. Furthermore, invasive devices for use in humans have a strict regulatory process that must be navigated, surgical procedures have to be performed by trained surgeons under aseptic conditions, anesthesia, and post-operative support are needed, all of which make this type of interface far more complicated to commercialize. From a signal processing standpoint, it is far less complicated and far more feasible derive useable control signals from an electrode placed in the brain as opposed to an electrode placed on the scalp. One of the factors that currently limits the useability of noninvasive interfaces is the difficulty in recovering usable control signals from the highly complex electrical activity recorded by surface detectors.

2.1.1 PERIPHERAL NERVOUS SYSTEM INTERFACES

The peripheral nervous system includes nerves which connect the senses, skin, muscles, and internal organs to the central nervous system (CNS). The CNS consists of the brain and spinal cord. A peripheral nerve contains multiple nerve fibers and is, in concept, like the trunk cable of an old style telephone system containing thousands of wires each carrying a different conversation. However, unlike the telephone cable each fiber in a nerve carries a unidirectional signal. Motor fibers carry signals from the brain to the muscles while sensory fibers communicate information on temperature, pain, pressure, vision, etc., to the brain. Some nerves contain all one kind of fiber (e.g., the optic nerve contains only sensory fibers) but many peripheral nerves are mixed. A mixed nerve complicates processing because in order to control a prosthesis we need to separate the signals originating from motor fibers from those of the sensory fibers. The nerve fibers are arranged in bundles (fascicles) which are contained within the nerve sheath (Fig. 2.4).

2.1.1.1 Cuff Electrodes

The general design for a cuff electrode is that of a flexible material containing wires on the inner surface which is then wrapped around a nerve. The edges of the material may be glued, sewn, or otherwise connected so as to form a tube and hold the bare wires in contact with the surface of the nerve. Cuff electrodes have also been fabricated from short lengths of silicone elastomer tubing (Fig. 2.5) as well as other materials. Cuff electrodes have the advantage of being relatively easy to fabricate from commonly available materials and are well tolerated for both acute and long-term experiments [52]. A self-sizing

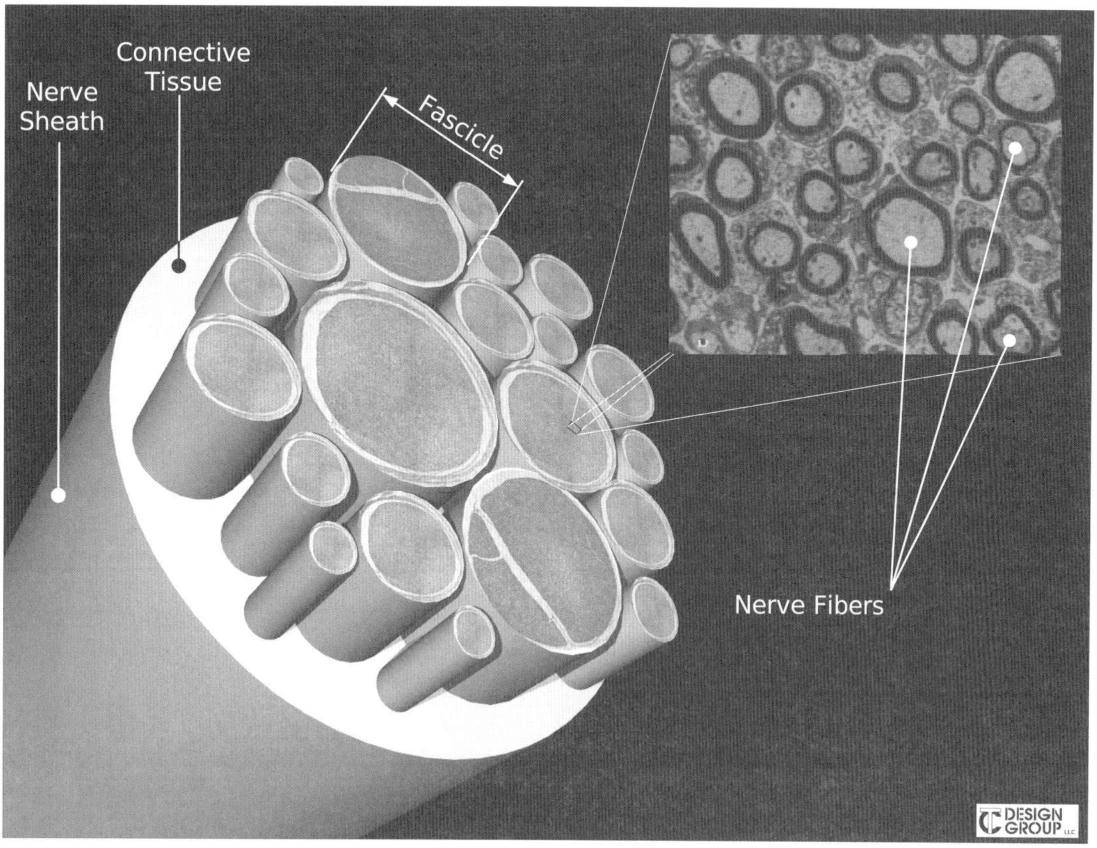

Figure 2.4: Diagram of the structure of a peripheral nerve. Nerve fibers are arranged in bundles (fascicles) which are stabilized by connective tissue and contained within the nerve sheath. Also contained within the connective tissue are blood vessels (not depicted).

cuff electrode [53, 54] can be fabricated from tubing by making a longitudinal cut which spirals from end to end. The tube is then un-coiled, wires glued into place following the spiral and the tube allowed to re-coil. When wrapped around a nerve, the cuff will expand as needed to fit. Spiral nerve cuffs have been used experimentally in humans with degenerative retinal disease as a means to directly stimulate the optic nerve [55].

2.1.1.2 Nerve Re-Shaping Electrodes

When a cuff electrode is placed around a nerve, each electrode detects the nearby electrical activity and, to a lesser extent, that occurring elsewhere in the nerve. This complicates the process of sorting out which fibers are active in the nerve at a given time. One way to address the problem of signal separation in a round nerve is to try to reconfigure the nerve to a more desirable shape [56]. That is, if the nerve were flat instead of round then the nerve fascicles would be arranged serially instead of bundled together. Dr. Dominique Durand of Case Western Reserve University and others have developed a type of cuff electrode that is

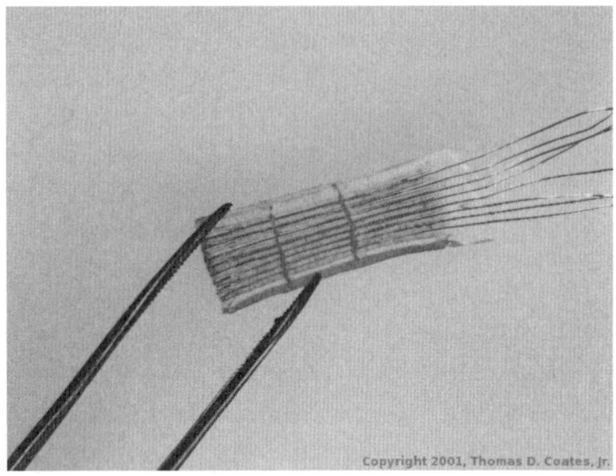

Figure 2.5: Photomicrograph of a cuff electrode [27]. The cuff is formed from a short length of silicone tubing which has wires affixed to the interior. The tube is slit parallel to the long axis to permit it to be opened and placed around a peripheral nerve. Here, the cuff is held open and flattened (using a pair of small tweezers) to reveal the wires running parallel to the length of the cuff. The cuff shown is approximately 1.5 cm in length.

capable of re-shaping the nerve. The flat interface nerve electrode (FINE) [57] is fabricated from silicone sheeting, iridium wire, and PLGA, a biocompatible polymer that slowly dissolves *in vivo*. The silicone sheeting is used to construct a flattened tube which represents the final state of the cuff. The tube is stretched into a round configuration and PLGA is applied. When the PLGA is cured, it forms a strong, rigid band around the flexible tubing which maintains it in the round configuration. Once placed around a nerve *in vivo*, the PLGA begins to dissolve over approximately a 1-month period. As the PLGA dissolves, the tube gradually resumes its original configuration and the nerve is safely flattened. Once flattened, the electrodes are able to detect activity originating from nearby sources with far better isolation than before. The FINE can also be used to stimulate nerves with better specificity than cuff electrodes [57, 58].

2.1.1.3 Nerve Regeneration Arrays

Regenerative Nerve Interfaces (RNIs), also known as sieve electrodes, capitalize on the limited ability of peripheral nerves to regenerate following injury. When a peripheral nerve is cut, the nerve fibers inside the fascicles will degenerate back to the cell bodies in the spinal cord (termed Wallerian degeneration). The cell body will then send out a new process which grows down the nerve fascicle inside the sheath toward the cut end. If the cut nerve is reconnected with the fascicles aligned many of the fibers will find their way back to their original connection sites in the limb. If a porous material is placed between the stumps, the nerve fibers will grow through it and into the distal stump (Fig. 2.6). The basic design of an a RNI consists of a thin, porous substrate (the sieve) with electrodes around some of the pores. The RNI is placed between the stumps of a cut nerve and the nerve allowed to regrow through the fenestrated substrate. The electrodes can then be used to record from and stimulate nearby nerve fibers. If sufficient electrode density could be achieved RNIs could provide fine control and feedback between a nerve and a neuroprosthesis.

Advanced RNI implementations have used silicon microchips containing perforating via holes, micro-fabricated electrodes, and on-chip amplifier/multiplexing circuitry. The early work in this field was

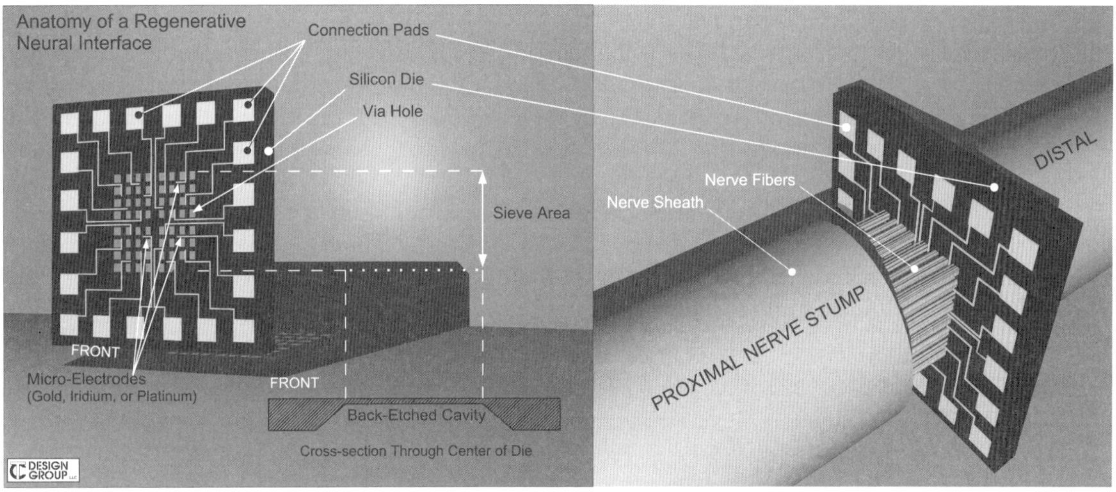

Figure 2.6: Illustration of a Regenerative Neural Interface. Left: An RNI similar to those developed in the U.S. and Europe. Right: Diagram of the RNI *in situ* (not to scale).

conducted by A. F. Marks [59] (1969) and R. Llinas [60] (1973) while trying to improve peripheral nerve regeneration following injury. Later, researchers at Stanford [61, 62], University of Michigan [63], and University of Stuttgart worked on advanced designs. Work on this type of interface still continues at The Fraunhofer Institute for Biomedical Engineering [64], University of Freiburg, and other sites in Europe and at the National Cardiovascular Research Institute in Osaka, Japan [65]. Continued progress on these types of interfaces is contingent upon advances in human nerve regeneration, biomaterials research, and micromachining techniques. Since this interfacing strategy requires nerve transection it is not likely that it would be useful for connecting devices to nerves in healthy, uninjured individuals. However, in cases of amputation (where the nerve is already cut) it could become practical in the relatively near term as an interfacing technique for prosthetic limbs.

2.1.2 CENTRAL NERVOUS SYSTEM

From the perspective of evolution, the primary purpose of the nervous systems of all animals is to allow the rapid response and adaptation to ever changing situations in their environments to increase the animal's chances of survival. The central nervous system, as already mentioned, consists of the brain and spinal cord. The spinal cord and lower brain are responsible for the reflexive responses, autonomic processes (like breathing), and lower-level processing functions. The higher brain, composed of the cerebral hemispheres and other structures, is responsible for high-level functions which gives rise to our cognitive processes and allows us to carry out voluntary actions. Interfaces designed for use in the CNS typically use arrays of penetrating microelectrodes, that is, microelectrodes that are inserted into the tissue. Microelectrode arrays can be constructed in a number of ways: using silicon micromaching techniques, using individually inserted wires, or by patterning conductive materials on other rigid or semi-rigid substrates which can then be implanted. One of the first micromachined-silicon, multi-electrode penetrating arrays was developed at the University of Utah.

2.1.2.1 "Hair Brush" Electrode Arrays

The Center for Neural Interfaces (CNI) was founded in 1995 by Richard Normann of the University of Utah. Among its goals are to develop multichannel neural interfaces, test the safety/efficacy of those devices, and develop new therapeutic approaches to treating nervous system disorders. As part of this mission new types of microelectrode arrays (Fig. 2.7) were constructed using three-dimensional microfabrication techniques [66, 67].

The present 3D microelectrode array design is the result of years of research by scientists at the University of Utah, Brown University, MIT, and Emory University. These high-density arrays have been developed to interface superficial cortical areas of the brain [68] and peripheral nerves to computers. They are used by several of the neuroprosthetic systems discussed later.

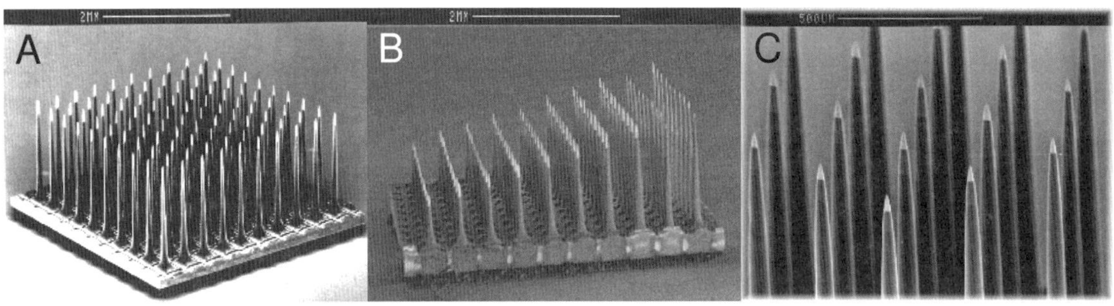

Figure 2.7: Photos of the (A) Utah Electrode Array, (B) Utah Slant Array, and (C) closeup of the microelectrode tips. *Photos courtesy of R.A. Normann.*

The Utah Electrode Array (UEA) is fabricated from silicon and contains an array of 100 penetrating microelectrodes (Fig. 2.7-A). The UEA is designed to be implanted in the cerebral cortex and is capable of stimulating and/or recording from large number of neurons simultaneously. The second type of array fabricated by Dr. Normann's team was the Utah Slant Array (USA) which is similar to the UEA except that it contains 100 penetrating microelectrodes of various lengths (typically from 0.5–1.5 mm in length). The USA is designed to be implanted in the peripheral nervous system and to provide a means for simultaneous stimulation and recording from a large number of nerve fibers.

In 2002, Dr. Kevin Warwick, a cybernetics professor at the University of Reading, U.K.[6], was implanted with an array similar to the UEA. The 100 electrode array was implanted into the median nerve of his left hand. Signals derived from the interface permitted him to control an electric wheelchair and a robotic arm. Warwick's use of himself as a research subject was, and still is, highly controversial. To date, the implants are functioning properly with no ill effects to the professor.

The UEA and USA have been used extensively in animals [69, 70, 71] and more recently in humans in conjunction with the BrainGate System®. The BrainGate® System[7] developed by Dr. John Donoghue's team at Brown University and marketed by Cyberkinetics is currently beginning clinical trials for individuals with severe motor impairment. If successful, the system will permit paralyzed individuals to interact with their environment through novel, computer-based communication interfaces.

[6] `http://www.kevinwarwick.org`
[7] `http://www.cyberkineticsinc.com`

2.1.2.2 Microwires

The use of fine wires for recording and electrical stimulation of the brain dates back to Walter Hess in the 1930's. Since then a number of researchers including José Delgado [72], C.A. Palmer [73], Westby [74], Richard Andersen, Joel Burdick [75], and others have refined the technique. Microwires can be placed at depth in the brain using stereotaxic equipment and a guide cannula or they can be coated with glass to stiffen them. Arrays containing hundreds of microwires can be constructed allowing detailed study of large areas of the brain.

One of the problems encountered with any chronically implanted microelectrode is that the electrode, neuron, or both can shift position over time resulting in loss of signal strength. Drs. Richard Andersen, Joel Burdick, and Yu-Chong Tai at Caltech are developing a micro-positioning unit using electrolysis-based diaphragm actuators [76] capable of repositioning the individual electrodes in a multiwire array. This MEMS device would enable the array to "intelligently" track the signal sources and reposition the electrodes as needed to maintain optimal contact. A larger version of the device [75] using piezo-electric positioners has already been constructed and the control algorithm tested *in vivo* with excellent results.

2.1.2.3 Silicon/Polymer Microelectrode Arrays

Silicon has a number of advantages over other materials for use in the fabrication of neural interface devices. Its long history of use in the electronics industry provides an extensive knowledge base of how to fabricate microelectronics in silicon including microprocessors, other digital circuitry and various types of analog devices. Research at Caltech, the University of Michigan (U. Mich), and other academic/industrial fabrication facilities around the world are constantly adding to the body of knowledge of how to construct novel devices out of one of the earth's most abundant minerals.

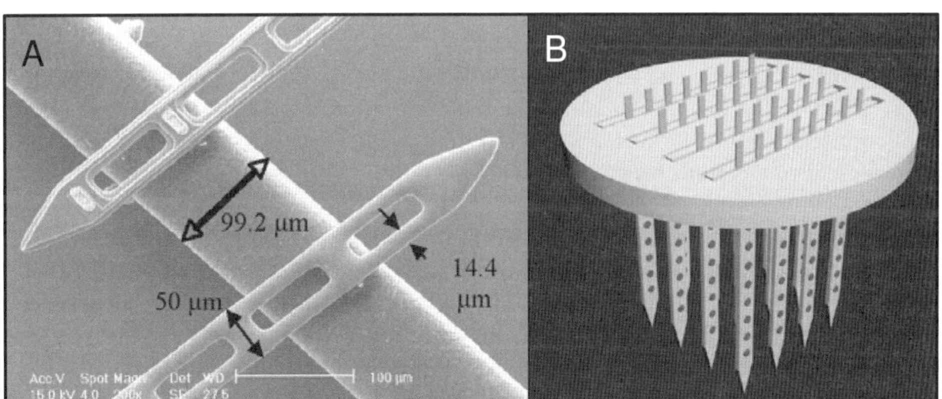

Figure 2.8: Examples of silicon-based technology developed at the University of Michigan's Center for Wireless Integrated MicroSystems. (A) lattice recording arrays compared to a human hair [77, 78]. (B) concept drawing of the lattice arrays used in a zero-rise recording array [77, 79]. *Photos courtesy of K. Wise.*

The brain and spinal cord is surrounded by cerebro-spinal fluid (CSF) contained in the arachnoid space located between the pia-mater[8] and the arachnoid[9]. The brain is not anchored to the skull at every point about its perimeter instead it is partially afloat in a sea of CSF. Thus, as we walk, run, and go about our everyday activities the CSF acts as a kind of shock absorber. This means that the brain is not absolutely stationary with respect to the skull. Micro-motion of the brain can cause problems with chronically implanted, rigid micro-electrodes that are anchored to the skull. While silicon is somewhat flexible when made into very thin structures, a more flexible substrate would be highly desireable. For example, a rigid microelectrode implanted in the brain might be integrated with a flexible micro-cable which connects it to the electronics attached to the skull. Toward this end, researchers are developing microelectrodes and cables constructed from flexible, biocompatible polymers such as parylene [80, 81, 82] and polyimide [83, 84].

At the University of Michigan's Center for Wireless Microsystems, researchers are developing a number of novel devices for use in neural interfacing applications. The center is headed by Dr. Kensall Wise who has over 30 years of experience with implantable microsystems. Among the center's most recent achievements are high-density cochlear electrodes for hearing prostheses, microelectrodes with integrated MEMS fluid handling systems, thin-film diamond probes for electrochemistry applications, and wireless implantable microsystems for multi-channel neural recording.

2.1.2.4 In Vivo Electrochemistry (IVEC)

There are basically two methods for detecting neurotransmitter release in living tissue: Microdialysis which relies on High Pressure Liquid Chromatography (HPLC) and In Vivo Electro-Chemistry (IVEC). Micro-dialysis provides excellent quantitative accuracy but poor temporal resolution due to the need to move fluid through tubing from the implant site to an HPLC apparatus for the analysis. Although it is an important clinical and experimental tool, microdialysis is too bulky (at present) to be useful as a self-contained, implantable neural interface. IVEC systems coupled with inexpensive, mass-manufactured microelectrode arrays have the potential to meet the criteria needed to produce small, implantable chemical detection systems.

The technique of in vivo electrochemical analysis was developed by Dr. Ralph Adams in the early 1970's [85] and has become a vitally important tool for the real-time detection and quantitation of neurochemicals in living tissue. Voltammetry refers to analytical methods in which a small amount of the analyte (the material being analyzed) is oxidized and/or reduced by an electrolytic reaction. The term"Oxidation" refers to the loss of electrons by a molecule or atom while the term "reduction" refers to a gain. Reduction/oxidation reactions are collectively referred to as redox reactions. During a redox reaction the movement of electrons sets up a measurable current that is proportional to the amount of the analyte that was electrolyzed.

In a voltammetry system there are two electrodes: the working electrode and the reference electrode. The working (a.k.a. recording) electrode is where the detection occurs. A voltage potential is applied to the reference electrode whose character and amplitude depends on the method of analysis (Fig. 2.9-A, CA, CV) and material being analyzed, respectively. An analyte that is electroactive will undergo a redox reaction in response to the applied voltage. The job of the reference electrode (and any auxiliary electrodes) is to allow the system to compensate for any drift that might occur in the potential applied to the reference electrode. The piece of equipment that monitors and compensates for any drift is the potentiostat. A simple diagram of a potentiostat and its connections are shown in Figure 2.9. There are a number of methods used in electrochemical detection and quantitation. The three that are the

[8] A thin highly vascular layer of connective tissue that is intimate with the folds of the brain.

[9] A delicate layer of connective tissue located under the dura-mater.

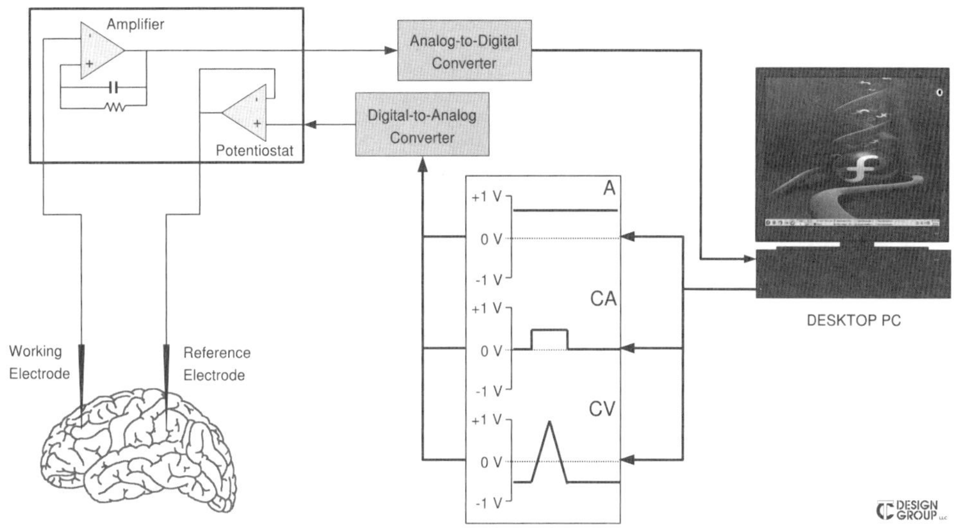

Figure 2.9: Block diagram of an IVEC system. A desktop PC is used to generate the waveform which is applied to the working electrode. The output of the D/A converter is sent to the potentiostat (shown here as a unity gain buffer). The current detected by the working electrode is amplified and converted into a proportional voltage (trans-resistance amplifier). The output waveform is digitized and processed in the PC. A: amperometry, CA: Chronoamperometry, CV: Cyclic Voltammetry.

most widely used for *in vivo* studies are: cyclic voltammetry, chronoamperometry, and amperometry. Differences between the three techniques include the shape of the waveform applied to the reference electrode (Fig. 2.9-A, CA, CV) and the way the output recorded at the working electrode is processed to compute the analyte concentration. Each technique has its own unique set of advantages and applications which are to lengthy to discuss here. Instead I will refer you to "Electrochemical Methods" by Bard and Faulkner [86] for a more detailed discussion on the subject.

The electrodes used for IVEC range from single carbon fibers to microelectrode arrays fabricated on a number of different kinds of substrates[10]. The ideal situation from a neural interfacing perspective would be to have a "lab-on-a-probe" which would be capable of detecting the simultaneous release of different neurotransmitters and the corresponding electrical activity. Realization of this goal depends on the fabrication of probes containing multiple microelectrodes and the ability to apply different coatings to each of these sites. Typically, the microelectrodes themselves tend to be fabricated from platinum although other materials have also been used. In the past, ceramic, and silicon were the most prevalent substrates for electrochemistry probes. While still widely used, more recent work has focused on flexible polymer substrates such as parylene, exotic silicon variants, diamond, saphire, etc. Generally speaking, the higher the dielectric constant of the substrate the better it will be for electrochemistry, although dielectric properties are not the only consideration. For example, microelectrode arrays fabricated on silicon, a semiconductor, have also been used for IVEC with good results (Fig. 2.10). Silicon substrates

[10]The base material from which the probe is fabricated.

offer the advantage of being able to fabricate micro-electronic devices directly on the probe using robust and widely available manufacturing techniques.

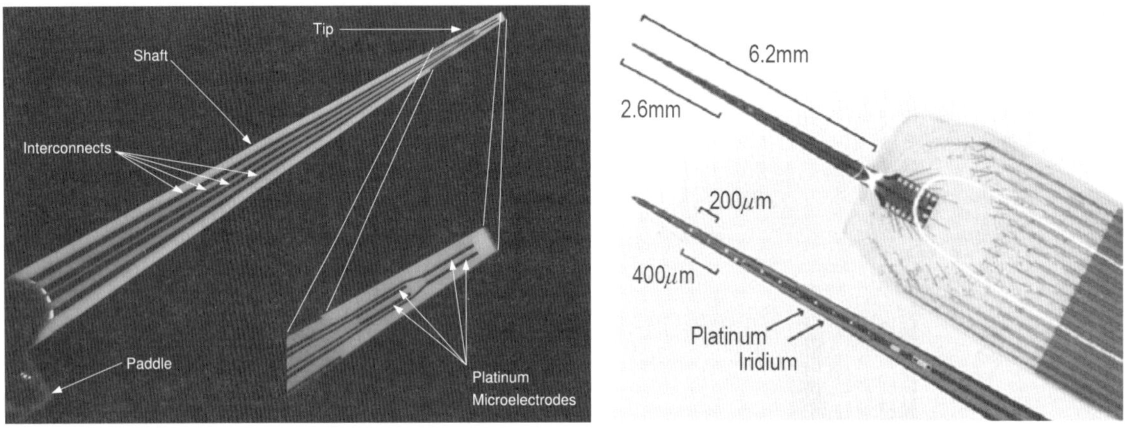

Figure 2.10: Examples of microelectrode arrays used to record *in vivo* electrochemical activity. (Left) 4 site, ceramic substrate array developed at the University of Kentucky's Center for Microelectrode Technology. (Right) 16 site, silicon substrate recording array developed at the University of Michigan's Center for Neural Communication Technology. *Left photo courtesy of P. Huettl; right photo courtesy of D.R. Kipke.*

Not all analytes are electroactive. That is, some molecules hold on to their electrons more tightly than others. In this case, microelectrodes must be "functionalized" with enzymes or other substances which convert the molecule of interest to a form that can be detected. One such example is the neurotransmitter Glutamate which requires the use of the enzyme Glutamate Oxidase to convert it into an electroactive form. As such, IVEC is a less direct method of detection than HPLC. In the case of an enzyme-assisted detection strategy, the output of the working electrode is a current that is proportional to the amount of the analyte that was converted to an electroactive form and subsequently oxidized/reduced. The redox current data is then used to compute the concentration of analyte present at the surface of the working electrode at the time the electrolytic reaction took place. Studies have demonstrated that the quantitation achieved with enzyme assisted IVEC is comparable to similar methods utilizing HPLC. As such, IVEC is becoming the method of choice for rapid, *in vivo* analysis of neurotransmitter release and is particularly well suited for neural interfacing applications.

At the time of this writing there are several companies offering IVEC systems for use in research. Quanteon, LLC was started by Greg Gerhardt (a former Ph.D. student of Ralph Adams) in 1995 to develop and market systems for *in vivo* electrochemisty. The system (Fig. 2.11) sold by Quanteon performs IVEC using amperometry and chronoamperometry. Mark Wightman (a former postdoc of Ralph Adams) of the University of North Carolina has developed IVEC systems which utilize cyclic voltammetry primarily for the detection of dopamine. IVEC equipment like that developed by Dr. Wightman is available from Cypress Systems[11]. Both Drs. Gerhardt and Wightman have dedicated

[11] A division of ESA Biosciences.

a large part of their careers to the development of the techniques and technology which make IVEC possible.

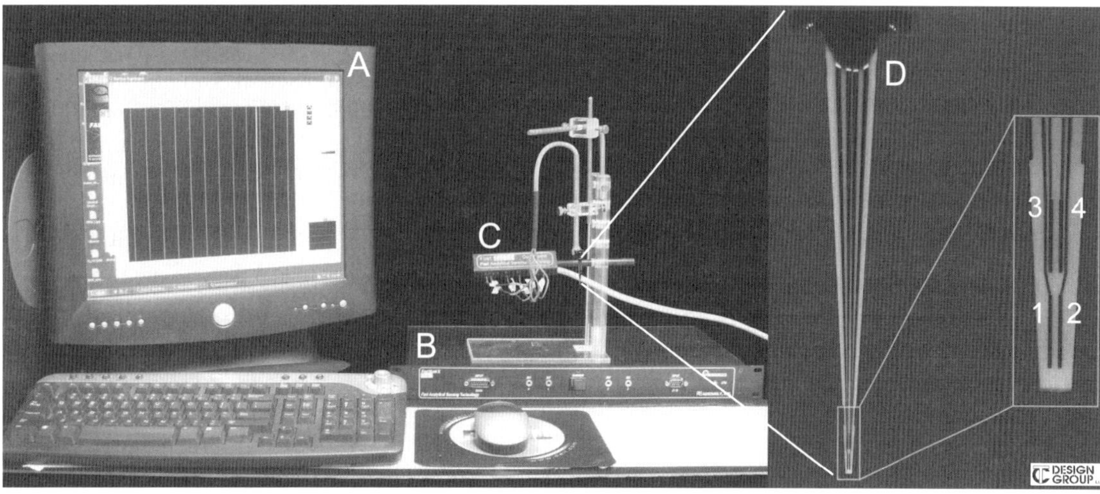

Figure 2.11: F.A.S.T. IVEC System by Quanteon LLC of Nicholasville, Kentucky shown with a typical microelectrode array attached. The system utilizes PC based data acquisition and measurement software (A). The system shown is capable of recording from up to 8 channels simultaneously *in vivo*, with a temporal resolution of less than 100 milliseconds. It is able to measure a variety of neurochemicals in living tissue such as those listed in Table 1.1 using either amperometry or chronoamperometry.

A relative newcomer to the IVEC scene is the University of Michigan's Center for Wireless Integrated Microsystems and the Center for Neural Communication Technology. There, a research effort lead by Drs. Daryl Kipke and Richard Brown is developing silicon-based, multi-channel probes (Fig. 2.10-B) and a system capable of simultaneously recording electrical and neurochemical (dopamine, choline, or serotonin) activity. Building upon the center's expertise with semiconductors, their current probes are fabricated on silicon substrates and will incorporate on-chip electronics. The group has plans to begin fabrication of high channel count devices using flexable polymers in the near future. If successful, theirs will be the first system capable of performing *in vivo* electrochemistry and electrophysiology using the same probe and a single software package.

2.1.2.5 Chemical Stimulation

The ability to stimulate a synapse or a small group of synapses using neurochemicals relies on the ability to deliver the chemical in question with extraordinarily high precision. At present chemical stimulation is accomplished using glass micropipettes. The pipettes are tapered to an extremely fine point using a pipette puller, filled with the desired solution, then piggy-backed on a microelectrode. The untapered end is fitted with flexable tubing connected to a Picospritzer[12] which provides very precise pressure to eject small quantities of the liquid from the tapered end of the pipette. When the micro-electrode/micro-pipette combination is placed in neural tissue it provides a means by which the electrical (or chemical)

[12]Picospritzer is a registered trademark of Parker Hannifin Corp, Pine Brook, NJ.

response to the ejected substance can be recorded. Compared to the size of a synaptic terminal which averages less than 1 micron, the probe/pipette combination is several orders of magnitude larger. Although this method is considered state-of-the-art at present, it is of course at best. Researchers at a number of universities around the world are developing Micro Electro Mechanical Systems (MEMS)[13] capable of handling nanoliter quantities of liquids. These miniature fluid handling systems, called micro-fluidics, hold the promise of one day allowing us to construct probes which can not only record chemical activity but release a variety of neurotransmitters on command. If micro-fluidics could be coupled with an on-board repository of living neuro-secretory cells the device could have an inexhaustible neurochemical reservoir. Otherwise, there will have to be an external means of refilling the device when its chemical supplies are exhausted.

There is presently no technology that will allow us to target a chemical stimulus with the precision of a single synapse. This is because we do not yet know how to fabricate (or place) devices on a scale which would permit such fine control. However, this fact does not negate the usefulness of less precise devices. Neuromodulators, which play a crucial role in neural processing, are released into the extracellular space and diffuse to the surrounding synapses. These neurochemicals effect the spiking activity of neurons proximal to the site of release. Dysfunction of this neuromodulatory system has been implicated in a number of conditions including certain forms of mental illness. Ergo, a device capable of even diffuse release of neurochemicals could prove highly useful for diagnosing and treating such disorders.

2.1.3 BIO-HYBRID CIRCUITS, NEUROCHIPS, AND NEUROPROBES

The ever-shrinking dimensions of what we can build will eventually allow the fabrication of arrays containing thousands of microelectrodes. These ultra-dense arrays will contain integrated processing circuitry to permit wireless communication with computers. Such arrays will permit the study of living neural networks with unprecedented detail. Cell biology is a field of study concerned with understanding cell growth, development, physiology, and proliferation. Cell culture refers to both a discipline and a body of knowledge regarding how cells can be maintined *ex vivo*, induced to proliferate and differentiate into specialized tissues in the laboratory. Through the combination of silicon micro-electrode arrays, cell biology, and cell culture, it is now possible to grow intact neural networks *in vitro* directly on silicon microchips. Development of these high-tech tools began in the Laboratories of Jerome Pine at the California Institute of Technology in the U.S. and Peter Fromherz at the Max Planck Institute of Biochemistry in Martinsried, Germany.

Dr. Fromherz, building upon his earlier work [87], developed a technique whereby the outgrowth of leech neurons could be guided along pre-determined pathways to form defined arborizations [88] (Fig. 2.12-A). These early experiments utilized glass coverslips as the substrate and extracellular matrix protein as the patterning agent. Neurons grown using this technique were then used to evaluate the hypothesis that dendritic arborizations were capable of performing nontrivial processing on their inputs. At the time the idea of dendritic processing was the subject of considerable debate.

To test the hypothesis that dendrites could be more than just passive conductors the XOR[14] function was chosen. Many artificial neural network models of the time relied on what is known as the McCulloch-Pitts neuron. This type of artificial neuron was modeled on the theory that dendrites of real neurons are passive structures which perform a simple summation of their inputs. McCulloch-Pitts neurons are not capable of implementing the XOR since to do so requires the ability to perform an

[13]MEMS research can be followed in the IEEE/ASME Journal of Microelectromechanical Systems (`http://www.asme.org`) and at the annual IEEE International Conference on MEMS (`http://www.ieee.org`).

[14]The output of the XOR function is false when the inputs are either all true or all false: otherwise the output is true.

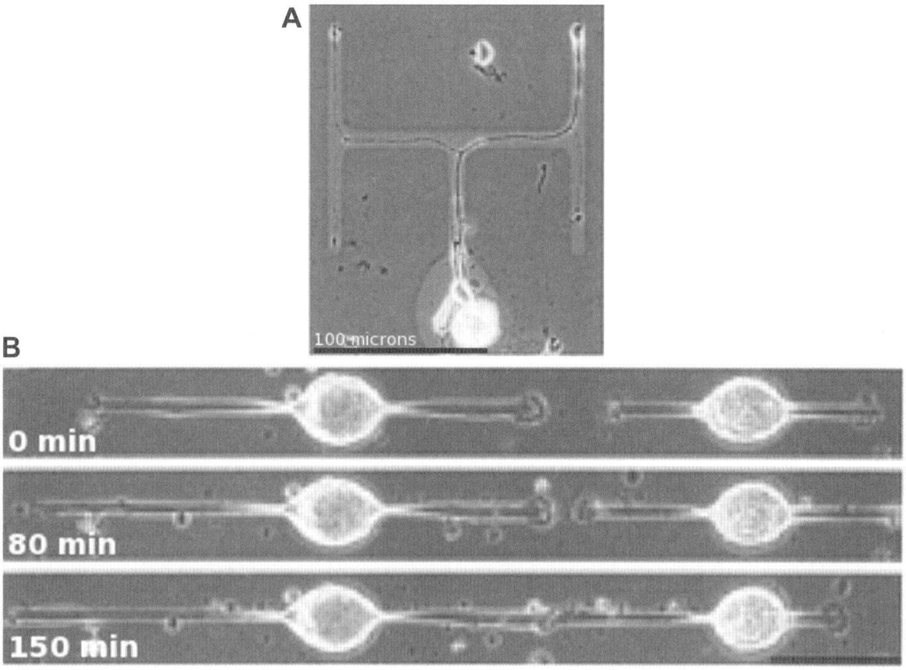

Figure 2.12: Results demonstrating patterned outgrowth of a leech neuron (A) [88] and a sequence of three photos demonstrating guided outgrowth and synapse formation between two neurons (B) [89]. Scale bars are 100 μm. *Photos courtesy of P. Fromherz.*

inversion[15] operation. Thus, the XOR was attractive as a test since it is one of the operations that cannot be performed by the theoretical "point neuron" of McCulloch-Pitts style artificial neural networks and should not be able to be performed by a real neuron if the current theory was correct. In 1993, the results of these experiments were published and demonstrated not one but four different ways that the XOR function could be implemented in the patterned arborizations [90] of living neurons.

In 1995, Fromherz and Stett demonstrated that leech neurons grown on specially designed silicon substrates could be capacitively stimulated using microscopic, metal-free, field effect transistors (FETs) [91]. By making the FETs metal-free the concern of leaking toxic metals into the extracellular medium and slowly poisoning the cells could be avoided. The neuron's cell membrane acts as one plate of a capacitor separated by the silicon dioxide insulating layer from the p-doped layer of the FET which acts as the other plate (Fig. 2.13). Stimulation occurs when the electric field coupled across the silicon-neuron junction achieves sufficient charge density to depolarize the voltage gated ion channels in the neuron and produce an action potential. In such an arrangement no faradic current flows which eliminates possible toxic effects of repeated stimulation. Stimulation using current can cause leaching (migration) of cytotoxic materials from the substrate into the surrounding extracellular space effecting the health and longevity of the neurons. Also, high currents can damage the cell directly. The following year both recording and stimulation were demonstrated using capacitive coupling of the silicon-neuron junction [92].

[15] An inverter is a basic logic element where the output is the opposite of the input.

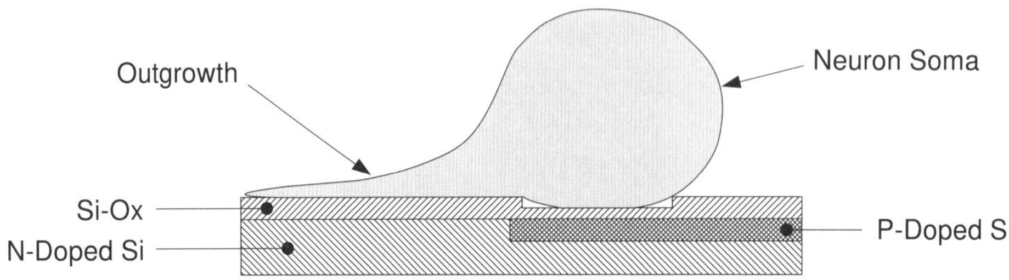

Figure 2.13: Illustration of the metal-free silicon-neuron junction pioneered by Peter Fromherz. *Based on [91, p. 1670, Fig. 1].*

The intervening years between 1995 and 1999 saw improvements in techniques, development of better methods for visualization of living neural networks, further characterization of the silicon-neuron junction, and development of new structures to retain the neurons on the silicon substrates. In 1999, this work produced the first demonstration of the formation of a functioning electrical synapse between two neurons using guided outgrowth (Fig. 2.12-B). Later in 2001, Dr. Fromherz's and Günther Zeck[16] published an article [93] describing a new methodology using a picket fence type of structure constructed from parylene to retain the neuron cell body in place over the active electronics. In previous experiments the cell bodies tended to migrate away from their initial positions over the FETs which caused problems with stimulation and recording. The new structure prevented migration of the soma while permitting the neural processes to grow un-hindered (Fig. 2.14). A later article [94] described the Electrolyte-Oxide-Silicon transistor which permits the electrical activity of the neuron to directly control the source-to-drain current of the FET.

Dr. Fromherz's has continued to advance the study of silicon-to-neuron devices and has added to his repertoire the construction of multi-transistor arrays which provides, "... time-resolved images of evoked field potentials and allows the detection of functional correlations over large distances"[95]. The arrays are capable of imaging electrical activity at exceptionally high resolution (16,384 pixels). Fromherz uses these devices to study patterns of activation in cultured hippocampal[17] slice preparations.

In the late 1970's, Dr. Jerome Pine of the California Institute of Technology (Caltech)[18] was developing a technique to record from cultured neurons using planar, extracellular microelectrode arrays. The arrays consisted of 32, 8μm x 10 μm thin film microelectrodes patterned on glass and positioned at the bottom of specially prepared petri dishes. Neurons obtained from rat superior cervical ganglia were cultured atop the arrays for 8 days after which electrical activity could be recorded using the array. Pine speculated in his 1980 paper on this work [96] that such cultures could, "...create neuronal networks which offer the potential for studies which are now technically difficult or impossible to perform *in vivo*." The technique also offered the ability to record from multiple neurons simultaneously without damaging the cells, thus enabling long-term studies of development and plasticity.

[16]Max Planck Institute for Biological Cybernetics in Tübingen.
[17]The hippocampus is a structure in the forebrain that is involved in memory and spatial navigation.
[18]While on leave at the Washington University School of Medicine.

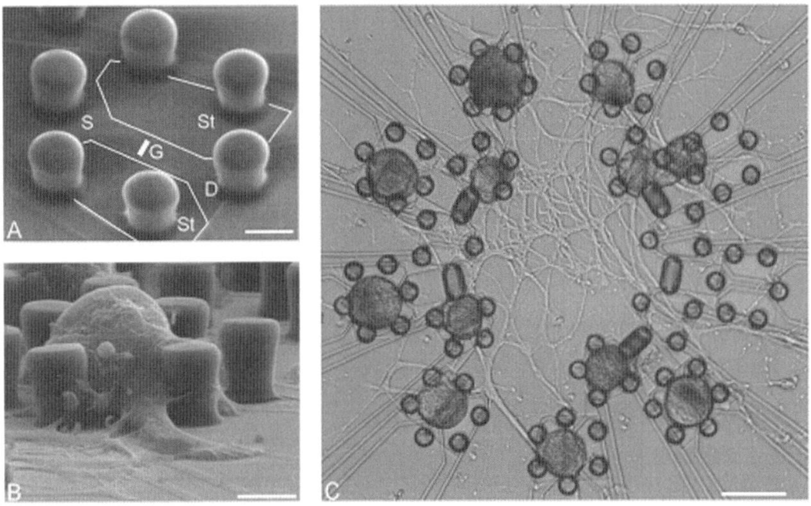

Figure 2.14: Biohybrid silicon-neuron technology from Dr. Fromherz's Laboratory. (A) EM of picket fence made of polyimide. Stimulator wings (St) and electrolyte-oxide-silicon (EOS) field-effect transistor (S source, D drain, G gate) are shown. Scale bar = 20 μm. (B) EM of snail neuron in a picket fence after three days in culture. Scale bar = 20 μm. (C) Micrograph (viewed from above) after 2 days in culture, showing neurons held in picket fences on a circle of two-way contacts. Neurites (bright threads) can be seen connecting the neurons forming a functioning neural network. Scale bar = 100 μm. *Photos courtesy of P. Fromherz* [93].

After returning to Caltech, Pine continued to develop this recording technique and began work with Dr. David Rutledge[19] and Wade Regehr[20] to develop a new silicon-based technology for use with cultured neural networks. In 1988 The trio published their work [97] describing a micromachined "diving board" electrode which extended from a pedestal and made contact with the top of the cell in the petri dish. While the work was successfully concluded with the demonstration of stimulation of- and recording from the cultured neurons, the technology needed improvement especially with respect to the seal obtained between the electrode and the cell.

In the early 1990's Dr. Pine formed a collaboration with Dr. Yu-Chong Tai of Caltech's Micromachining Laboratory to continue the work on applying microelectrode arrays and silicon substrates to the *in vitro* study of living neural networks. The technology developed from this collaboration would later come to be known as the "Neuro-probe" and the "Neuro-chip." The basis of both devices were microfabricated "neurowells" into which immature neurons were placed. Grillwork at the top of the well prevented the soma from migrating out but allowed the axons and dentrites to grow out of the well onto the top of the chip where they could synapse with processes from other "caged" neurons. A gold electrode at the bottom of each well-permitted recording and stimulation.

[19]California Institute of Technology.
[20]Ph.D. candidate, Caltech.

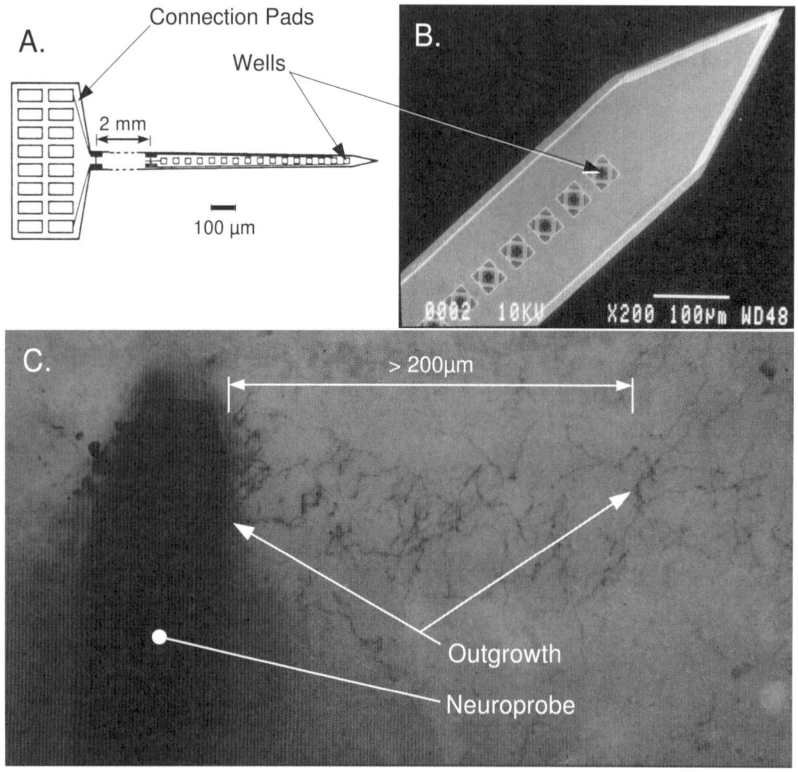

Figure 2.15: The Neuroprobe. (A) diagram showing the layout of the neuroprobe. Circuitry (not shown) connects each of the 16 pads on the paddle with their corresponding microelectrode at the bottom of each of the 16 wells on the shank. (B) electron micrograph of the tip of a completed neuroprobe prior to loading with neurons. (C) *In vivo* results: Light micrograph of stained [98] neurons showing neurite outgrowth from the wells into rat brain tissue. *Diagram and photos courtesy of J. Pine* [99].

The neuroprobe (Fig. 2.15) was a dagger-like, linear array of neurowells with a paddle shaped connection area at the top. The shape of the neuroprobe allowed it to be loaded with neurons then the shaft inserted into brain tissue. Over the course of the next few days the neurons grew their processes out of the well and into the surrounding neural tissue. Experiments were performed first with cultured brain slices and then *in vivo* in rat hippocampus. Both studies demonstrated that the cells in the neuroprobe were capable of surviving, growing, and sending out processes that are incorporated into surrounding brain tissue. All the probes used in the studies lacked electrodes and as such were not capable of recording or stimulation. The primary goal of the study was to test the effectiveness of the well design for containing the cells, demonstrate survival of the neurons and outgrowth of neural processes into the surrounding tissue. All of the projects initial goals were met and the next step was to fabricate probes with electrodes for testing. Unfortunately, this work was not continued much past the mid 1990's due to factors unrelated to the efficacy of the technology. However, work on the Neurochip has progressed significantly since this time.

The first version of the Neurochip utilized the same well design used on Neuroprobe. The neurochip is an *in vitro* device intended to serve as a substrate on which networks of cells could be grown and studied much like Pine's earlier work with thin-film microelectrodes. Early versions of the neurochip featured wells etched into bulk silicon [100] and enabled Pine's group to demonstrate viability of the cells and show proof-of-concept of the design. The neurochip allowed them to engage in the long-term study of both the development and activity of live neural networks. Despite these successes, the early neurochip suffered from several problems, primarily: the wells were difficult to fabricate in large numbers and the neurons sometimes lost connection with the electrodes as they grew out of the top of the well.

The second generation of the neurochip was implemented using cages constructed of parylene (Fig. 2.16-A), a flexable polymer. The new design incorporates advances in materials, fabrication techniques, and lessons learned from the first generation chips. These improvements include changes in the cage design to better contain the cell body and to maintain electrical contact. Rather than have the soma sit at the bottom of the well and its processes grow out the top, the second generation uses tunnels extending from the side of the cage. As the neuron sends out processes through these tunnels they help retain the soma in place over the microelectrodes. This ingenious design along with the method of fabricating parylene structures on silicon was arrived at though years of trials by Pine and Tai. The cages themselves are anchored into the silicon using vertical holes that extend deep into the substrate. The process by which the holes are constructed was pioneered in Dr. Tai's lab.

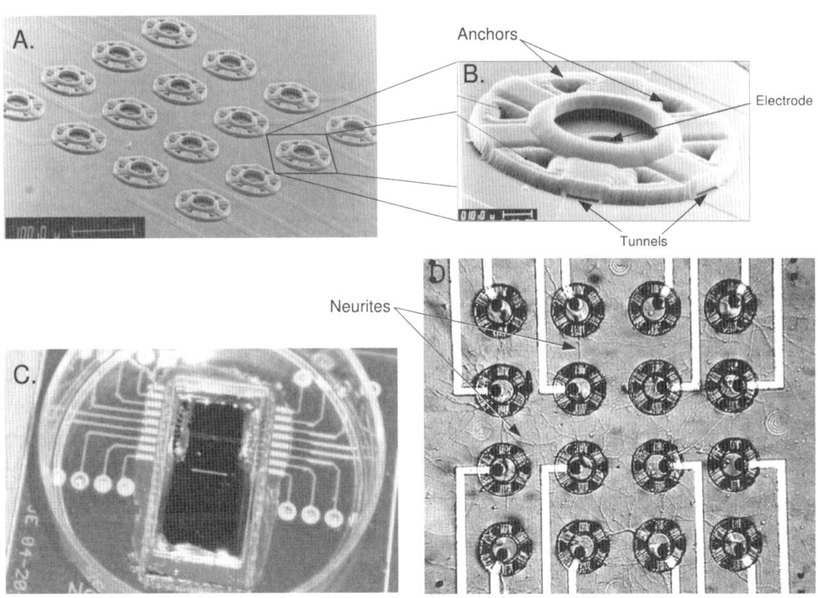

Figure 2.16: The Neurochip 2007. (A) SEM of a 4 x 4 array of parylene wells fabricated on silicon substrate. (B) Close-up of one of the wells. (C) the assembled, mounted and connected neurochip shown positioned at the bottom of a petri dish. (D) Photo of neurites growing from the wells and forming synapses on the surface of the silicon [101]. *Photos courtesy of J. Pine.*

One of the problems encountered with the use of the neurochip was the time it took to load the neurons into the cages. The second generation neurochip contained a 4 x 4 array of neurocages but the plan was to fabricate much larger arrays to permit far more complex networks to be grown and studied. A new method had to be found which would permit rapid and perhaps semi-automatic loading. This required a new approach to cell handling and placement. The solution to this problem came in the form of optical tweezers built by Gary Chow [101] to pick up and place the neurons into the neurocages.

Whenever light strikes an object and is refracted (changes direction) it imparts a small momentum to the object; optical tweezers make use of this phenomena to pick up and move cells. When a laser beam is focused onto a transparent refractile sphere (the neuron), the beam is bent as it enters the cell due to differences in the refractive indices of the cell and surrounding medium, and bent again as it exits on the other side. Due to the opposing forces exerted on the cell by each bend in the beam, the cell is held at the beam's point of focus. By moving the laser beam (and the cell along with it) the "captured" cell can be maneuvered into a neurocage. Optical tweezers have been used in a number of other applications [102] including manipulation of other type of cells [103], airborne particles [104], and protein translation [105]. To minimize the chance of damage to the cells, an infrared[21] laser was used. Future Neurochip designs will consist of 60 (or more) cages and permit the study of large *in vitro* neural networks. The optical tweezer technology enables the rapid loading of these high density neurochips.

The work being conducted by Drs. Pine and Fromherz is creating new avenues of research that will help us to better understand the functioning of biological neural networks. The ability to grow neurons on active substrates capable of recording and stimulation will permit the study of biological neural networks in ways that would never be possible in a living organism. It is difficult to study the details of such networks *in vivo* for a number of reasons including:

- The physical placement of electrodes into living brain tissue may disrupt fine connections we wish to study.
- Networks in the brain are highly interconnected making isolation and study of smaller interconnected groups of cells impossible.
- We presently lack tools able to visualize synaptic morphology *in vivo*[22]. Current research indicates that the size of the synapse may have a role in information processing and changes in size may have a role in age-related mental degradation [106].

Tools like those being developed by Drs. Pine and Fromherz combined with other technologies could overcome this problem by permitting the detailed electro-chemical study of both the function of the network as a whole and the individual neurons of which it is comprised. Furthermore, networks grown in petri dishes can be periodically placed under a microscope and their physical properties observed; something that is not possible with a living brain.

With regard to brain-machine interfaces, if the theory that each brain is truly unique with respect to how we process and store perceptual experiences, an adaptive interface will be required. Some future device like the Neuroprobe that uses biological neurons to make the connection between the brain and the electronics, might enable the requisite adaptive, direct neural interface. Using some training paradigm the cells could, as they grow and connect, learn to interpret the activity of the surrounding neural network into a form that could be used by the connected computer system. There are a number of advantages to this strategy (if it could be made to work) over using microelectrodes:

- Neurons are capable of making far more precise and well-defined connections than are possible now or for the foreseeable future with MEAs, microwires, or other inorganic means.

[21] Living cells are semi-transparent to infrared radiation.
[22] The average size of a synapse is approximately 0.5 μM compare that to a human hair which is about 100μM.

- Neurons have the capacity to adapt and learn, thus providing the possibility of signal processing and interpretation built into the substrate of the interface.
- Neurons are not susceptible to electrical noise like microelectrodes.
- The connections made by the neurons might be better tolerated over the lifetime of the person than nonbiological/inorganic materials.
- Future advances in other fields such as genetics and cellular/molecular biology may allow the engineering of a type of "interface neuron" specifically designed to connect human-made circuitry to *in vivo* neural networks.

Neural interfaces of the future will probably look far more "organic" than those envisioned today. Such interfaces may contain a core of electro-optical processing circuitry surrounded by human-engineered cells. The hybrid device would then be implanted into a damaged brain area where it would grow and integrate with the remaining tissue. The organic and inorganic components would work together, interacting with the existing brain tissue, to "learn" how to replace the lost functionality. Such a device is far in the future indeed but the technologies needed to realize it are already in the early stages of development. The creation of such a device relies on expertise from many disciplines. The work of Drs. Fromherz and Pine are but two examples of the highly interdisciplinary nature of neural interfacing/enginering.

2.1.4 FUNCTIONAL ELECTRICAL/NEUROMUSCULAR STIMULATION

Functional Electrical Stimulation or FES is useful when the muscles are intact but signals cannot reach them due to damage to the nerves and/or the spinal cord [107]. FES systems control muscle movements using stimulating electrodes located either on the skin above the muscle [108] or implanted directly in the muscle tissue [109]. Present systems such as the Parastep®[23] use a remote keypad to select pre-defined muscle activation patterns to, for example, stand from a sitting position and/or walk using a walker.

Dr. Gerald Loeb of the A.E. Mann Institute for Biomedical Engineering and pioneer of the BION®system has been studying electrode construction, spinal reflexes, EMG[24], and FES for decades. His laboratory is developing wireless, implantable micro-stimulators which can be injected into muscles using a large diameter needle. These tiny implants are inductively powered and controlled from a small computer worn by the user. Another research group headed by Drs. Ranu Jung and Jimmy Abbas at the University of Arizona's Center for Adaptive Neural Systems, is developing systems and methodologies using FNS (Functional Neuro-muscular Stimulation) and other paradigms to re-animate paralyzed limbs. Their current efforts include the development of a MEMS-based neural clamp that can be reversibly attached to spinal roots. The clamp would be used as the mechanism to hold micro-electrodes in place for the purpose of stimulating peripheral efferents to produce muscle contraction.

FES/FNS systems not only help to restore limited movement to paraplegics, but help prevent muscle atrophy which occurs due to disuse. Some day we will be able to build the "ultimate" system which bypasses the injured spine and permits direct mental control of the muscles. This might be done using the combination of a brain-to-machine interface and a system to translate detected brain activity into movement commands. The commands would then be converted into the correct sequence of stimuli to contract the muscles and carry our the persons intended movement.

[23]Parastep is a registered trademark of Sigmedics Inc., `http://www.sigmedics.com`.
[24]ElectroMyoGraphy: recording the electrical activity in muscle using surface electrodes.

As I read about these systems I cannot help but be reminded of the old Star Trek episode "Spock's Brain." In this episode the brain of the ill-fated Vulcan is stolen and his friends must bring his body along as they search for the missing organ. Instead of carrying his body on a stretcher the Doctor constructs a device (controlled by a wireless, handheld remote) to move Spock's muscles and make him walk. I remember joking as I watched that episode years ago, "Gee, it's the 23rd century and the best they can do is have him move like Mr. Roboto? They should have him break dancing by then." A "few" years and a Neuroscience Ph.D. later, I realize that having a brain-less body do anything at all would be pretty amazing.

2.2 NON-INVASIVE NEURAL INTERFACES

This category of interfacing strategies does not require surgery, injection or any other type of procedure which "invades" the internal spaces of the body. These technologies (if perfected) would be useable by any person without bodily alteration. This includes devices or electrodes that can be worn, adhered to the skin, or otherwise nonpermanently attached to ones person. Such devices have the potential advantage of not requiring specific licensing or approval for use by normal individuals other than ensuring electrical and mechanical safety. Thus, noninvasive neural interfaces, in all likelyhood, will be the first type of interface that shows up on the store shelves for you and I to use. These technologies are not currently fast enough to replace a computer keyboard and mouse. So, unless you are a quadriplegic and have no other option it is not likely that you would want to use them in their present state of development; In the future this will surely change.

2.2.1 ELECTROENCEPHALOGRAPHY (EEG) AND MAGNETOENCEPHALOGRAPHY (MEG)

Experimental, EEG-based neural interfaces often utilize scalp electrodes affixed to a type of snug-fitting, fabric cap that can be worn by the user. The location and spacing of the electrodes is usually according to one of the standard systems of placement such as the "10-20" system [110]. The electrodes are used to detect fluctuations in the electric field produced by the brain. The field is a composite of all the electrical activity emanating from all the neurons at a given instant in time. Deconvolution of this activity into something useful is similar to the "cocktail party" problem discussed earlier where each electrode sees a different "view" of the composite activity.

Use of EEG as the basis of a neural interface depends on some very sophisticated and computationally intensive calculations to extract meaningful control information from the composite activity. EEG BMIs have improved remarkably over the years both in accuracy and speed. This increasing rate of success is directly tied to the increase in processing "horsepower" available on desktop PC's and their decreasing cost. Deconvolution of EEG is a computationally bound problem, that is, a problem which relies on the ability to perform many calculations at a high rate of speed. As computer technology progresses and the systems become smaller, faster, and less expensive, EEG-based BMI will likely become practical as a way for everyone to interact with machines.

Drs. Jonathan Wolpaw and Dennis McFarland are two well-known names in the field of electroencephalography (EEG). Their work on building an EEG based brain-computer interface [5] has spanned nearly 20 years, produced scores of journal articles, and led to the open-source BCI-2000 system [111][25]. In Europe and Asia there are several efforts to develop EEG-based brain-computer interface systems. The Berlin Brain-Computer Interface project at the Fraunhofer Institute directed by Drs. Klaus-Robert

[25] http://bci2000.org

Müller, Gabriel Curio and Benjamin Blankertz [112, 113] has been on-going since 2000. They have made a number of advances in EEG-BCI especially in producing systems which require minimal user training. Also in Germany at the Institute of Medical Psychology and Behavioral Neurobiology at the University of Tübingen, researchers are working to decode hand movement from EEG signals to allow control of a robotic limb [114]. In Switzerland at the IDIAP Research Institute, Dr. José Millán and his team have developed an EEG-based system which permits a user to mentally guide a robot around an indoor maze [115]. At the Brain Science Institute in Japan the work of Dr. Andrzej Cihcocki is helping to improve the speed and accuracy of EEG-BCIs by developing novel algorithms to extract pertinent features from human EEG signals [116].

Magnetoencephalography (MEG) provides similar information to EEG except the detected magnetic field is 90° rotated from the electric field. That is, the electric field detected by the EEG is perpendicular to the skull while the magnetic field is parallel to the skull. The equipment needed for MEG is very bulky owing to the super-cold temperatures needed by the array of ultra-sensitive, superconducting quantum interference devices (SQuIDs). SQuIDs are capable of detecting the tiny magnetic fields produced by an action potential. Since EEG can deliver essentially the same information but with far less bulk and expense it is currently preferred over MEG as the method of choice for noninvasive interface systems.

2.2.2 ELECTROMYOGRAPHY

Electromyography or EMG involves the placement of electrodes on the skin over a muscle whose activity one wishes to monitor. The contractile state of the muscle is reflected by the detected electrical activity and can be used to control a prosthesis. EMG has been used to control various type of myo-electric prosthetics with good results. However, it cannot deliver the fine motor control and sensory input of the natural limb. The first prototype of the prosthetic arm for the DARPA RP-2009 project (discussed in Sec. 3.3) used this type of control scheme. There have been a number of research efforts to use EMG to interact with virtual environments including a hand model [117], neck muscle EMG to track head position [118], a system for gesture recognition [119] and the just-mentioned DARPA project.

2.3 TECHNICAL CHALLENGES TO IMPLANTABLE DEVICES

One of the significant challenges to neural interface development is designing technologies which can withstand years of implantation in the body. The internal environment of the body is highly ionic (salty) and will wreak havoc on most materials commonly used in the production of semiconductors and electrodes. Not only does the device have to survive this corrosive environment but it has to remain chemically inert to prevent damage to the surrounding tissue. There is a need to find coatings for present materials or alternative materials which will allow implants to withstand the harsh environment of the body for decades. Much of the neural interfacing work conducted thus far has utilized microfabricated components constructed from silicon. Since it is highly desirable from both a cost and R&D standpoint to utilize microchip manufacturing techniques, methods for protecting the fragile electronics must be found. Additional issues for chronically implanted devices include:

- Heat: implanted, active devices generate a small amount of heat during powered operation. As devices become more complex, smaller and circuitry density increases, more heat will be generated. The surface temperature of the device should not exceed body temperature to prevent possible undesirable effects to the surrounding tissues.

- Size: the devices must be small enough so as not to interfere with movement or other activities.
- Aesthetics: scarring, devices visible as lumps under the skin, etc.

2.4 POWER AND COMMUNICATION

The final topic to be covered in this section is how implanted devices are powered and how they communicate with the "outside" world. In the past, these two items were accomplished using a hardwired connection. That is, a multi-conductor electrical socket mounted to the skull or elsewhere that was connected to the external equipment. Though this type of arrangement is still used due to its low cost and simplicity, wireless technologies are gaining in popularity for a number of reasons:

- Less chance of infection since everything is contained under the skin.
- Decreasing cost: always a big factor with limited research budgets.
- Increased availability: a number of companies now offer out-of-the-box solutions so researchers no longer have to spend the time to develop their own systems.
- Decreased size.
- Decreased power requirements: in the past, large battery packs have limited the usefulness of un-tethered systems.
- In awake, behaving subjects it is difficult to study natural behaviors if the subject is in the unnatural situation of being tethered to a rack of equipment by a cable.

Long-distance, wireless transmission of data is possible utilizing a number of standard methods including Amateur (HAM) Radio, WiFi, cellular, and/or satellite communication channels. In the U.S., the band of frequencies utilizable for long-range communication is strictly regulated and a license may required to operate on certain frequencies. Short-range communication devices are less strictly regulated and the frequencies assigned for these devices are (in general) available for un-licensed use. Many of the custom designed devices available for laboratory use, utilize short range transmitter/receivers which operate in the bands assigned for consumer products such as bluetooth and cordless telephones. These devices usually have a ranges from 2 to approximately 100 meters.

Implanted devices such as cardiac pacemakers are often powered using batteries that must be replaced every five years or so. Neural interfaces which operate continuously and transmit/receive data have much higher power requirements and cannot be powered solely from an implanted power cell. These types of devices are usually powered either one of two ways: (1) by a cord which extends through the skin to plug into a power supply, or (2) by an implanted rechargeable battery with inductive charging circuitry. In the case of the latter, the transmission of power occurs between two wire coils, one implanted under the skin the other lying on the surface of the skin. The two coils form an air-gap (or in this case a skin-gap) transformer where the magnetic field generated by the alternating current flowing through the exterior coil induces a current to flow in the internal coil. The current is then rectified into DC, regulated to the proper level and fed to the battery's charging circuit.

There are certain species of animals that are capable of generating natural electricity. The best known is *Electrophorus electricus* or the electric eel found in the Amazon. The electric eel can produce up to 800 watts of energy per discharge, enough to light a small Christmas tree. Perhaps some day a way will be found to produce an implantable biological power source based on this or other examples found

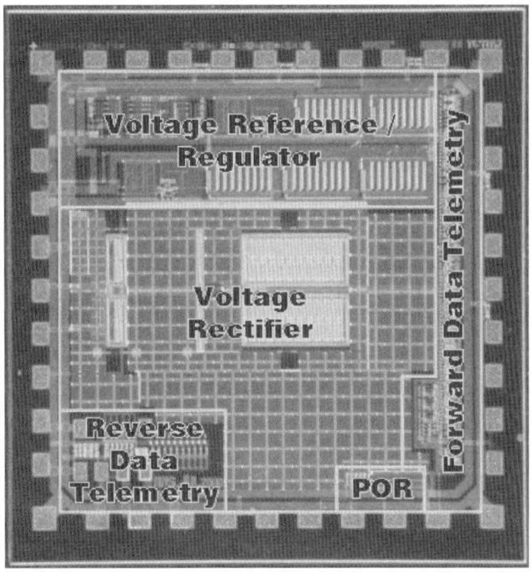

Figure 2.17: An example of the degree of miniaturization that can be presently be attained in wireless circuitry. The General telemetry module developed at University of Michigan's Center for Wireless Integrated MicroSystems provides both two way communication and power regulation. The module measures 2mm x 2mm [120]. *Photo courtesy of A. Sodagar.*

in nature. Such a device would resemble an artificial organ, connected to the blood supply and utilizing the nutrients found there to feed the cellular mechanisms that produce bioelectricity. There have been a few investigations into how bioelectricity might be harnessed for similar applications but (as far as I know) no practical methodologies for its use have yet emerged.

Present interface designs which have no trans-cutaneous (through the skin) connections use a combination of an impanted wireless transmitter/receiver and inductively rechargeable batteries. Systems such as these have proven very useful in research since they permit the subject complete, natural freedom of movement. Wireless systems have been used in both humans and animals since the 1960's and have been refined and miniaturized over the intervening years.

CHAPTER 3

Current Neuroprosthesis Research

"The technology for nonsensory communication between brains and computers through intact skin is already at our fingertips, and its consequences are difficult to predict. In the past the progress of civilization has tremendously magnified the power of our senses, muscles, and skills. Now we are adding a new dimension: the direct interface between brains and machines."

- Prof. José M.R. Delgado, MD/DSc [39, p. 95] -

Neuroprosthetics represent the amalgamation of many technologies and disciplines to create systems that, for example, permit the direct connection of a prosthetic arm to the brain. At the time of this writing, there are a large number of groups conducting neuroprosthetic research. This work is most often done in conjunction with work on neural interfacing since you need the interface for the prosthesis function. This chapter features highlights from some of those groups but should in no way be construed as the final word on the field as that changes from year to year. I have included contact information for the labs mentioned in this and the previous chapter at the end of the book. If you are interested by what you read here you should follow up by visiting the lab's web site or contacting them directly for more information. In the previous chapter, I discussed the various means by which the interface between the excitable tissue (muscle, nerve or brain) and a machine may be accomplished. This chapter focuses on current systems which use those interfaces.

3.1 VISION PROSTHESES

The entire process by which the visual world is converted to neural impulses, interpreted by the brain, and integrated into our perceptual experience is not well understood. However, there are parts of this process that are better understood than others and because of ongoing research more is being learned each year. The aim of artificial vision is to replace the damaged portion of the visual system with man-made components thus restoring some form of sight to the blind. There are three points in the human visual system where a neural interface might be established for the purpose of connecting a vision prosthesis: the retina, optic nerve, and the visual areas of the brain; prosthetics applicable to each area are discussed below.

The human eye functions much like a camera; rays of light pass through lenses (the cornea and lens) and strike the photo-sensitive film (the retina). Unlike the film in a camera, the retina is comprised of several interconnected layers of cells. The bottom-most layer contains the light-sensitive photoreceptors which convert an incident image into neural impulses. The layers above the photoreceptor layer (the light passes through the other layers to get to the photoreceptors) contain neurons which enhance/process the image. Thus, the light strikes the photoreceptors and is processed through the other cell layers until finally the neural data is conveyed to the brain via the optic nerve. The region of the retina which contains the highest density of photoreceptors (thus the highest resolution) is called the macula and is of significant importance in vision. The human eye contains two type of photoreceptors: cones which are involved in color vision and rods which are not sensitive to color but are more light sensitive than cones. The macula [leutea] contains a higher density of photoreceptors than the surrounding areas. It appears as a yellow spot on the retina owing to the large number of ganglion cells that can be found there. At the center of the macula is a slight depression called the fovea [centralis] which contains the highest density of cones and is responsible for the central part of our vision. As you move from the center of the visual field toward the periphery, the ratio of rods to cones increases while the overall density of photoreceptors decreases. Thus, we have the best sensitivity to light (night vision) with lower resolution in our periphery but the highest resolution and best color perception in the central part of our vision. This fact can be easily demonstrated in a near-dark situation by looking directly at an object then looking slightly off to the side while keeping the object in view. You will find that the object is more easily visualized by <u>not</u> looking at it directly. In

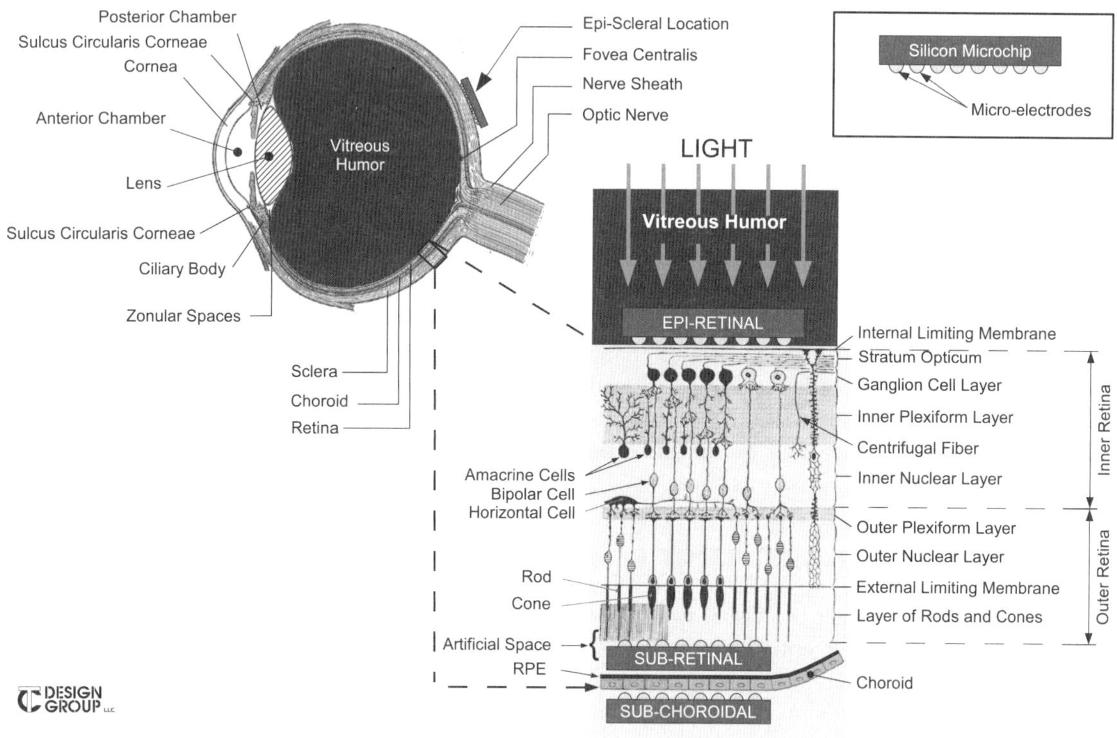

Figure 3.1: Basic anatomy of the eye and approximate locations for retinal implants. *Portions of this illustration were adapted from [7, Figs. 869 and 882].*

some individuals the photoreceptors in the macula begin to degenerate (termed macular degeneration or MD) leading to progressive loss of central vision. In others, the photoreceptors of the periphery (around the macula) begin to degenerate resulting in tunnel vision. This condition is termed Retinitis Pigmentosa (RP) and although it begins in the peripheral vision, it does eventually destroy the photorecptors of the macula as well. The end stage of both MD and RP is blindness. It is presumed that in individuals afflicted with these conditions that the other cell layers of the retina containing neurons remain intact and could continue to relay information to the brain if only there were a source of stimulation. In actuality, once the source of input (the photoreceptors) to the neural layers is lost, changes to those layers begin. Since a neuron's connectivity with other neurons is, in part, determined by its (and its neighbors) inputs, loss of those inputs results in undesirable changes in connectivity. Current research seems to indicate that artificial stimulation of those layers early in the course of the disease may help to maintain functional connectivity within the neural layer, protect the neurons from loss, and perhaps slow the degeneration of the photoreceptors [121, 122]. Since the artificial retina schemes discussed below rely on there being a functioning neural layer, preservation of the neurons is imperative.

3.1.1 SUB-RETINAL IMPLANTS

Sub-retinal implants are positioned in the eye in an artificially created space between the inner retina (containing the neurons) and outer retina (containing the photoreceptors). They have a slightly more complicated implantation procedure than epi-retinal implants but allow the stimulating portion of the retinal prosthesis to be placed in closer proximity to the inputs of the neural cell layer of the retina. Also, as the retina heals from the surgery, the implant is held in place by the surrounding layers of tissue. Thus, external means of restraint like tacks or a bio-adhesive are not needed (see epi-retinal implants below).

The first design for a totally self-contained sub-retinal implant was the Artificial Silicon Retina (ASR) conceived by Dr. Alan Chow in the mid 1980's and patented in August of 1989 [123]. The technology was further developed by Chow et al. and was the subject of numerous peer-reviewed articles [124, 125, 126, 127] and subsequent patents. The device consisted of a small array (2mm disc x 25 microns thick) of 5,000 micro-photodiodes which directly convert light into electrical pulses for stimulating existing retinal cells (Fig. 3.2). The early work on the feasibility of sub-retinal electrical stimulation and the refinement of the ASR design was carried out by Dr. Chow while at Loyola University of Chicago and his brother Vincent Chow, an Electrical Engineer [128]. In these experiments, strip electrodes were placed in the subretinal space of three adult Dutch Belted rabbits. Each electrode was connected to an external photodiode that, when illuminated, produced a current sufficient to stimulate the neural tissue of the retina. The results of these early experiments demonstrated that stimulation in the subretinal space could produce a response in the visual system. Furthermore, it provided support to the, "...idea that implantation of a photodiode based device into the subretinal space may provide a possible route for the restoration of visual function in patients blinded by photoreceptor degeneration" [128, p. 15]. Dr. Chow's work continued next in cats where the first prototype of the ASR microchip was tested. These experiments demonstrated that the microchip-based device would produce a response to light that could be measured in the neural tissue of the retina by recording electrodes placed on the eye. Some of the implants were allowed to remain in the cats and functioned for over three years, demonstrating the durability of the technology. Around the time of these experiments Dr. Chow and others formed Optobionics Corporation to further develop and commercialize the technology.

In January of 2000, the ASR entered clinical trials to assess the safety and feasibility of the implant. As part of this study, a total of 6 human patients suffering from retinitis pigmentosa were implanted. Since the photodiodes comprising the ASR were sensitive to both visible and infra-red light and the human eye is sensitive to only visible light, this provided a way to verify function of the ASR. That is, if a subject could perceive the sensation of light when the researcher illuminates the eye with an infra-red light source then the perception must be due to stimulation of the retinal cells by the ASR. Having the device implanted in human subjects who could directly report their experiences allowed the group to collect more definitive data on the functioning of the implant as a visual prosthesis.

In initial testing following implantation, four of the six patients reported the sensation of seeing light in response to infrared illumination of the ASR. This response slowly exhausted and became less noticeable by some patients in subsequent months. It was believed that certain characteristics of the

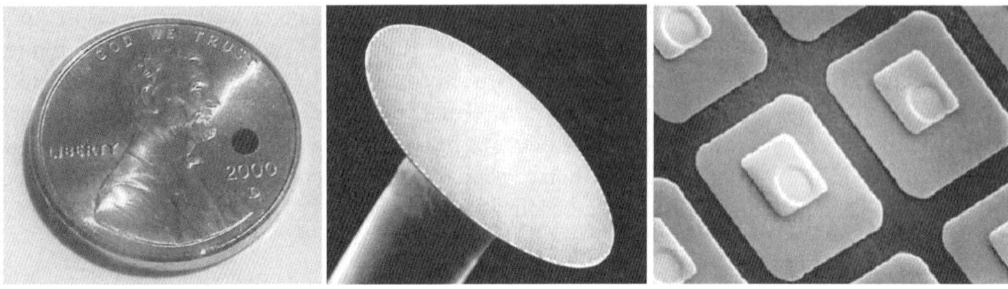

Figure 3.2: The Artificial silicon retina (ASR) pioneered by Dr. Alan Chow, MD and Vincent Chow, BSEE. The model shown measures 2 mm in diameter and 25 μm thick. It contains approximately 5,000 microphotodiode pixels each measuring 20 x 20 μm square. Left: ASR microchip shown atop a penny; Center: EM of the ASR microchip; Right: EM of the surface of the ASR microchip showing detail of pixels. The electrodes are the light colored, dimpled rectangles which are fabricated on top of the photodiodes (larger rectangles) [127]. *Photos courtesy of A. Chow.*

stimulation pulse caused undesirable changes to the tissue surrounding the ASR electrodes. This resulted in the stimulation current no longer being adequate to produce a response in the retinal cells. However, as the patients were monitored over the next 18 months they all showed remarkable improvement in vision, even in areas of the retina some distance from the implant. In an article published by the group in 2004, they indicated that, "Visual function improvements occurred in all patients and included unexpected vision improvements in retinal areas distant from the implant" [127, p. 469]. This led the group to the conclusion that, "...the observation of retinal visual improvement in areas far from the implant site suggests a possible generalized neurotrophic-type rescue effect on the damaged retina caused by the presence of the ASR" [127, p. 460]. In other words, the presence of the ASR in the sub-retinal space itself might be stimulating the damaged cells to produce neurotrophins (chemicals that stimulate growth in neural tissue) which lead to the rescue and repair of some of the retinal cells. Also, it has been shown by other researchers that certain types of electrical fields can affect cell growth, nerve repair, and retinal tissue preservation [129, 122]. This may mean that even though the microchip wasn't generating enough current to stimulate the retinal cells, it could be producing an electric field that had a positive influence on repair.

The data from the clinical trial complicated the picture of the efficacy of the ASR since it could not be clearly determined if the improvement was due to some neurotrophic rescue effect or a low-level electric field generated by the ASR or both. This result prompted the group to launch another study, this time in RCS Rats[1]. The results of the study suggested that in the rat, "...subretinal electrical stimulation provides temporary preservation of retinal function..." and "...implantation of an active or inactive device into the subretinal space causes morphological preservation of photoreceptors..."[121]. Later work [130] published in 2007 by a group working closely with Dr. Chow concluded that direct activation of the retina by the ASR induces activity in the superior colliculus[2] of RCS rats; indicating that it was indeed functioning as a visual prosthesis. Furthermore, the study demonstrated that there was, "... evidence for

[1]A breed of rat that develops a condition similar to RP in humans.
[2]An area of the brain responsible for orienting the eyes toward an intended visual target.

implant induced neurotrophic effects as a consequence of both it's presence and its activity in the retina." These findings put to rest the question of the ability of the ASR to function as a retinal prosthesis and indicate that it has additional efficacy as a means to slow or partially reverse photoreceptor degeneration.

The visionary work of Dr. Chow and his associates has inspired subsequent work by others on retinal prosthetics and demonstrated the safety, feasibility and durability of sub-retinal implants. While the story of the ASR continues to be written the work thus far bodes well for the treatment of photoreceptor degeneration in humans and it seems to only be a matter of time before a totally self-contained retinal prosthesis, powered only by incident light, will be realized.

Eberhart Zrenner, MD of the University Eye Hospital of Tübingen (Germany) and his colleagues are building on the work of Dr. Chow and others to design an externally powered, sub-retinal implant. The device is being developed in conjunction with his spin-off company Retinal Implant, AG[3]. The implant is a square microchip approximately 3mm x 3mm and about 50μm thick. It is fabricated on a flexable polyimide substrate and contains 1500 photocells connected to a 4 x 4 array of stimulating electrodes. The entire device is encapsulated in silicone elastomer. Unlike the ASR, the photocells do not produce the current used to stimulate the retinal tissue. Instead they gate the external power which is used for stimulation. This scheme insures that even in low light there will be sufficient power available for stimulation. Beginning in 2005, seven totally blind patients have undergone surgery for implantation of the Retina Implant device with promising results. Much was learned from these early human trials including several important technical modifications that needed to be made to the device. All patients were able to detect the location of light sources and sometimes identify simple shapes when presented with high foreground-to-background contrast (e.g., white on black) [131].

3.1.2 EPI-RETINAL IMPLANTS

Second Sight Medical Products, Inc.[4] was founded by Alfred Mann in 1998 to develop implants designed to help those blinded by retinal degeneration. Their first product, the Argus 16, entered clinical trials in February 2002 and to date a total of six subjects have been implanted with the technology. This experimental device consists of a 4 x 4 epi-retinal (Fig. 3.1)[5] array of electrodes that are attached directly to the retina using a tack (Fig. 3.3). The microelectrode array is connected to the external system via a flexible cable. When used in conjunction with a small video camera (usually mounted on a pair of glasses) and an external processing system, it provides a rudimentary form of sight to the implanted subjects [132, 133, 134]. The next generation of the device, the Argus II, will contain 60 electrodes which will provide the subject with much higher resolution images. The Argus II entered clinical trials at four hospitals in the U.S. in Jan 2007 and was expanded to three European sites (London, Switzerland, and France) in early 2008. At the time of this writing, Second Sight is the only manufacturer with an actively powered permanently implantable retinal prosthesis under clinical study in the United States. The Argus II is the highest electrode count epi-retinal device being made anywhere in the world.

Intelligent Medical Implants[6] of Switzerland is developing a similar device called the Learning Retina Implant System. The device is fabricated on a flexible polymer substrate and is connected to a wireless telemetry/power module. The external receiver/transmitter and camera is mounted on the earpiece of a pair of glasses. The first chronic study of this device began in the fourth quarter of 2005. Results included the identification of simple patterns and the location of light sources. The second chronic

[3] `http://www.eye-chip.com/`
[4] `http://www.2-sight.com/`
[5] Epi-Retinal implants sit on the surface of the retina and stimulate the underlying cells.
[6] `http://www.intmedimplants.de/`

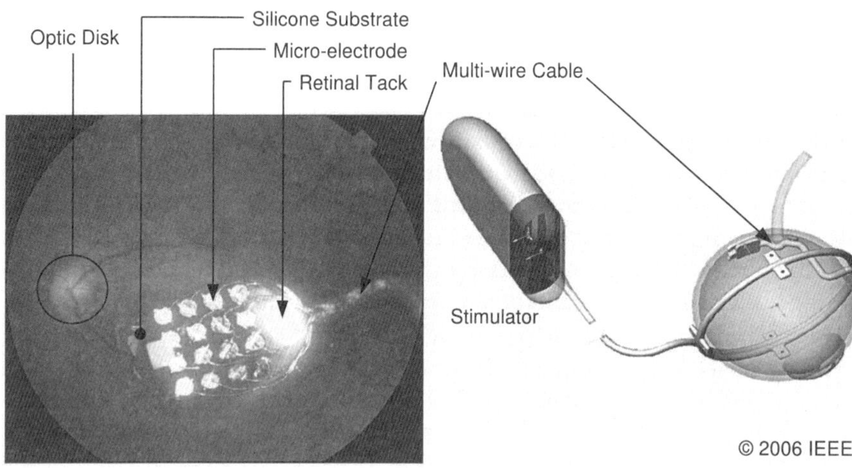

Figure 3.3: The Argus I epi-retinal implant. Left, a photomicrograph of the Argus array *in-situ*. Right, illustration of the components of the Argus system showing the external stimulation unit, multi-wire cable and implanted components [135]. *Reprinted with permission of IEEE EMBS.*

study is on-going and involves the use of the full unit with its wearable camera and pocket sized processing unit.

3.2 COCHLEAR (HEARING) IMPLANTS

Cochlear implant development began in the 1950's when researchers demonstrated that stimulating the auditory nerve could produce auditory sensations. In the sixties scientists learned that stimulation of specific nerve fibers in the auditory nerve via the cochlea was needed to reproduce sound. During this era, three patients were implanted with experimental devices with positive results. Research continued to rapidly advance during the period of the 1970's through the 1980's. In December 1984, cochlear implants were granted FDA approval for nonexperimental implantation into adults. Now systems are FDA approved, safe, and effective. Over 60,000 recipients including children and adults have received these devices to have their hearing at least partially restored.

There is considerable debate in the deaf community regarding the ethics of cochlear implants. The primary argument centers around the deaf communities objection to the mainstream view that deafness is a disability and that deaf people need to be "fixed" through cochlear implants or other means. They argue that there is little or no understanding in the general population that deaf individuals can lead successful, well-adjusted lives without such implants. The counter argument is that while deaf individuals can lead successful lives within the deaf community, it can be very challenging when the individual tries to leave that community and live in the world at large. There have been many tools developed over the years to assist the deaf and hard of hearing: Braille, closed captioning, text-telephones, telecommunications relay services, hearing aids, etc. The cochlear implant is the newest of these tools but is fundamentally different because it is invasive and actually attempts to restore hearing itself rather than providing a "work-around." However, it should be noted that the current technology does not restore hearing per-se,

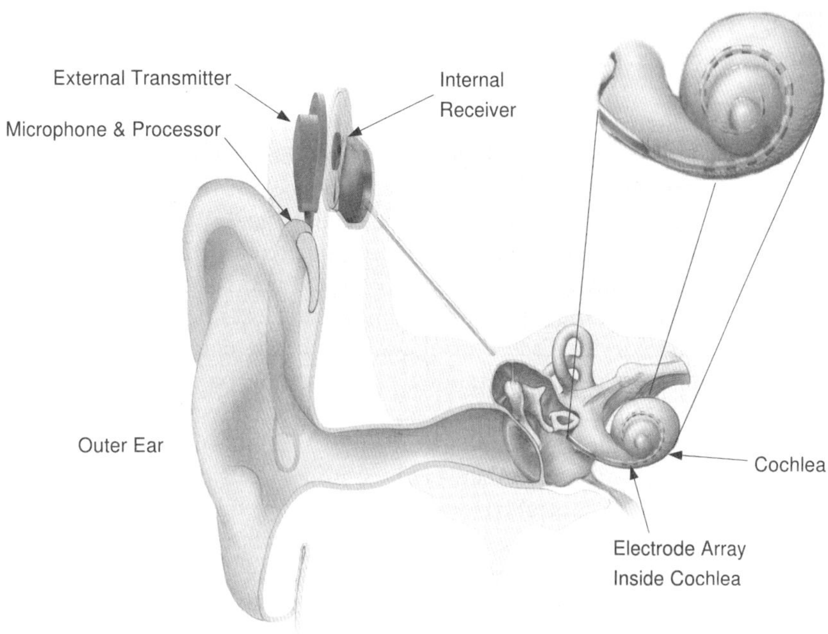

Figure 3.4: Illustration showing the key parts of a cochlear implant and its placement in the inner ear. *Adapted from an illustration by NIH Medical Arts and provided by the NIDCD* [136].

rather a hearing-like sensation described as an electrical buzzing. A further ethical complication is that the implants work best when they are implanted in the very young, ideally around two years of age before language acquisition has begun. This means that the decision to implant is in the hands of the parents who are likely not deaf and have little or no exposure to the deaf community. Thus, the decision to implant is often made without an understanding of what the deaf community has to offer and with probable bias against any option other than the attempt to restore the child's hearing. Like sign language and braille, it requires extensive training to learn to interpret the information from the implant in a useful way and not all individuals are equally successful at this task. Regardless of which side of the ethical argument you find yourself on, cochlear implant technology continues to evolve and will (probably) be able to restore full, "natural" hearing at some point in the not-too-distant future. Despite its controversy and, for the moment, its shortcomings it still provides another tool in the arsenal to help level the playing field between those who can hear and those who cannot. The National Association for the Deaf has conditionally endorsed cochlear implants as one choice among many choices of assistive tehcnologies for the deaf and hard of hearing.

3.3 COGNITIVE-BASED NEURAL PROSTHETICS: THE RP-2009 INITIATIVE

In the previous chapter, I discussed some of the neural interface technologies used to connect neural prostheses to the nervous system. Cognitive-based neural prosthetics are experimental systems designed to utilize the electrical activity in the brain (via a neural interface) for the purpose of controlling a prosthesis. There are a number of systems under development whose goal is to restore function to amputees, paraplegics, and quadraplegics. One of the most advanced (that I know of) and certainly one of the most ambitious, is DARPA's Revolutionizing Prosthetics (RP)—2009 initiative.

Limb prosthetics have been around probably as long a humans, from simple wooden arms, hooks, or peg-legs to the next-generation device discussed below. Advances in materials, robotics, neural interfaces, and computers have enabled the development of natural looking prosthetics with functionality that approaches that of the natural limb. At DARPATech 2005 (the Defense Advanced Research Projects Administration's annual research meeting), Col. Geoff Ling, MD, outlined his vision to restore function to combat upper limb amputees. He stated that: "This is not a far-off dream... In 4 years, we anticipate having a prosthetic arm that will be controlled identically to the way that we control our biological arms." This large, collaborative program is being carried out at over 30 sites throughout the U.S., Canada, and Europe including the Johns Hopkins University Applied Physics Laboratory (APL) in Laurel, Maryland, which is the project lead. The goal of the project is to have a completed, functioning prototype of a neural-interfaced, prosthetic arm by the end of 2009. Dr. Brett Giroir, Director of DARPA's Defense Sciences

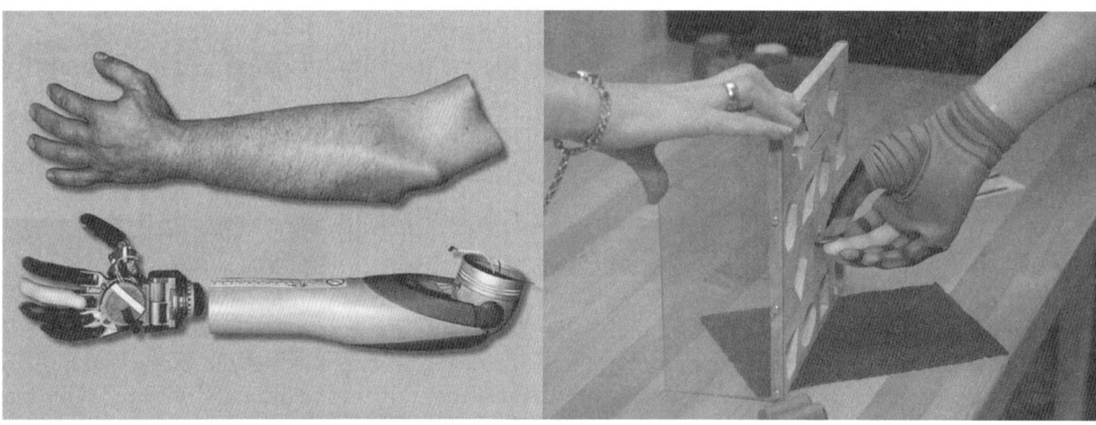

Figure 3.5: DARPA's RP-2009 initiative promises to revolutionize prosthetic limbs for soldiers returning from war; Left: Prototype I of the DARPA arm (below) shown with cosmesis (above); Right: Demonstration of hand dexterity by one of the patients in the study. *Left photo courtesy of the Rehabilitation Institute of Chicago and the Johns Hopkins University APL; Right photo courtesy of the Johns Hopkins University APL.*

Office, stated at the DARPA 25th Systems and Technology Symposium in 2007 that, "...if warfighters could play the piano before their amputation, they would again play the piano with their new prostheses." While Giroir's vision may seem more science fiction than fact, the RP-2009 team is turning it into reality.

The term "degrees of freedom" or DOF when applied to artificial limbs refers to how many ways the limb can move (a simple hinge has 1 degree of freedom, a universal joint has 2). While the degree of movement (DOM) is the number of axes on which movement can occur relative to a reference plane. Thus, a limb may have a higher number for the DOF than the DOM since some of the movements (relative to a given reference plane) permitted by the joints are redundant. For example, how many different ways can you position your arm which result in the fingers of your hand touching your shoulder? Because of its high number of DOF, the joints of the arm permit several possibilities of positioning to move the hand along the same axis of motion.

The prototype-1 system utilizes a Cobot drive [137, 138] located in the forearm to control the movements of the fingers, hand, and wrist. It provides 25+ degrees of freedom and 18 degrees of movement, making it a very good approximation of the natural limb. This prototype provided a testbed for many aspects of the design, construction, testing, and actual use of the arm using control signals derived from muscles. The first prototype did not incorporate direct neural control and sensory feedback. Instead, it is controlled using skin electrodes placed over the pectoral muscle on the same side of the body as the prosthesis. Following the loss of the natural limb the amputee undergoes a surgical procedure which subdivides the pectoral muscle. The nerves that once served the arm are re-directed into the newly created subdivisions. Activation (contraction) of different parts of the muscle control different movements of the prosthesis. After some training the amputee is able to use the prosthesis much as they would their natural limb. Unfortunately, this method of control is not adequate for the prototype-2 system which will have more degrees of movement than the first prototype.

The prototype-2 system has 22 degrees of movement and incorporates over 80 tactile elements for sensing pressure, vibration, heat flow, and proprioception. The prosthesis will be connected to the users nervous system via either a nerve implant or a brain implant. The hand is capable of producing a 9 kg pinch and a 32 kg grasp which is more than most people can manage with their natural limb. The neural interface component is being developed at the University of Utah and will be the "hair brush" type of array discussed earlier. These arrays have been used in humans as part of several other experimental systems. The Utah slant array technology will be used to create an interface to one of the large nerves carrying signals to the arm. After training, nerve activity corresponding to arm movement would be translated into control signals to guide the prosthesis or a computer generated limb in a virtual environment. The real advantage of the neural interface, versus using surface electrodes, will be that the system can stimulate nerve fibers to create the sensation of touch, hot, cold, or even pain in response to the limb's tactile sensors. A direct prosthesis-to-brain interface using the Utah electrode array would be utilized when a suitable nerve is not accessible or there is damage to the spinal cord. In this case, the brain would be directly connected to the computer system controlling the prosthesis thus bypassing the damaged or nonfunctional area.

3.4 CARDIAC PACEMAKERS

Cardiac pacemakers have been in use for almost 50 years and have helped countless people, but are often overlooked in the discussion of neuroprosthetics. Defibrillators, also known as implantable cardioverter-defibrillators or ICDs, are not only capable of pacing the heart[7] but can intervene in cases of abnormal heart rhythm. When an abnormal heart rhythm occurs the heart can no longer effectively pump blood and may result in death. Defibrillators can deliver an electric shock allowing the heart to hopefully resume a normal rhythm. This neural-prosthetic technology has had a profound impact on human lives.

[7] Pacing refers to the process of artificially regulating the rate and intensity of the heart's contractions.

CHAPTER 4

Conclusion

"The only way of discovering the limits of the possible is to venture a little way past them into the impossible."

- Author C. Clarke -

Before diving into the the ethical questions of the technology itself, I would like to say a few words about how bio-technology makes it's way from the drawing board to clinical application in humans. When a device that is destined for use in humans is developed it is required to undergo extensive testing for safety, tolerability, efficacy, and feasibility. Safety testing a material destined for implantation in the human body often begins by placing it in a cell culture. If the culture dies or becomes abnormal in some way then the material is probably not bio-compatible. Next, a device might be fabricated from the material then implanted in a living specimen. While an animal that is low on the phylogenetic tree (e.g., a crustacean) can sometimes be used, work that is destined for use in humans will often begin in a mammalian species. Rats and mice are commonly used for experiments because of their low cost of procurement and maintenance. Also, there are a number of breeds which exhibit disorders similar to those found in humans making them a good testbed for experimental therapies. Usually extensive testing is done in rats or mice prior to moving to a higher mammal or directly to humans. The use of non-human primates (NHPs) in research is highly controversial, they are expensive to procure, and their cost of maintenance is high. In general, if a device is safe enough to use in a NHP, chances are that it is safe enough to begin limited use in humans. However, it is important to understand that this is not always the case and it is still necessary in certain instances to use NHPs for testing. In these cases there are very strict guidelines regulating the care and use of these intelligent animals. These guidelines dictate practices that are very similar to those required for human patients.

Often there are several cycles of design-test-modify-test before obtaining a prototype device which is a candidate for use in humans. During the prototyping period implanted devices are tested in animals. Researchers work with their institution's Internal Review Board (IRB) which oversees the use of animals (and humans) in research and insures that researchers follow established ethical practices and humane use guidelines. When the prototyping period is complete and the device is deemed ready for human testing, the next step may be to enter pre-clinical trials with a limited number (usually less than 10) of human subjects or apply to the country's regulatory body[1] to begin limited experimental use in humans. If approved, the devices are implanted in a small number (~20–80) of the most needy patients. For example, experimental cancer drugs are often tested only in patients with terminal prognoses for whom conventional treatments have failed or cannot be used. The primary aims of early testing are usually limited to assessing the safety and tolerability of the device. This type of clinical testing of a treatment on a small group of patients to determine safety and side effects is referred to as a *Phase I* clinical trial[2].

Following successful outcome of Phase I trials the device may be tested on a larger number of patients (~100–300) to further evaluate its safety and determine its effectiveness. This is referred to as *Phase II* clinical trials. *Phase III* testing is performed in potentially thousands of patients to further assess

[1] E.g., The Food and Drug Administration in the United States.
[2] `http://www.clinicaltrials.gov/ct2/info/understand`

effectiveness, side effects, and outcomes across a large patient population. Finally, *Phase IV* trials use post-marketing studies to evaluate the device's risks, benefits, and additional uses that were not part of the original study. At each step of the testing process of the device information is collected which will guide designers in making refinements and clinicians in its optimal use. There is also a fast-track approval process for treatments which demonstrate safety and high efficacy. The process of developing a device for use in humans takes years of work and the cooperative efforts of a large number of professionals and patients alike. Bringing a device to market represents a huge investment both in terms of taxpayer dollars in grants that funded the basic science and in terms of money spent by companies to fund clinical testing and product development.

Ideally, we would like to have a highly detailed computer simulation of the human body which could be tweaked many different ways to test the devices under different use scenarios, individual physiological variations, etc. Unfortunately, computers are not nearly powerful enough, our understanding of the body woefully inadequate and there does not exist a modelling language or simulation environment that could even remotely handle the complexity. As technology and our knowledge of the body's systems advances this will change. There are fledgling efforts underway to simulate various body systems in software. Such simulations help us to validate our present knowledge (when the simulation performs as the natural system would), show us our errors and underscore our lack of understanding. I have no doubt that some day a large part of the safety testing of drugs and implantable materials/devices will be performed on computers. Unfortunately, the present state-of-the-art leaves us with the imperfect and often ethically complicated alternative of using intact, animal systems that closely resemble our own for the early stages of testing.

From a financial standpoint, animal testing is sub-optimal since animals do not respond to drugs or implanted materials exactly as humans would. Companies sink billions of dollars into the development of new drugs only to find out in clinical trials that what seemed to work in animals either doesn't work in humans or produces undesirable side-effects. From a materials standpoint, animals heal better, have stronger resistance to infection, and tolerate implants better than humans. So, if an implant is not tolerated well in the animal it will almost certainly not be tolerated in the human. However, just because it is well tolerated in the animal does not necessarily mean the same for the human patient. There has been tremendous change in the past 50 years in the way animals are used in research thanks to various efforts to increase awareness and advance the guidelines for their ethical use. Hopefully, with the continuing advances in cell culture, computers, and the fledgling field of organ culture, we will gradually be able to replace the use of animals with engineered human tissue. Assessing drugs and devices in human tissues grown in the lab should prove far more useful and cost effective. Not that ethical choices should be reduced to financial terms but for large companies what is best for the bottom line is definitely a factor in the decision.

I believe that the ethical use of animals in research is necessary at this point in time, but only insofar as alternatives are not available. I feel that there should be a concerted effort to develop systems which provide a better testbed for drugs and devices than what is currently available using animal models. In my opinion, the best place for animals is in their natural habitat, not caged in the lab or zoo. While there are certain organizations and individuals who believe that *any* use of animals is unacceptable and the research which uses them should not be sanctioned, one must consider the alternatives. If we were to halt the use of animals in research we would be left with testing using cell culture. While cell culture can give us a rough idea regarding biocompatibility of implanted materials, it cannot test immune responses, complex tissue reactions, glial scaring and a host of other things that can happen in the intact animal. Thus, we would have one of two choices:

1. Enter clinical trials with minimally tested drugs and devices, using humans as the initial test subjects.

2. Halt research into the development of new therapies.

In my opinion both of these scenarios are ethically unacceptable from a human standpoint. The use of humans as guinea pigs is not only unethical but illegal and the cessation of research is unethical from the standpoint of alleviating human suffering. I am reminded of a scenario posed in class by one of my ethics professors. It goes something like this:

> You are stranded in a raft adrift on the ocean. Among your number are yourself, two other humans, and a pet chimpanzee. You have water which you collect when it rains, but no food and no way to obtain food from the sea. You have been stranded now for nearly two weeks and you are all on the brink of starvation. It is believed that you will be rescued within the next week if you can just hold out a little longer. *What would you do to survive?*

I was surprised by how much debate this scenario elicited (an entire 50 minute class period as I recall) over what seemed to be a relatively straightforward survival decision. So, I posed the same scenario to my 4 year old son and 7 year daughter who arrived at the response, "eat the monkey" in 15 seconds and 2 minutes, respectively. The debate over the use of animals in research, though analogous, is not quite so cut-and-dried. While I understand the viewpoint of animal-rights activists and respect the strength of their convictions, I cannot in good conscience agree with all of their tenets. I feel that the goal of alleviating human suffering trumps the rights of animals. So, for the time being, we have three alternatives: to use humans as the proverbial "guinea pigs," to stop the development of new therapies, or to use animals in our research. In other words, it's the choice of the lesser of three evils and I choose the latter.

4.1 NEURAL INTERFACING ETHICS

As mentioned earlier, a nervous system provides organisms with the unparalleled ability to adapt to their dynamic environments; something that computers are a long way from achieving. This will almost certainly change given enough time and the continued progress of technology. For the time being, we are very far from the day when human-like machines will be able to match the adaptability and creativity of their creators. Some might say that the best way to achieve this end is the merger of humans with the machines they might create, to others this is a horrific notion that is best left on the pages of science fiction.

While restoration of function is a noble, high-priority research goal, one cannot ignore the potential of this technology for another more far-reaching application, the augmentation of human abilities. In this context the realization of a complete human-machine neural interface could potentially be one of the most important events in human evolution since our distant ancestors evolved opposable digits. Neural prosthetics could not only allow the crippled to walk, the blind to see, and the deaf to hear, but could eventually augment/expand the human mind in ways we can only begin to imagine. Direct Neural Interface (DNI) augmented intelligence would permit users to instantly tap into global databases containing the store of human knowledge. Abstract ideas and thoughts for which language is an inadequate medium could be communicated directly, mind to mind, through a wireless link. The precise number-crunching abilities of computers could be merged with the creative ability of the human mind permitting users to perform complex calculations "in their head." An architect could, for example, imagine a new design for a bridge while simultaneously calculating load factors, material costs, etc. A musician could imagine a musical composition and then through the use of computer interfaced instruments have it performed. Technical ability and manual dexterity would no longer limit the ability of a person to bring to fruition whatever their mind can imagine. Computer-augmented senses could allow scientists and engineers to directly experience with their own virtual eyes any portion of the electromagnetic spectrum for which a sensing modality exists. Astronomers could peer into the furthest reaches of the universe and doctors could look into patients using their own cybernetically enhanced vision.

Immersive virtual reality (IVR) refers to totally computer generated environments in which the user has complete motor control and sensory feedback just as in the real world. The current technologies of body position tracking devices, headgear, haptic feedback, etc., which attempt to approximate IVR are as primitive as stone knives when compared to DNI. Neural interface based IVR could put the user in a virtual environment that seems as real as the physical world. One of the many promises of DNI-IVR is to bring people living in physically distant locations together in virtual environments where they can interact with the same richness of experience as a face to face meeting. Thus, a person living in the Australian outback could instantly be in the heart of downtown "Cyberopolis" where they could interact with millions of other connected users from around the world. Likewise, virtual schools and colleges could bridge the education gap between rural and urban areas. People might go to work by "plugging in" and joining their co-workers in a virtual office.

The word "avatar" is taken from Hindu mythology and refers to the reincarnation of an advanced or enlightened soul on earth. In virtual reality parlance an avatar is a 3D construct which is used to represent a user in a virtual environment and may or may not look anything like the user it serves. When a robot is used as an avatar it becomes a tool for immersive tele-presence. Through the use of DNI-IVR and an avatar a person could be instantly anywhere in the real or virtual world. For example, a business person in Nome, Alaska would rent an android avatar in Beijing, China. Then, using DNI-IVR, have his or her sensorium mapped onto the body of the avatar in order to conduct business in the real world far away from their physical location. Likewise, an astronaut could explore another world without ever leaving the safety of earth. Others who want to share the experience could participate as observers, feeling everything the astronaut feels but not controlling the avatar. Recordings of the event could be stored allowing future generations to directly re-experience it in a first person sense. Other events could likewise be recorded allowing people to later experience history through the eyes and minds of the people who were there.

A complete interface to the brain might permit a user to copy part of their knowledge/personality into an avatar which then could act as an agent working on the user's behalf to perform some duty. While performing its task the avatar would maintain communication with the user and exchange information as needed. This is analogous to a computer program forking processes then communicating with those processes through a pipe[3]. Using a similar mechanism, and prior to the expiration of a user's physical body, their knowledge might be uploaded into a global repository. This global repository might exist as a type of generalized expert system or "counsel of elders" composed of millions of incorporeal intelligences which could be consulted whenever their particular expertise is needed. Future scientific advances might allow a new body to be grown for an incorporeal intelligence which could then be downloaded into it to become corporeal again.

The use of neuroprosthetics beyond what is strictly required for rehabilitation is fraught with both practical and ethical concerns. To use a more familiar example, cosmetic surgery is widely available to anyone with the motivation and financial resources. One can safely and easily be shaped and enhanced to look younger, slimmer, or better endowed. With a bit more effort and financial commitment, one can even change their sex or become something in between male and female. All of this has come about in the last 50 years (give-or-take) with the advance in surgical techniques, implantable materials and varying degrees of social acceptance. What will the next 50 bring? Aside from the regulation of medical standards

[3] A communication channel used to exchange data.

of practice, we do not prohibit people from altering their bodies. We would not say to an actor or actress "you can't have any more cosmetic surgery because you look too good and are taking jobs from other less attractive professionals." Likewise, what would be the rationale for limiting and/or preventing people from enhancing themselves with safe and effective cybernetic implants should they one day be available?

Sci-fi author and father of the "Cyberpunk" genre, William Gibson, wrote extensively on a fictitious, future society whose denizens often spent a great deal of their time connected to the vast virtual world he termed "Cyberspace." As with many things which promise great benefits there is the potential for abuse. The dark side of DNI-IVR is that it could become a "drug" more powerful and addictive than any present day pharmaceutical. Persons looking for an escape could become lost in a world of anything-goes synthetic reality that is seemingly as real as the physical world. A whole new class of mental disorders could arise from people who spend too much time in IVR environments, who are unable to cope with what is presented in those environments, and/or lose the ability to distinguish them from the "real" world.

A user plugged into a DNI which completely disconnects them from all the physical sensation of their real body could die of thirst or hunger if nutrition is not provided. Furthermore, our sensory and motor systems don't respond on nanosecond time scales. When we first learn to grasp an object, crawl, and eventually walk, our brains have learned all these tasks while compensating for built in delays, errors, or irregularities unique to a person's own natural system. A user's brain directly connected to a virtual environment could experience instantaneous feedback without the delays imposed by their natural sensory/motor systems. New forms of physical disability could result from the disparity between the responsiveness of virtual limbs and the users own limbs. A user who spends too much time in an ultra-responsive virtual world could find their coordination, motor control, and sensory perception severely affected when they are "unplugged." A person could literally forget how to use their own limbs or process their own natural sensory input.

The ultimate neural interface would seamlessly integrate cybernetic and biological components. With advances in the fields of molecular biology, genetics, and nanotechnology, new types of interfacing paradigms may become possible. For example, presently there is an extensive effort to create materials which can be implanted in the body without fear of adverse or undesirable reactions. In the future it may be possible to genetically modify a person's body to more readily accept implants. Alternately, cultured implants containing both electronic and neuronal components would be grown in labs, then placed in the brain or body where they continue to grow and make connections. Such an implant might be custom made for a specific individual with the biological components created from their own cells or DNA. Further genetic modification could enhance strength, intelligence, or possibly re-engineer the brain to make it physically easier to access for cybernetic implantation. Might a human so radically modified by genetic alterations and implants be considered a new species, perhaps even a next step in our evolution? What would this mean for the future of the human race, particularly those who have not been modified?

Thus far we have discussed human enhancement using neuroprosthetics, but what of using biological elements to enhance computers? Before continuing I should point out that there many technical hurdles to overcome with using living neural networks as components in computers. However, based upon some of the work discussed earlier, in the *distant* future it may become possible to build arrays of biological processing elements containing specially grown neural networks. These networks could be incorporated into electronic/photonic systems, endowing them with more human-like abilities. These biological components could provide the essential front-end for natural language recognition, speech synthesis, vision processing, and other functions where humans and animals presently outperform computers. It is possible that the next paradigm shift in computer technology will be "wet." That is, computer systems of the future might consist of organic and in-organic components with the organic components sustained by a nutrient and oxygen-rich solution not unlike blood. Since biological systems are (at present) far smaller and more portable than computers, organo-synthetic processors could become the technology that paves the way for the "android" so often written about in science fiction. That is, a true synthetic being comprised of biological and technological components; having blood, needing air to breathe, and consuming nutrients to power its systems.

This brings to mind the larger ethical question which is, "would such systems have rights?" Would they be subject to the same treatment before the law as humans? If not, what would be the rationale for denying them these protections? If one were to use the "sentience" argument then how do we judge sentience? If the argument simply is, "machines don't have rights," then aren't we just machines of a different type? We, having evolved to our present state through natural processes, instead of being built by another race of beings. Fortunately for society and jurisprudence, such systems are unlikely to exist even in our great grandchildren's lifetimes, but the ethical dilemma posed by the emergence of engineered beings makes for lively academic debate as well as interesting fiction.

Science is many years away from producing the kind of technologies discussed above (Tab. 4.1), however, given enough time and effort it is easy to imagine a day when such things will become possible. There are those like Dr. José Delgado who believe the type of neural interface hoped for by the transhumanists[4] and feared by the mind-control, conspiracy theorists is simply not realizable [139]. Others believe that producing technologies such as hybrid silicon-neuron computer components is so fraught with technical challenges that it may never be practical. While few would argue against using neuroprosthetics to help the disabled, what about the question of enhancing otherwise normal individuals? Banning particular uses of cybernetic enhancement or lines or research through legislation will prove no more effective than bans on cloning or stem cell research. Someone, somewhere in the world will conduct the research, develop the product and perform the surgery because there is money to be made. In these situations the law of supply and demand has more effect than the laws of some body of government.

The difficulty is that once the body of knowledge pertaining to a particular technology becomes known it cannot be withdrawn again from public knowledge. Once that knowledge is realized in a deliverable therapy, device, or treatment no one government will be able to regulate its use. If you can't obtain the desired enhancement in the U.S. then one could simply travel to Thailand, China, Europe or elsewhere. At that point the ethical discussion will be rendered largely academic and we will find ourselves dealing with the ramifications of a two tier society, those who are enhanced and those who are not. The proper way to address the ethical issues of neural interfacing is to begin discussion well ahead of the availability of such technologies, bringing as many scientists, business people, and countries to the table as possible. A strong international regulatory structure may help insure the safety of therapies that are developed around neural interfacing technologies and moderate potential deleterious social effects of wide spread non-therapeutic use.

The ethical and social ramifications of producing a society of cybernetically or genetically enhanced individuals should not be ignored. Almost certainly these types of enhancements will be expensive (especially at first) and as such only available to a wealthy few. Employer sponsored cybernetic enhancement could become both a benefit and condition of employment. Such "benefits" could also become a means for unscrupulous employers to make it difficult for employees to quit; especially if those enhancements are proprietary and not easily undone. Normal, unenhanced individuals who cannot afford, or do not want to be upgraded could find themselves relegated to menial jobs; becoming the untouchables of a technologically imposed caste system.

The formation of a total, direct neural interface between humans and machines holds both great promise and potentially dire consequences. However, neural interfacing is one of a few technologies that has the potential to completely change the face of human society by bridging the language, cultural, edu-

[4] http://www.transhumanism.org

Table 4.1: Some examples of neural-based technologies discussed in this book and "guesses" as to when they might be fully realized (assuming continuing research efforts)

	CA	CT	NF	FF	?
Auditory Neuroprosthesis (Ear)			•		
Whole-Eye Neuroprosthesis				•	
† Artificial Retina		•			
Olfactory Neuroprosthesis (Nose)			•		
‡ Cochlear Implant	•				
† EEG Computer Interface	•				
† DNI Computer Interface		•			
DNI Based Remote Telepresence			•		
Total DNI based Immersive VR				•	
Mind Reading (EEG or some other technology)					•
Mind/Memory Uploading/Downloading					•
Mind Linking & Cognitive Expansion					•
Hybrid Silicon-neural Computer Components				•	

CA=Commercially Available Now, CT=Presently in Clinical Trials,

NF=Near Future (50-100 yrs), FF=Far Future (100-200 yrs), ?= Very distant future, if ever.

† Indicates a technology in the early stages of development

‡ Indicates an advanced but still evolving technology

cational, and physical barriers that have traditionally divided humans along so many lines. The realization of the technology needed to accomplish a DNI has begun. So as the Cyborg makes its way from the pages of science fiction into the pages of scientific journals it will be up to both present and future generations to decide how humanity takes what could amount to a significant, next step in its evolution.

CHAPTER 5

Resources for Students

This Chapter contains an abbreviated listing of groups conducting neural-interfacing related research. This listing is based on the work discussed in this book and *is not* a comprehensive listing of all the work occurring in the field. Of the work discussed, only the contact information for those researchers known to be conducting neural engineering research at the time of this writing are included. A more complete and up-to-date list of researchers, universities, groups, and companies is available at `http://www.neuropunk.org`.

If you are contemplating a career in neural engineering there are several things you may want to consider. First, as I have already mentioned, the field is highly interdisciplinary so a broad education is often advantageous. A general knowledge of the gross anatomy, cytoarchitecture, and functional organization of the nervous system (well beyond what is covered in this book) is essential for persons interested in pursuing a career in Neural Engineering. At the very least you should have some familiarity with the organization of the nervous system and what functions its various structures are thought to perform. A good, senior- or graduate-level neuroscience course should fulfill this requirement. Other knowledge/skills you might want to accumulate during your undergraduate or graduate training may include:

- Technical Writing and the ability to illustrate your ideas: an absolute must—you have to be able to communicate your ideas and concepts with other professionals. If drawing is not your thing, learn to use Computer Aided Drafting and Design (CADD) software.
- Computer programming: at the very minimum learn C/C++ or Java, Neuro-prosthetics are run by software, even if you are not the one writing the program a knowledge of computer programming will help you in your interactions with other team members.
- Electronics:
  - A working knowledge of electronics and basic principles such as resistance, impedance, and capacitance. This is often taught as part of college physics.
  - The ability to design and fabricate basic analog circuitry (e.g., an amplifier) and basic digital logic circuitry. If your university does not offer this, check the local community college.
- Biology: Freshman- and sophomore-level possibly including cellular/molecular biology.
- Mathematics through introductory Calculus.
- College-level physics.

Remember any advice is just that, advice. Don't take the opinions of one person (including myself) as the final word on what you should be doing. If I had relied on the advice of my high school guidance counselor I would never had gone to college, earned a Ph.D. and gone on to do other things like write this book. Ultimately, it's your decision and you are the one that has to live with the choices you make. To make infomed decisions about your future you need more than just sage advice, you need the one thing you may not have at this point, experience. Many universities and laboratories offer summer internships for high school/college students and lab rotations for graduate students. Internships and rotations will allow you to explore your interests, get a taste of what working in a lab is really like. More to the point, it will put you in contact with those who can best advise you.

In addition to the subjects mentioned above there are several unrelated but important things you should consider. We live in a global economy where modern communication and transportation are shrinking the distances between countries. It is important that you have some knowledge of cultures other than the one you in which you were raised. The best way to learn about other cultures is through foreign language courses and foreign exchange programs. We as Americans tend to have a myopic view of the world since we live in a large, more-or-less homogeneous country which "officially" recognizes only a single language[1]. The world is a large diverse place filled with many different cultures, languages and peoples. Financially, most of the market capitalization is vested in the rest of the world, not the US. Additionally, there are numerous neural engineering research efforts in Europe, Japan, and China. To give youself the most options for having the career of your choice, commit to learning a second language to the point of fluency.

If possible, work some business management courses into your schedule or take the extra time needed to get a minor (or a second degree) in business. In today's environment graduate students become post-docs then become professors only to discover that they now have to run a lab, manage personnel, obtain funding or even spin off businesses from their research. While at the same time trying to publish their work and navigate through a university's tenure and promotion process. They have little or no experience delegating tasks, managing people, writing business proposals, etc. As a student, you only have to worry about completing your own work. Research professionals not only have their own work but often have to manage a team. A good business management curriculum will provide you with the basic skills you need so you won't be starting "cold" when confronted with managing a lab or a business. Furthermore, if you decide to go into industry instead of academia you will find some business management knowledge invaluable.

Completing a degree is not the end of your studies but the beginning. One purpose of college is to transition you from high school where you are told what assignments to complete, how they are to be completed and when they are due, to working and learning independently. Technology is constantly changing, our knowledge of the brain grows and is revised frequently and new tools are always being developed. We scientists must commit to a lifelong journey of self education. College and graduate school are essential stepping stones along the way designed to give you the foundation you need, but the foundation is only the beginning of any well-made structure.

5.1 CONTACT INFORMATION

Please visit the website listed for the professors/research groups prior to contacting them. Many of the groups have information on their site regarding educational opportunities. When making intital contact by e-mail please check the group's homepage(s) for the current address.

[1] Although Spanish is the un-official, second language in the U.S.

5.1.1 JAPAN

Cardiovascular Research Institute

`http://www.ncvc.go.jp/english/res/rese.html`

Dr. Masaru Sugimachi, M.D.
National Cardiovascular Center Research Institute
Department of Cardiovascular Dynamics
5-7-1 Fujishiro-dai
Suita, Osaka 565-8565
Phone:+81 06 6833 5012
Fax: +81 06 6835 5403

Laboratory for Advanced Brain Signal Processing

`http://www.bsp.brain.riken.jp`

Dr. Andrzej Cichocki, Ph.D.
RIKEN Brain Science Institute
Laboratory for Advanced Brain Signal Processing
2-1 Hirosawa
Wako, Saitama 351-0198
Phone:+81 48 467 9668
Fax: +81 48 467 9686

5.1.2 EUROPE

Intelligent Data Analysis Group

`http://ida.first.fraunhofer.de/homepages/ida/`

Prof. Dr. Klaus-Robert Müller, Ph.D.
Kekuléstr. 7
D-12489 Berlin
Germany
Phone: +49 0 30 6392 1860
Fax: +49 0 30 6392 1805

The Fraunhofer Institute for Biomedical Engineering

`http://www.ibmt.fraunhofer.de/fhg/ibmt_en/biomedical_engineering/medical_engineering_neuroprosthetics`

Prof. Dr. Klaus-Peter Hoffmann, Ph.D.
Fraunhofer-Institut für Biomedizinische Tech. IBMT
Dept. of Medical Engineering and Neuroprosthethics
Ensheimer Straße 48
66386 St. Ingbert
Germany
Phone: +49 6894 980 401
Fax: +49 6894 980 400

Institute of Microsystem Technology - IMTEK

`http://www.imtek.uni-freiburg.de/index_en.php`

Prof. Dr. Thomas Stieglitz, Ph.D.
Laboratory for Biomedical Microtechnology
Georges-Koehler-Allee 102
D-79110 Freiburg
Germany
Phone: +49 0 761 203 7471
Fax: +49 0 761 203 7472

Institut de recherche IDIAP

`http://www.idiap.ch/current_projects.php?project=27`

Centre du Parc
Av. des Prês-Beudin 20
Case Postale 592
CH-1920 Martigny
Switzerland
info@idiap.ch
`http://www.idiap.ch/jobs.php`
Phone: +41 27 721 77 11
Fax: +41 27 721 77 12

Max Planck Institute of Biochemistry
http://www.biochem.mpg.de/mnphys/

Prof. Dr. Peter Fromherz, Ph.D.
Max Planck Institute of Biochemistry
Department of Membrane and Neurophysics
Am Klopferspitz 18
D-82152 Martinsried
Germany
Phone: +49 89 8578 2821
Fax: +49 89 8578 2822

Warwick Biosensors Group
http://template.bio.warwick.ac.uk/staff/ndale

Prof. Dr. Nicholas Dale, Ph.D.
University of Warwick
Department of Biological Sciences
Gibbet Hill Road
Coventry CV4 7AL
United Kingdom
Phone: +44 0 24 7652 3729
Fax: +44 0 24 7657 2594

5.1.3 UNITED STATES

The Andersen Lab
http://vis.caltech.edu/

Dr. Richard A. Andersen, Ph.D.
California Institute of Technology
Division of Biology
MC 156-29
Pasadena, CA 91125
Phone: +1 626.395.4951
Fax: +1 626.449.0756

The Applied Physics Laboratory
http://www.jhuapl.edu

Dr. Stuart Harshbarger, Ph.D.
The Johns Hopkins University
Applied Physics Laboratory
11100 Johns Hopkins Rd
Laurel MD 20723-6099
Phone: +1 240 228 5000
Fax: +1 240 228 1093

Biomimetic MicroElectronic Systems Engineering Research Center.
http://bmes-erc.usc.edu

Dr. James D. Weiland, Ph.D.
Biomimetic MicroElectronic Systems Eng. Research Ctr.
1450 San Pablo Street
DVRC 130
Los Angeles, CA 90033-1035
Phone: +1 323 442 6786
Fax: +1 323 442 6790

Biomimetic MicroElectronic Systems Engineering Research Center
http://bmes-erc.usc.edu

Dr. Gerald E. Loeb, M.D.
University of Southern California
Denney Research Building B12, MC1112
1042 Downey Way
Los Angeles, CA 90089
Phone: +1 213 821 1112
Fax: +1 213 821 1120

Brain-Computer Interface Research Project
http://www.bciresearch.org/

Dr. Jonathan R. Wolpaw, M.D., Ph.D.
New York State Department of Health
Wadsworth Center
P.O. Box 509
Albany, New York 12201-0509
http://www.wadsworth.org/docs/education.shtml
Phone: +1 518 473 3631
Fax: +1 518 486 4910

Center for Adaptive Neural Systems
`http://ans.asu.edu/`

Dr. Ranu Jung, Ph.D.
Dr. Jimmy Abbas, Ph.D.
Arizona State University
Fulton School of Engineering
PO Box 874404 — Phone: +1 480 965 9489
Tempe, AZ 85287-4404 — Fax: +1 480 727 7624

Center for Microelectrode Technology
`http://www.mc.uky.edu/cenmet`

Dr. Greg Gerhardt, Ph.D.
University of Kentucky
306 Davis Mills Bldg.
800 Rose Street — Phone: +1 859 323 4531
Lexington, KY 40536-0098 — Fax: +1 859 257 5310

Center for Neural Engineering
`http://www.neural-prosthesis.com/`

Dr. Theodore W. Berger, Ph.D. — **info@neural-prosthesis.com**
Department of Biomedical Engineering
164 Denney Research Building
University of Southern California — Phone: +1 213 740 8017
Los Angeles, CA 90089-1111 — Fax: +1 213 821 2368

Center for Wireless Integrated Microsystems ERC
`http://www.wimserc.org`

Dr. Kensall D. Wise
The University of Michigan
Dept. of Electrical Eng. and Computer Science
2401 EECS Bldg.
1301 Beal Avenue — Phone: +1 734 764 3346
Ann Arbor, MI 48109-2122 — Fax: +1 734 763 9324

Computational NeuroEngineering Laboratory
`http://www.cnel.ufl.edu/`

Dr. José Carlos Príncipe, Ph.D.
University of Florida
Electrical & Computer Engineering Department
NEB 451, Bldg #33 — Phone: +1 352 392 2662
Gainesville, FL 32611 — Fax: +1 352 392 0044

The Donoghue Laboratory
`http://donoghue.neuro.brown.edu`

Dr. John Donoghue, Ph.D.
Brown University
Department of Neuroscience
185 Meeting Street — Phone: +1 401 863 9524
Providence, RI 02912 — Fax: +1 401-863-1074

Neural Engineering Center
`http://nec.cwru.edu/`

Dr. Dominique M. Durand, Ph.D.
Case Western Reserve University
Department of Biomedical Engineering
Wickenden 309 — Phone: +1 216 368 3974
Cleveland, OH 44106 — Fax: +1 216 368 4872

Neural Engineering Laboratory
`http://nelab.engin.umich.edu`

Dr. Daryl R. Kipke, Ph.D.
1101 Beal Ave.
Ann Arbor MI 48109-2110
Phone: +1 734 764 3716
Fax: +1 734 647 4834

Neuroprosthetics Research Group
`http://nrg.mbi.ufl.edu/`

Dr. Justin C. Sanchez, Ph.D.
University of Florida
Department of Pediatrics
P.O. Box 100296, JHMHC
Gainesville, Florida, 32610
Phone: +1 352 846 2180
Fax: +1 352 392 2515

Nicolelis Lab
`http://www.nicolelislab.net/NLnet_Load.html`

Dr. Miguel A. L. Nicolelis, M.D., Ph.D.
Duke University Medical Center
Department of Neurobiology
Box 3209, DUMC
Durham, NC 27710
Phone: +1 919 684 4580
Fax: +1 919 684 4431

The Pine Lab
`http://www.its.caltech.edu/$\sim$pinelab`

Dr. Jerome Pine, Ph.D.
California Institute of Technology
Dept. of Physics, Mathematics and Astronomy
MC 256-48
Pasadena, CA 91125
Phone: +1 626 395 6677

Wightman Research Group
`http://www.chem.unc.edu/people/faculty/wightmanrm/rmwgroup/`

Dr. R. Mark Wightman, Ph.D.
The University of North Carolina at Chapel Hill
Department of Chemistry
Campus Box 3290
Chapel Hill, NC 27599-3290
Phone: +1 919 962 1472
Fax: +1 919 962 2388

5.2 LINKS TO FREE RESOURCES ON THE INTERNET

Table 5.1:

Location	Name and URL
US	**NINDS - National Institute for Neurological Disorders and Stroke** Neural Prosthesis Program, `http://www.ninds.nih.gov/funding/research/npp/index.htm`
US	**National Science Foundation** `http://www.nsf.gov/funding/pgm_summ.jsp?pims_id=5147\&from=fund`
US	**Neuropunk.Org - Online Encyclopedia of Neural Engineering** `http://www.neuropunk.org`
DE	**BCI-Info - International Portal for Brain Computer Interfaces** `http://www.bci-info.tugraz.at`
US	**Scholarpedia.Org - The free peer reviewed encyclopedia** `http://www.scholarpedia.org`

5.3 FURTHER READING

- *From Neuron to Brain*, John G. Nicholls et al., ISBN: 0-87893-439-1.
- *Brain-Machine Interface Engineering*, by José Principe and Justin Sanchez, ISBN: 9781559820349.
- *Toward Replacement Parts for the Brain*, by Theodore Berger and Dennis Glanzman, ISBN: 0262025779.
- *Cybernetics: or Control an Communication in the Animal and Machine*, by Norbert Wiener, ISBN: 026273009-X.
- *Electrochemical Methods: Fundamentals and Applications*, by Allen J. Bard and Larry R. Faulkner, ISBN: 0-47104-372-9.

ACKNOWLEDGMENTS

I would like to thank the following individuals without whom this book would not have been possible. My wife for her patience and support. Fred Prior, Linda Larson-Prior, and Greg Gerhardt for their guidance and mentoring. Jerry Pine and Peter Huettl for their feedback. Last, but not least, a special thanks to all of those who contributed graphics material.

ABOUT THE AUTHOR

A native of Virginia, Dr. Coates received his undergraduate degree in biology from Virginia Commonwealth University in 1995, his Ph.D. in Neuroscience from the Pennsylvania State University in

2001 [27], and completed postdoctoral work at the University of Kentucky's Center for Microelectrode Technology. He is presently a private research consultant who specializes in issues surrounding neural interface construction. His Company (TC Design Group, LLC) is the sponsor of Neuropunk.Org (http://www.neuropunk.org), a nonprofit educational project dedicated to the free dissemination of quality information relating to neural engineering. When not building Cyborgs he enjoys exploring planet Earth, SCUBA diving, and pursuing his interests in music and art.

Commentary and discussion on this book can be found at:
http://thomcoates.net/cms/index.php?blog=1\&cat=37

Address for Correspondence
Thomas D. Coates, Jr.
PO Box 554
Frederick MD 21705-0554
E-mail: ni-book@thomcoates.net

This book was written entirely using open-source tools running on the Linux operating system.

Bibliography

[1] A. Abbott. In search of the sixth sense. *Nature*, 442(13):125–127, Jul. 2006. DOI: 10.1038/442125a

[2] J. G. Nicholls, R. A. Martin, B. G. Wallace, and P. A. Fuchs. *From Neuron to Brain*, vol. 1. Sinauer Associates, Inc., Sunderland, MA, 4th ed., 2001.

[3] K. H. Pribram. *Brain and Perception: Holonomy and Structure in Figural Processing*. John M Maceachran Memorial Lecture Series. Lawrence Erlbaum, Mahwah, NJ, 1st ed., 1991.

[4] K. H. Pribram. *Rethinking Neural Networks: Quantum Fields and Biological Data*. Lawrence Erlbaum, Mahwah, NJ, 1st ed., 1993.

[5] J. R. Wolpaw, N. Birbaumer, D. J. McFarland, G. Pfurtscheller, and T. M. Vaughan. Brain-computer interfaces for communcation and control. *Clin. Neurophysiol.*, 113:767–791, 2002. DOI: 10.1016/S1388-2457(02)00057-3

[6] The stanford encyclopedia of philosophy. World Wide Web.

[7] H. Gray. *Anatomy of the Human Body*. Henry Gray (1825 - 1861). Lea and Febiger, Philadelphia, 20th ed., 1918.

[8] S. Finger. *Origins of Neuroscience: A History of Explorations into Brain Function*. Oxford University Press, 1st ed., Oct. 11, 2001.

[9] G. Perea and A. Araque. Synaptic information processing by astrocytes. *J. Physiol. Paris*, 99:92–97, 2006. DOI: 10.1016/j.jphysparis.2005.12.003

[10] E. R. Kandel, J. H. Schwartz, and T. M. Jessell. *Principles of Neural Science*. McGraw-Hill Medical, New York, 4th ed., Jan. 2000.

[11] H. B. Barlow. *Sensory Communication*, Chapter 13, pp. 217–234. The MIT Press, Cambridge, MA, 1st ed., 1961. Possible principles underlying the transformations of sensory messages.

[12] F. Rieke, D. Warland, R. de Ruyter van Steveninck, and W. Bialey. *Spikes: Exploring the Neural Code*. The MIT Press, Cambridge, MA, 1st ed., 1997.

[13] E. D. Adrian and Y. Zotterman. The impulses produced by sensory nerve endings, part ii: The response of a single end organ. *J. Physiol.*, 61(2):151–171, Apr. 23, 1926.

[14] D. O. Hebb. *The Organization of Behavior: A Neuropsychological Theory*. John Wiley & Sons Ltd., New York, 1st ed., 1949.

[15] G. J. Stuart and B. Sakmann. Active propagation of somatic action potentials into neocortical pyramidal cell dendrites. *Nature*, 6:69–72, Jan. 1994. DOI: 10.1038/367069a0

[16] H. Markham, M. H. Lübke, M. Frotscher, and B. Sakmann. Regulation of synaptic efficacy by coincidence of postsynaptic aps and epsps. *Science*, 275(5297):213–215, Jan. 10, 1997. DOI: 10.1126/science.275.5297.213

[17] G. J. Stuart, S. G. Spruston, B. Sakmann, and M. Häusser. Action potential initiation and backpropagation in neurons of the mammalian cns. *Trends Neurosci.*, 20(3):125–131, Mar. 1997. DOI: 10.1016/S0166-2236(96)10075-8

[18] I. R. Winship and T. H. Murphy. In vivo calcium imaging reveals functional rewiring of single somatosensory neurons after stroke. *J. Neurosci.*, 28:6592–6606, Jun. 2008. DOI: 10.1523/JNEUROSCI.0622-08.2008

[19] T. J. Sejnowski and C. R. Rosenberg. Parallel networks that learn to pronounce english text. *Complex Syst.*, 1:145–168, 1987.

[20] H. Wang. *Reflections on Kurt Gödel.* MIT Press, Cambridge, MA, 1st ed., Mar. 1990.

[21] S. M. Ulam. Reflections on the brain's attempts to understand itself. *Los Alamos Science*, pp. 283–287, Oct. 1987.

[22] H. Moravec. *Robot: Mere Machine to Trancendent Mind*, vol. 1, Oxford University Press, New York, 1st ed., 1999.

[23] C. A. Skarda and W. J. Freeman. How brains make chaos to make sense of the world. *Behav. and Brain Sci.*, 10:161–195, 1987.

[24] R. Linsker. Local synaptic learning rules suffice to maximize mutual information in a linear network. *Neural Computat.*, 4:691–702, 1992. DOI: 10.1162/neco.1992.4.5.691

[25] A. J. Bell and T. J. Sejnowski. An information-maximization approach to blind separation and blind deconvolution. *Neural Computat.*, 7:1129–1159, 1995. DOI: 10.1162/neco.1995.7.6.1129

[26] C. E. Shannon. A mathematical theory of communication. *Bell Syst. Tech. J.*, 27:379–423, 623–656, Jul. - Oct. 1948. DOI: 10.1145/584091.584093

[27] T. D. Coates. *A Neural Network Based Cybernetic Interface for Identification of Simple Stimuli Based on Electrical Activity in an Intact Whole Nerve.* Ph.D. thesis, Pennsylvania State University, 2001.

[28] P. Pajunen, A. Hyvärinen, and J. Karhunen. Nonlinear blind source separation by self-organizing maps. In *Proc. of the 1996 International Conference on Neural Information Processing (ICONIP'96)*, vol. 2, pp. 1207–1210. Springer-Verlang, Hong-Kong, 1996.

[29] J. Karhunen, E. Oja, L. Wang, R. Vigario, and J. Joutsensalo. A class of neural networks for independent component analysis. *IEEE Trans. Neural Netw.*, 8:486–504, 1997. DOI: 10.1109/72.572090

[30] Y. Tan, J. Wang, and J. M. Zurada. Nonlinear blind source separation using a radial basis function network. *IEEE Trans. Neural Netw.*, 12:124–134, 2001.

[31] C. H. Zheng, D. S. Huang, K. Li, G. Irwin, and Z. L. Sun. MISEP method for postnonlinear blind source separation. *Neural Comput.*, 19:2557–2578, Sept. 2007. DOI: 10.1162/neco.2007.19.9.2557

[32] M. Imai. *Kaizen: The Key To Japan's Competitive Success.* McGraw-Hill/Irwin, 1st ed., Nov. 1, 1986.

[33] N. Wiener. *Cybernetics: or Control and Communication inthe Animal an the Machine.* MIT Press, Cambridge, MA, 2nd ed., 1999.

[34] M. Clynes and N. S. Kline. Cyborgs and space. *Astronautics*, p. 27, Sept. 1960.

[35] D. M. Durand. Editorial by Dominique M. Durand. *J. Neu. Eng.*, 4(4), Dec. 2007.

[36] D. Schmorrow. Augmented cognition: Building cognitively aware computational systems. Technical report, Defense Advanced Research Projects Administration (DARPA), Anahiem, CA, Jul. 30, Aug. 7, 2002.

[37] G. S. Brindley and W. S. Lewin. The sensations produced by electrical stimulation of the visual cortex. *J. Physiol.*, 196:479–493, 1968.

[38] J. M. R. Delgado, V. Mark, W. Sweet, F. Ervin, G. Weiss, G. Bach-y Rita, and R. Hagiwara. Intracerebral radio stimulation and recording in completely free patients. *J. Nerv. Ment. Dis.*, 147:329–340, 1968.

[39] J. M. R. Delgado. *Physical Control of the Mind*, vol. 1. Harper and Row, New York, 1st ed., 1969.

[40] J. M. R. Delgado. Personal communication, May 2008.

[41] O. Foerster. Beitriige zur pathophysiologie der sehbahn und der sehsphare. *J. Psychol. Neurol.*, 39:463–485, 1929.

[42] F. Krause and H. Schum. *Neue deutsche Chirurgie*, vol. 49a, pp. 482–486. Enke Publishers, Stuttgart, Germany, 1931.

[43] G. Holmes. The organization of the visual cortex in man. *Proc. Roy. Soc. London Ser. B (Biol.)*, 132:348–361, 1945. DOI: 10.1098/rspb.1945.0002

[44] G. S. Brindley, P. E. K. Donaldson, M. D. Falconer, and D. N. Rushton. The extent of the region of occipital cortex that when stimulated gives phosphenes fixed in the visual field. *J. Physiol.*, 225:57–58, 1972.

[45] G. S. Brindley. *Sensory effects of electrical stimulation of the visual and paravisual cortex in man*, volume vnI/3B of *Handbook of Sensory Physiology*. Springer-Verlag, New York, 1973.

[46] G. S. Brindley and P. E. K. Donaldson. Striate cortex stimulator. *US Patent #3,699,970*, Oct. 24, 1972.

[47] W. H. Dobelle and M. G. Mladejovsky. Phosphenes produced by electrical stimulation of human occipital cortex, and their application to the development of a prosthesis for the blind. *J. Physiol.*, 243(2):553–576, Dec. 1974.

[48] E. M. Schmidt, M. J. Bak, F. T. Hambrecht, C. V. Kufta, D. K. O'Rourke, and P. Vallabhanath. Feasibility of a visual prosthesis for the blind based on intracorticai microstimulation of the visual cortex. *Brain*, 119:507–522, 1996. DOI: 10.1093/brain/119.2.507

[49] W. H. Dobelle. Artificial vision for the blind by connecting a television camera to the visual cortex. *ASAIO J.*, 46(1):3–9, Jan. - Feb. 2000. DOI: 10.1097/00002480-200001000-00002

[50] R. A. Normann, E. M. Maynard, P. J. Rousche, and D. J. Warren. A neural interface for a cortical vision prosthesis. *Vision Res.*, 39(15):2577–2587, Jul. 1999. DOI: 10.1016/S0042-6989(99)00040-1

[51] E. J. Tehovnik and W. M. Slocum. Phosphene induction by microstimulation of macaque v1. *Brain Res. Rev.*, 53(2):337–343, Feb. 2007. DOI: 10.1016/j.brainresrev.2006.11.001

[52] J. J. Struijk, M. Thomsen, J. O. Larsen, and T. Sinkjaer. Cuff electrodes for long-term recording of natural sensory information. *IEEE Eng. Med. Biol. Mag.*, 18(3):91–98, May - Jun. 1999. DOI: 10.1109/51.765194

[53] G. G. Naples, J. T. Mortimer, A. Scheiner, and J. D. Sweeney. A spiral nerve cuff electrode for peripheral nerve stimulation. *IEEE Trans. Biomed. Eng.*, 35(11):905–916, Nov. 1988. DOI: 10.1109/10.8670

[54] M. Sahin, M. A. Haxhiu, D. M. Durand, and I. A. Dreshaj. Spiral nerve cuff electrode for recordings of respiratory output. *J. Appl. Physiol.*, 83(1):317–322, Jul. 1997.

[55] C. Veraart, C. Raftopoulos, J. T. Mortimer, J. Delbeke, D. Pins, G. Michaux, A. Vanlierde, S. Parrini, and M. C. Wanet-Defalque. Visual sensations produced by optic nerve stimulation using an implanted self-sizing spiral cuff electrode. *Brain Res.*, 813(1):181–186, Nov. 30, 1998. DOI: 10.1016/S0006-8993(98)00977-9

[56] J. Perez-Orive and D. M. Durand. Modeling study of peripheral nerve recording selectivity. *IEEE Trans. Rehabil. Eng.*, 8(3):320–329, Sept. 2000.

[57] D. J. Tyler and D. M. Durand. Functionally selective peripheral nerve stimulation with a flat interface nerve electrode. *IEEE Trans. Neural Syst. Rehabil. Eng.*, 10(4):294–303, Dec. 2002. DOI: 10.1109/TNSRE.2002.806840

[58] D. K. Leventhal and D. M. Durand. Subfascicle stimulation selectivity with the flat interface nerve electrode. *Ann. Biomed. Eng.*, 31(6):643–652, Jun. 2003. DOI: 10.1114/1.1569266

[59] A. F. Marks. Bullfrog nerve regeneration into porous implants. *Anatom. Rec.*, 162:226, 1969.

[60] R. Llinas, C. Nicholson, and K. Johnson. *Implantable monolithic wafer recording electrodes for neurophysiology*, Chapter 7, pp. 105–111. Brain Unit Activity During Behavior. Thomas, Springfield, IL, 1st ed., 1973.

[61] G. T. Kovacs, C. W. Storment, and J. M. Rosen. Regeneration microelectrode array for peripheral nerve recording and stimulation. *IEEE Trans. Biomed. Eng.*, 39(9):893–902, 1992. DOI: 10.1109/10.256422

[62] G. T. Kovacs, C. W. Storment, M. Halks-Miller, C. R. Belczynski, C. C. Della Santina, E. R. Lewis, and N. I. Maluf. Silicon-substrate microelectrode arrays for parallel recording of neural activity in peripheral and cranial nerves. *IEEE Trans. Biomed. Eng.*, 41(6):567–77, 1994. DOI: 10.1109/10.293244

[63] A. F. Mensinger, D. J. Anderson, J. Buchko, M. A. Johnson, D. C. Martin, P. A. Tresco, R. B. Silver, and S. M. Highstein. Chronic recording of regenerating viiith nerve axons with a sieve electrode. *J. Neurophys.*, 83(1):611–616, 2000.

[64] A. Ramachandran, M. Schuettler, N. Lago, T. Doerge, K. P. Koch, X. Navarro, K. P. Hoffmann, and T. Stieglitz. Design, in vitro and in vivo assessment of a multi-channel sieve electrode with integrated multiplexer. *J. Neural Eng.*, 3:114–124, 2006. DOI: 10.1088/1741-2560/3/2/005

[65] T. Kawada, C. Zheng, S. Tanabe, T. Uemura, K. Sunagawa, and M. Sugimachi. A sieve electrode as a potential autonomic neural interface for bionic medicine. *Proc. IEEE Eng. Med. Biol. Soc. 26th Annu. Int. Conf.*, 2:4318–4321, Sept. 2004.

[66] P. K. Campbell, K. E. Jones, and R. A. Normann. A 100 electrode intracortical array: structural variability. *Biomed. Sci. Instrum.*, 26:161–165, 1990.

[67] K. E. Jones, P. K. Campbell, and R. A. Normann. A glass/silicon composite intracortical electrode array. *Ann. Biomed. Eng.*, 20(4):423–437, 1992. DOI: 10.1007/BF02368134

[68] E. M. Maynard, C. T. Nordhausen, and R. A. Normann. The utah intracortical electrode array: a recording structure for potential brain-computer interfaces. *Electroencephalogr. Clin. Neurophysiol.*, 102(3):228–239, Mar. 1997.

[69] C. T. Nordhausen, E. M. Maynard, and R. A. Normann. Single unit recording capabilities of a 100 microelectrode array. *Brain Res.*, 726(1-2):129–140, Jul. 8, 1996. DOI: 10.1016/0006-8993(96)00321-6

[70] P. J. Rousche and R. A. Normann. Chronic intracortical microstimulation (icms) of cat sensory cortex using the utah intracortical electrode array. *IEEE Trans. Rehabil. Eng.*, 7(1):56–68, Mar. 1999. DOI: 10.1109/86.750552

[71] E. M. Maynard, N. G. Hatsopoulos, C. L. Ojakangas, B. D. Acuna, J. N. Sanes, R. A. Normann, and J. P. Donoghue. Neuronal interactions improve cortical population coding of movement direction. *J. Neurosci.*, 19(18):8083–8093, Sept. 15, 1999.

[72] J. M. R. Delgado. Permanent implantation of multilead electrodes in the brain. *Yale J. Biol. Med.*, 24:351–358, 1952.

[73] C. Palmer. A microwire technique for recording single neurons in unrestrained animals. *Brain Res. Bull.*, 3(3):285–289, May - Jun. 1978. DOI: 10.1016/0361-9230(78)90129-6

[74] G. W. Westby and H. Wang. A floating microwire technique for multichannel chronic neural recording and stimulation in the awake freely moving rat. *J. Neurosci. Meth.*, 76(2):123–133, Oct. 3, 1997. DOI: 10.1016/S0165-0270(97)00088-5

[75] J. G. Cham, E. A. Branchaud, Z. Nenadic, B. Greger, and J. W. Burdick. Semi-chronic motorized microdrive and control algorithm for autonomously isolating and maintaining optimal extracellular action potentials. *J. Neurophys.*, 93(1):570–579, Jan. 2005.

[76] C. Pang, Y. C. Tai, J. W. Burdick, and R. A. Andersen. Electrolysis-based diaphragm actuators. *Nanotechnology*, 17:S64–S68, 2006. DOI: 10.1088/0957-4484/17/4/010

[77] WIMS-ERC. Wims erc annual report 2007. Technical report, University of Michigan, Ann Arbor, 2007.

[78] K. D. Wise, A. M. Sodagar, Y Yao, M. N. Gulari, G. E. Perlin, and K. Najafi. Microelectrodes, microelectronics, and implantable neural microsystems. *Proc. IEEE, Special Issue on Implantable Biomimetic Microelectronics Systems*, Jul. 2008.

[79] G. E. Perlin and K. D. Wise. A compact for three-dimensional neural microelectrode arrays. *IEEE Int. Engr. In Med. Biol. Conf. Proc.*, Aug. 2008.

[80] J. P. Seymour and D. R. Kipke. Fabrication of polymer neural probes with sub-cellular features for reduced tissue encapsulation. *Conf. Proc. IEEE Eng. Med. Biol. Soc.*, 1:4606–4609, 2006. DOI: 10.1109/IEMBS.2006.260528

[81] Y. Kato, I. Saito, T. Hoshino, T. Suzuki, and K. Mabuchi. Preliminary study of multichannel flexible neural probes coated with hybrid biodegradable polymer. *Conf. Proc. IEEE Eng. Med. Biol. Soc.*, 1:660–663, 2006. DOI: 10.1109/IEMBS.2006.259978

[82] J. Wang, M. N. Gulari, and K. D. Wise. A parylene-silicon cochlear electrode array with integrated position sensors. *Conf. Proc. IEEE Eng. Med. Biol. Soc.*, 1:3170–3173, 2006.

[83] S. Kisban, S. Herwik, K. Seidl, B. Rubehn, A. Jezzini, M. A. Umiltà, L. Fogassi, T. Stieglitz, O. Paul, and P. Ruther. Microprobe array with low impedance electrodes and highly flexible polyimide cables for acute neural recording. *Conf. Proc. IEEE Eng. Med. Biol. Soc.*, 2007:175–178, 2007.

[84] E. Patrick, M. Ordonez, N. Alba, J. C. Sanchez, and T. Nishida. Design and fabrication of a flexible substrate microelectrode array for brain machine interfaces. *Conf. Proc. IEEE Eng. Med. Biol. Soc.*, 1:2966–2969, 2006. DOI: 10.1109/IEMBS.2006.260581

[85] P. T. Kissinger, J. B. Hart, and R. N. Adams. Voltammetry in brain tissue. *Brain Res.*, 55(1):209–213, 1973. DOI: 10.1016/0006-8993(73)90503-9

[86] A. J. Bard and L. R. Faulkner. *Electrochemical Methods: Fundamentals and Applications*. John Wiley & Sons Ltd., New York, 2nd ed., Dec. 18, 2000.

[87] P. Fromherz, H. Schaden, and T. Vetter. Guided outgrowth of leech neurons in culture. *Neurosci. Lett.*, 129:77–80, 1991. DOI: 10.1016/0304-3940(91)90724-8

[88] P. Fromherz and H. Schaden. Defined neuronal arborizations by guided outgrowth of leech neurons in culture. *Eur. J. Neurosc.*, 6:1500–1504, 1994. DOI: 10.1111/j.1460-9568.1994.tb01011.x

[89] A. A. Prinz and P. Fromherz. Electrical synapses by guided growth of cultured neurons from the snail lymnaea stagnalis. *Biol. Cybern.*, 82:L1–L5, 2000. DOI: 10.1007/PL00007969

[90] P. Fromherz and V. Gaede. Exclusive-OR function of single arborized neuron. *Biol. Cybern.*, 69:337–344, 1993. DOI: 10.1007/BF00203130

[91] P. Fromherz and A. Stett. Silicon-neuron junction: Capacitive stimulation of an individual neuron on a silicon chip. *Phys. Rev. Lett.*, 75(8):1670–1673, Aug. 21, 1995. DOI: 10.1103/PhysRevLett.75.1670

[92] A. Stett, B. Müller, and P. Fromherz. Two-way silicon-neuron interface by electrical induction. *Phys. Rev. E.*, 55(2):1779–1782, Feb. 1997. DOI: 10.1103/PhysRevE.55.1779

[93] G. Zeck and P. Fromherz. Noninvasive neuroelectronic interfacing with synaptically connected snail neurons immobilized on a semiconductor chip. *Proc. Natl. Acad. Sci. U S A*, 98(18):10457–10462, Aug. 28, 2001. DOI: 10.1073/pnas.181348698

[94] M. Voelker and P. Fromherz. Signal transmission from individual mammalian nerve cell to field-effect transistor. *Small*, 1(2):206–210, Feb. 2005. DOI: 10.1002/smll.200400077

[95] M. Hutzler, A. Lambacher, B. Eversmann, M. Jenkner, R. Thewes, and P. Fromherz. High-resolution multitransistor array recording of electrical field potentials in cultured brain slices. *J. Neurophysiol.*, 96:1638–1645, Sept. 2006. DOI: 10.1152/jn.00347.2006

[96] J. Pine. Recording action potentials from cultured neurons with extracellular microcircuit electrodes. *J. Neurosci. Meth.*, 2(1):19–31, Feb. 1980. DOI: 10.1016/0165-0270(80)90042-4

[97] W. G. Regehr, J. Pine, and D. B. Rutledge. A long-term in vitro silicon-based microelectrode-neuron connection. *IEEE Trans. Biomed. Eng.*, 35(12):1023–1032, Dec. 1988. DOI: 10.1109/10.8687

[98] Scanning Microscopy International. *Neural Transplant Staining with DiI and Vital Imaging by 2-Photon Laser-Scanning Microscopy*, P.O. Box 66507, AMF O'Hare (Chicago), IL 60666, Aug. 1995.

[99] J. Pine, M.P. Maher, S. Potter, Y.C. Tai, S. Tatic-Lucic, J. Wright, G. Buzsaki, and A. Bragin. A cultured neuron probe. *Proc. IEEE-EMBS 18th Ann. Mtg.*, 5(31):2133–2135, Oct. 31, Nov. 3, 1996.

[100] M. P. Maher, J. Pine, J. Wright, and Y. C. Tai. The neurochip: a new multielectrode device for stimulating and recording from cultured neurons. *J. Neurosci. Meth.*, 87:45–56, 1999. DOI: 10.1016/S0165-0270(98)00156-3

[101] A. Tooker, J. Erickson, G. Chow, Y. C. Tai, and J. Pine. Parylene neurocages for electrical stimulation on silicon and glass substrates. *28th Annu. Conf. IEEE Eng. Med. Biol. Soc.*, pp. 4322–4325, New York, 2006. IEEE-EMBS.

[102] J. R. Moffitt, Y. R. Chemla, S. B. Smith, and C. Bustamante. Recent advances in optical tweezers. *Annu. Rev. Biochem.*, Feb. 28, 2008. DOI: 10.1146/annurev.biochem.77.043007.090225

[103] H. Zhang and K. K. Lui. Optical tweezers for single cells. *J. R. Soc. Interface*, Apr. 1, 2008. DOI: 10.1098/rsif.2008.0052

[104] L. Mitchem and J. P. Reid. Optical manipulation and characterisation of aerosol particles using a single-beam gradient force optical trap. *Chem. Soc. Rev.*, 37(4):756–69, Apr. 2008. DOI: 10.1039/b609713h

[105] J. D. Wen, L. Lancaster, C. Hodges, A. C. Zeri, S. H. Yoshimura, H. F. Noller, C. Bustamante, and I. Tinoco. Following translation by single ribosomes one codon at a time. *Nature*, 452(7187):598–603, Apr. 3, 2008. DOI: 10.1038/nature06716

[106] M. L. Hayashi, S. Y. Choi, B. S. Rao, H. Y. Jung, H. K. Lee, D. Zhang, S. Chattarji, A. Kirkwood, and S. Tonegawa. Altered cortical synaptic morphology and impaired memory consolidation in forebrain- specific dominant-negative PAK transgenic mice. *Neuron*, 42:773–787, Jun. 2004. DOI: 10.1016/j.neuron.2004.05.003

[107] G. E. Loeb and R. Davoodi. The functional reanimation of paralyzed limbs. *IEEE Eng. Med. Biol. Mag.*, 24(5):45–51, 2005. DOI: 10.1109/MEMB.2005.1511499

[108] E. Chaplin. Functional neuromuscular stimulation for mobility in people with spinal cord injuries. The parastep i system. *J. Spinal Cord. Med.*, 19(2):99–105, Apr. 1996.

[109] G. E. Loeb, F. J. Richmond, and L. L. Baker. The bion devices: injectable interfaces with peripheral nerves and muscles. *Neurosurg. Focus*, 20(5):1–9 (E2), May 15, 2006.

[110] E. Niedermeyer and F. Lopes da Silva. *Electroencephalography: Basic Principles, Clinical Applications, and Related Fields.* Lippincott Williams & Wilkins, New York, 1st ed., 2004.

[111] G. Schalk, D. J. McFarland, T. Hinterberger, N. Birbaumer, and J.R. Wolpaw. BCI2000: a general-purpose brain-computer interface (BCI) system. *IEEE Trans. Biomed. Eng.*, 51:1034–1043, Jun. 2004. DOI: 10.1109/TBME.2004.827072

[112] K. R. Müller, M. Tangermann, G. Dornhege, M. Krauledat, G. Curio, and B. Blankertz. Machine learning for real-time single-trial EEG-analysis: from brain-computer interfacing to mental state monitoring. *J. Neurosci. Meth.*, 167:82–90, Jan. 2008. DOI: 10.1016/j.jneumeth.2007.09.022

[113] B. Blankertz, G. Dornhege, M. Krauledat, K. R. Müller, and G. Curio. The non-invasive Berlin Brain-Computer Interface: fast acquisition of effective performance in untrained subjects. *Neuroimage*, 37:539–550, Aug. 2007. DOI: 10.1016/j.neuroimage.2007.01.051

[114] S. Waldert, H. Preissl, E. Demandt, C. Braun, N. Birbaumer, A. Aertsen, and C. Mehring. Hand movement direction decoded from MEG and EEG. *J. Neurosci.*, 28:1000–1008, Jan. 2008. DOI: 10.1523/JNEUROSCI.5171-07.2008

[115] J.d.e.l. R. Millán, F. Renkens, J. Mouriño, and W. Gerstner. Noninvasive brain-actuated control of a mobile robot by human EEG. *IEEE Trans. Biomed. Eng.*, 51:1026–1033, Jun. 2004.

[116] Y. Washizawa, Y. Yamashita, T. Tanaka, and A. Cichocki. Extraction of steady state visually evoked potential signal and estimation of distribution map from EEG data. *Conf. Proc. IEEE Eng. Med. Biol. Soc.*, 2007:5449–5452, 2007.

[117] S. M. Sukthankar and N. P. Reddy. Virtual reality of "squeezing" using hand EMG controlled finite element models. In *Proc. of the 15th Annual International Conference of the IEEE*, Engineering in Medicine and Biology Society, pp. 972–972, 1993.

[118] Y. Barniv. Using electromyography to predict head motion for virtual reality. Technical report, National Aeronautics and Space Administration, Ames Research Center, 2001.

[119] L. J. Trejo, K. R. Wheeler, and C. C. Jorgensen. Multimodal neuroelectric interface development. Technical report, National Aeronautics and Space Administration, Ames Research Center, 2001.

[120] A. M. Sodagar, K. D. Wise, and K. Najafi. General telemetry chip for power and bidirectional data telemetry in implantable microsystems. *WIMS World*, p. 4, Summer 2007.

[121] M. T. Pardue, M. J. Phillips, H. Yin, B. D. Sippy, S. Webb-Wood, A. Y. Chow, and S. L. Ball. Neuroprotective effect of subretinal implants in the RCS rat. *Inv. Opthalmo. Vis. Sci.*, 46(2):674–682, Feb. 2005. DOI: 10.1167/iovs.04-0515

[122] T. Morimoto, T. Fujikado, J-S Choi, H. Kanda, T. Miyoshi, Y. Fukuda, and Y. Tano. Transcorneal electrical stimulation promotes the survival of photoreceptors and preserves retinal function in royal college of surgeons rats. *Inv. Opthalmo. Vis. Sci.*, 48(10):4725–4732, Oct. 2007. DOI: 10.1167/iovs.06-1404

[123] A. Y. Chow. Artificial retina device. *US Patent #5,016,633*, Aug. 8, 1989.

[124] G. Peyman, A. Y. Chow, C. Liang, V. Y. Chow, J. I. Perlman, and N. S. Peachey. Subretinal semiconductor microphotodiode array. *Ophthalmic Surg. Lasers*, 29:234–241, Mar. 1998.

[125] M. T. Pardue, S. L. Ball, J. R. Hetling, V. Y. Chow, A. Y. Chow, and N. S. Peachey. Visual evoked potentials to infrared stimulation in normal cats and rats. *Doc. Ophthalmol.*, 103:155–162, Sept. 2001. DOI: 10.1023/A:1012202410144

[126] A. Y. Chow, M. T. Pardue, J. I. Perlman, S. L. Ball, V. Y. Chow, J. R. Hetling, G. A. Peyman, C. Liang, E. B. Stubbs, and N. S. Peachey. Subretinal implantation of semiconductor-based photodiodes: durability of novel implant designs. *J. Rehabil. Res. Dev.*, 39:313–321, 2002.

[127] A. Y. Chow, V. Y. Chow, K. H. Packo, J. S. Pollack, G. A. Peyman, and R. Schuchard. The artificial silicon retina microchip for the treatment of vision loss from retinitis pigmentosa. *Arch. Ophthalmol.*, 122(4):460–469, 2004. DOI: 10.1001/archopht.122.4.460

[128] A. Y. Chow and V. Y. Chow. Subretinal electrical stimulation of the rabbit retina. *Neurosci. Lett.*, 225:13–16, Mar. 1997. DOI: 10.1016/S0304-3940(97)00185-7

[129] T. Morimoto, T. Miyoshi, T. Fujikado, Y. Tano, and Y. Fukuda. Electrical stimulation enhances the survival of axotomized retinal ganglion cells in vivo. *Neuroreport*, 13(2):227–230, Feb. 2002. DOI: 10.1097/00001756-200202110-00011

[130] P. J. DeMarco Jr., G. L. Yarbrough, C. W. Yee, G. Y. McLean, B. T. Sagdullaev, S. L. Ball, and M. A. McCall. Stimulation via a subretinally placed prosthetic elicits central activity and induces a trophic effect on visual responses. *Inv. Opthalmo. Vis. Sci.*, 48(2):916–926, Feb. 2007. DOI: 10.1167/iovs.06-0811

[131] E. Zrenner. [Initial success in retinitis pigmentosa patients. Retinal chip brings hope to the blind (interview by Dr. med. Thomas Meissner)]. *MMW Fortschr Med.*, 149:16, Apr. 2007.

[132] M. S. Humayun. Intraocular retinal prosthesis. *Trans. Amer. Ophthalmol. Soc.*, 99:271–300, 2001.

[133] R. R. Lakhanpal, D. Yanai, J. D. Weiland, G. Y. Fujii, S. Caffey, R. J. Greenberg, E. de Juan, and M. S. Humayun. Advances in the development of visual prostheses. *Curr. Opin. Ophthalmol.*, 14:122–127, Jun. 2003. DOI: 10.1097/00055735-200306000-00002

[134] M. Javaheri, D. S. Hahn, R. R. Lakhanpal, J. D. Weiland, and M. S. Humayun. Retinal prostheses for the blind. *Ann. Acad. Med. Singap.*, 35:137–144, Mar. 2006.

[135] J. D. Weiland and M. S. Humayun. Intraocular retinal prosthesis. Big steps to sight restoration. *IEEE Eng. Med. Biol. Mag.*, 25:60–66, 2006. DOI: 10.1109/MEMB.2006.1705748

[136] Cochlea implants: National institute on deafness and other communication disorders (nidcd), 2008. `http://www.nidcd.nih.gov/health/hearing/coch.asp`.

[137] J. E. Colgate and M. A. Peshkin. Passive robotic constraint devices using non-holonomic transmission elements. *US Patent #5,923,139*, Jul. 13, 1999.

[138] J. E. Colgate and M. A. Peshkin. Cobots. *US Patent #5,952,796*, Sept. 14, 1999.

[139] J. Horgan. The work of Jose Delgado, a pioneering star. *Sci. Amer.*, pp. 65–78, Oct. 2005.

Quantitative Neurophysiology

Joseph V. Tranquillo
Bucknell University

MORGAN & CLAYPOOL PUBLISHERS

ABSTRACT

Quantitative Neurophysiology is supplementary text for a junior or senior level course in neuroengineering. It may also serve as an quick-start for graduate students in engineering, physics or neuroscience as well as for faculty interested in becoming familiar with the basics of quantitative neuroscience. The first chapter is a review of the structure of the neuron and anatomy of the brain. Chapters 2-6 derive the theory of active and passive membranes, electrical propagation in axons and dendrites and the dynamics of the synapse. Chapter 7 is an introduction to modeling networks of neurons and artificial neural networks. Chapter 8 and 9 address the recording and decoding of extracellular potentials. The final chapter has descriptions of a number of more advanced or new topics in neuroengineering. Throughout the text, vocabulary is introduced which will enable students to read more advanced literature and communicate with other scientists and engineers working in the neurosciences. Numerical methods are outlined so students with programming knowledge can implement the models presented in the text. Analogies are used to clarify topics and reinforce key concepts. Finally, homework and simulation problems are available at the end of each chapter.

KEYWORDS

Neurophysiology, neuroengineering, neuroscience, electrophysiology, biomedical engineering, neuron, neural signal processing, neural anatomy, brain, CNS

Contents

Preface

The function of the brain and connection to the mind is perhaps one of the most enigmatic pursuits in all of science. Some have even dubbed the study of the brain the final frontier. Others claim that we will never know exactly how our brain works. We do, however, understand a great deal about the brain, as evidenced by nearly 100 years of Nobel Prizes

Year	Winners	For
1906	Golgi and Cajal	Structure of the nervous system
1920	Nernst	Membrane potentials
1932	Sherrington and Adrian	The synapse
1936	Dale and Loewi	Chemical propagation across a synapse
1944	Erlanger and Gasser	Differentiation of nerve fibers
1949	Hess	Role of interbrain in regulation
1950	Kendall, Reichstein and Hench	Hormones in the adrenal cortex
1963	Eccles, Hodgkin and Huxley	Ionic mechanisms of action potentials
1970	Katz, Euler and Axelrod	Function of neurotransmitters
1977	Guillemin and Schally	Neurohormones in the brain
1981	Sperry, Hubel and Wiesel	Specialization of cerebral hemispheres
1991	Neher and Sakmann	Function of single ion channels
1994	Gilman and Rodbell	G-proteins in signal transduction
1997	Prusiner	Prions as a principle for infection
2000	Carlsson, Greengard and Kandel	Dopamine signal transduction
2003	Lauterbur and Mansfield	Magnetic Resonance Imaging (MRI)
2003	Agre and MacKinnon	Ion channel selectivity

The objective of "Quantitative Neurophysiology" is to provide an overview of the theory underlying electrical impulses in the neuron. The material presented is purposely distinct from a traditional neuroscience text in that it is geared toward engineers and applied scientists. As such the focus will be place more on neurons in the central nervous system, where a deep quantitative understanding has been developed, and less on the large-scale functions of the brain. Despite this narrow focus you will find some classic neuroscience woven into the text.

For the student, it is hoped that you will see concrete applications of fundamental engineering and science concepts in the description of a real biological system. It will be most helpful if you have already taken differential equations as well as a basic circuits course. The philosophy of the text is that by understanding the mathematical models of neuronal function (Chapters 2-6), you will gain a deeper understanding of how neurons work together to create complex behavior (Chapter 7) and how we might listen in on the chatter of neurons to unravel their language (Chapters 8-9). Finally, the theory presented should prepare you to read the latest neuroengineering literature and move on to more advanced topics (Chapter 10).

For the instructor, it is hoped that this text will provide a resource that can aid you in your instruction. I have found that students are capable of digesting one chapter each week ($\approx$3 hours of class time, supplemented by homework problems). At the end of each chapter are homework and simulation problems which can be assigned as is, or modified to better suit the abilities of the students. This schedule leaves ample time to cover more advanced topics at the instructor's discretion.

This text would not be possible without the support of two important groups. First, I wish to thank the Bucknell students who have taken my Neural Signals and System course. Second, I wish to thank my family for their patience and keeping me focused on what is really important in life.

Joseph V. Tranquillo
Lewisburg, Pennsylvania

CHAPTER 1

Neural Anatomy

The brain may be studied at many scales, ranging from molecular to gross anatomy. In this text, the focus will be primarily on quantifying cellular behavior for two reasons. First, much more is known about the function of individual neurons. Second, an underlying assumption of neuroscience is that a complete understanding of brain function will require deep insights at the cell and molecular level. In this chapter, the anatomy of the neuron is explored to the depth needed to understand future chapters. It is very important, however, to not lose site of the goal of understanding the brain at the larger scales. Therefore, this chapter also outlines some of the gross level structures of the brain.

1.1 THE NEURON

The human brain is composed of more than 10^{12} neurons. The table below is a summary of some types of neurons.

Table 1.1:

Bipolar	Found in eye, transduction of light to neural impulse
Somatosensory	Found in skin, tactile senses of pain and proprioception
Motor	Found in spinal cord, projects to muscles
Pyramidal cell	Found in cortex, relay information within brain
Purkinje Cell	Found in cerebellum, motor skills
Association cell	Found in thalamus, connect neurons together

Although the anatomy and function of the individual neuron can vary throughout out the brain, Fig. 1.1 shows the features of a generic neuron. Like other cells, the neuron separates space inside from space outside by a $5nm$ bi-lipid membrane that acts as an electrical insulator. Typically, the neuron is divided into the dendritic tree, soma and axon.

The *dendritic tree* is a complex web of branching structures that range in diameter from $1-20\mu m$. One interesting feature of some dendrites is the presence of *spines* that play a somewhat unclear electrical or chemical role in neuronal function.

The *soma*, or cell body, is approximately $20\mu m$ in diameter and contains most of the organelles, including the nucleus, Golgi apparatus, mitochondria, microtubuals, and endoplasmic reticulum. It is here that the cell generates ATP, packages neurotransmitters, houses the genetic material, and assembles proteins for the cell.

The *axon* branches off of the soma at the *axon hillock*. Axons can vary greatly in length ($1mm - 1m$) and diameter ($1-25\mu m$). The intracellular space of the axons are covered in small specialized proteins called *legs kinesin* which mechanically transport ATP, vesicles filled with neurotransmitter and enzymes to and from the soma. Some axons have offshoots called *axon collaterals*.

The *synapse* is the site where the dendrite of on neuron interfaces with the axon of another neuron. It was first described in 1897 by Sherrington who coined the term from the Greek, Syn (together) +

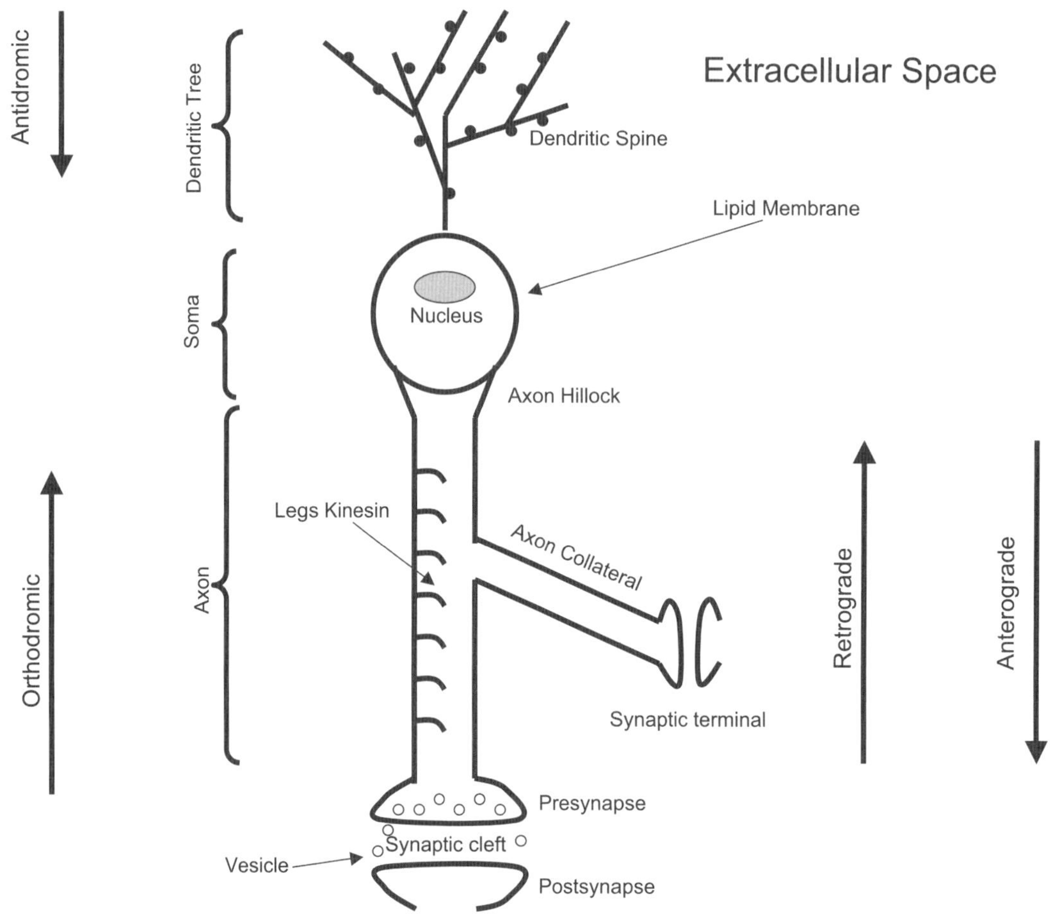

Figure 1.1: Schematic of a typical neuron.

Haptin (to fasten). There are approximately 10^{15} synapses in the brain, so each neuron has on average 1,000 synapses. The number of synapses in an individual neuron, however, can vary greatly. It is important to note that the synapse is not a singular structure, but rather the combination of three structures. The *pre-synapse* is the very end of an axon and houses *vesicles*, or small spheres of membrane, that contain *neurotransmitters*. The *post-synapse* is on the very end of a dendrite. The $20nm$ space between the pre-synapse and post-synapse is called the *synaptic cleft* and is technically outside of both neurons.

Neurotransmitters are specialized molecules that are packaged into vesicles in the soma, transported to the end of the pre-synaptic axon by the legs kinesin, and released into the synaptic cleft in response to an electric impulse. Neurotransmitters diffuse across the synaptic cleft and reach the post-synapse where they either excite or inhibit electrical impulses in the dendrite of a new neuron. There is enormous diversity in the molecules that function as neurotransmitters as well as the mechanisms by which they affect the post-synapse.

To indicate the direction of ionic or molecular movement in a neuron, two sets of terms have been defined. When discussing motion within the axon, *anterograde* is movement from the soma to the end of the axons and *retrograde* is movement from the axon to the soma. To discuss movement in the entire neuron, *orthodromic* is in the direction from the axon to the dendrites, while *antidromic* is from the dendrites to the axon.

1.2 GLIAL CELLS

Although often overlooked, *glial cells* in the brain outnumber neurons nearly two to one. Historically, glial cells were thought to only impact electrical properties indirectly by maintaining extracellular ion concentrations and by speeding electrically propagation through the growth of the *myelin sheath*. Recent evidence, however, suggests that glial cells may play a more direct role in the propagation of impulses. Some of the more common types of glial cells are listed in the table below.

Table 1.2:

Astroglia	Structural support and repairs
Oligodendroglia	Insulate and speed transmission in CNS
Schwann Cells	Insulate and speed transmission in PNS
Microglia	Perform phagocytosis of damaged or dead cells
Ependymal cells	Line the ventricles and produce cerebrospinal fluid

1.3 THE BRAIN

While our primary focus will be on quantifying the function of nerve cells, it is important not to lose sight of the fact that our brain regulates nearly all voluntary and involuntary actions. It is often reported that we only use 10% of our brain, with the implication that there is 90% of the brain that we do not understand. The 10% number is the percentage of the brain that is active *at any one time*. Each structure of the brain is known to perform some function. What is not clear is how these structures use neurons to perform their functions. Below is a brief overview of the structure and function of the most significant regions of the brain.

1.3.1 From Neuron to Nuclei

Typically, cells that are close to one another are more densely connected by synapses and perform similar functions. These groupings of cells form a *nucleus*. In histological slices, the dense tangle of cell bodies, dendrites, and capillaries have a grayish color and so have been given the name *grey matter*. Several nuclei may be connected together to form larger structures in the brain. The connections between nuclei are made by bundles of long axons, called *tracts*, that may carry large amounts of information. Long axons require fast propagation and so are covered in a fatty coating of myelin. In histological slices, these dense regions of axons have a white color and so have been given the name *white matter*.

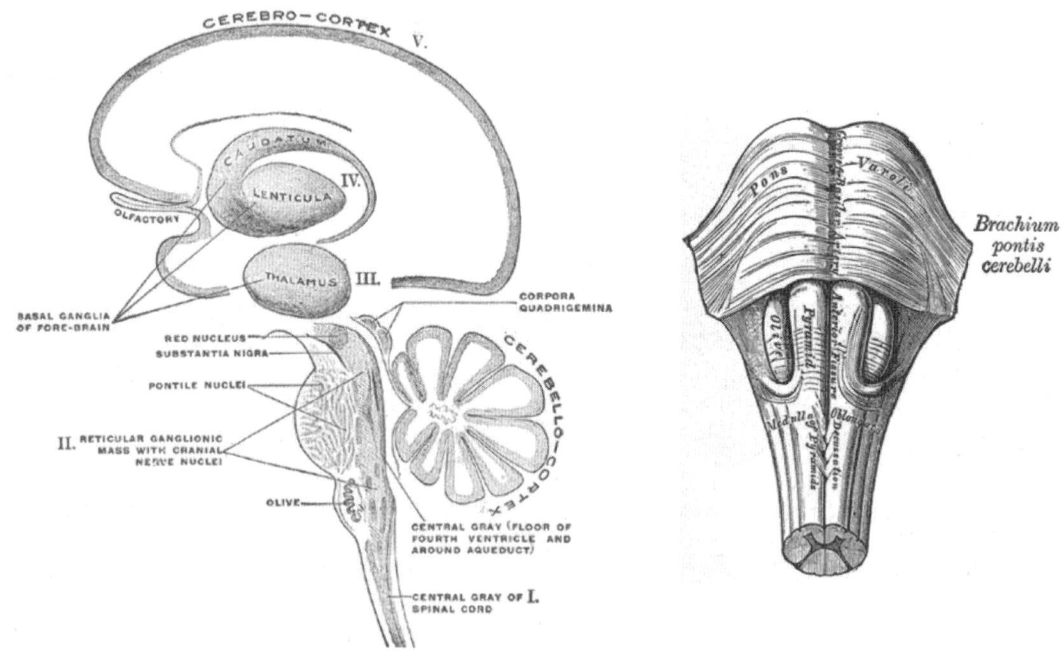

Figure 1.2: Gross anatomy of the brain.

1.3.2 Brain Systems

The gross anatomy of the brain consists of both tissue (neurons and glia) as well as chambers called ventricles that are filled with cerebral-spinal fluid. The tissue is organized into three regions based upon embryonic development.

The *hindbrain*, or Rhombencephalon, is an offshoot of the spinal cord and was inherited from the reptile brain. It is primarily involved in involuntary control of basic functions such as breathing and heart rate. In addition to forming the lower portion of the brain stem, it also includes the cerebellum, a relatively large outcropping at the base of the brain that controls motor movements and relays signals to the spinal cord.

Table 1.3:

Myelencephalon		Medulla oblongata
Metencephalon		Contains 4th Ventricle
	Pons	Relay sensory information
	Cerebellum	Integration of sensory and motor
	Reticular Formation	Rudimentary eye movements

The *midbrain*, or Mesencephalon, forms the upper part of the brain stem. Its primary function is to connect the lower brain stem with higher-level brain structures. In size, it is much smaller than the other two embryonic regions. It is involved in involuntary motor control and sensation.

The *forebrain*, or Prosencephalon, forms the bulk of the brain tissue in mammals. It contains the limbic system that regulates drives, hunger, hormones, emotions, and memory, as well as the cerebral cortex. The regions of the forebrain are tightly connected together by tracts in complex feedback networks.

Table 1.4:

Diencephalon		Lateral and 3rd ventricles, limbic system
	Epithalamus	Pineal body, wake/sleep patterns
	Thalamus	Relay information to cerebral cortex
	Hypothalamus	Link to endocrine system, metabolism, hunger
	Pituitary Gland	Master endocrine gland
Telencephalon		Also called the cerebrum
	White Matter	Axons that connect cortex to other structures
	Amygdala	Process memory and emotions
	Hippocampus	Process memory and spatial navigation
	Rhinencephalon	Olfaction (smell)
	Cerebral Cortex	Higher level tasks

The *cerebral cortex* occupies the top $2-4mm$ of the forebrain and is composed of grey matter organized into six layers. The cortex is split into a number of *lobes* which are roughly mirrored on left and right *hemispheres* of the brain.

Table 1.5:

Structure	Location	Notes
Frontal Lobe	Behind forehead	Impulse control, language
Parietal Lobe	Crown of head	Sensory, numbers
Temporal Lobe	Behind temples	Auditory, speech, and vision
Occipital Lobe	Back of head	Primary visual center
Insula	Between temporal and parietal	Emotions, pain, addition
Cingulate Cortex	Surrounds the corpus collosum	Emotions
Corpus Collosum	Connection between hemispheres	Millions of axons

1.3.3 Blood Brain Barrier

In the body, capillaries are lined with endothelial cells that contain small gaps, allowing for easy diffusion of chemicals and gasses to and from the blood. In the brain, a similar membrane, called the *blood brain barrier*, is present but the cells are more packed tightly so that most chemicals can not diffuse through. The only chemicals that can pass are either lipid soluble (e.g., O_2, CO_2, ethanol, steroid hormones) or carried by a specific channel (e.g., sugars, amino acids). The result is that the brain is able to maintain a high rate of metabolism but is protected from potentially dangerous chemicals and infections. In addition, glial cells surround capillaries in the brain, forming a second barrier. There are, however, a few entry points that allow the brain to sense the concentrations of hormones in the blood. The downside of the blood brain barrier is that it is difficult to design drugs that will work directly on the brain.

1.3.4 Directions in the Brain

Directions in the brain are the same as in the body. Anterior refers to the front of the brain while posterior is toward the back. Medial is toward the center of the brain while lateral is away from the center. Superior (also dorsal) is toward the top of the brain while interior (also ventral) is toward the base of the brain.

Two additional terms are used in the brain that trace a curved line from the bottom of the spinal cord to the nose (or frontal lobe). Moving in a *rostral* direction is moving toward the nose. Moving in a *caudal* direction is moving toward the bottom of the spinal cord.

1.3.5 Inputs and Outputs

The focus of the text is on neurons in the central nervous system. The mathematical models, however, apply equally well to neurons in the peripheral nervous system. We therefore briefly mention that the brain interacts with neurons in the body. Neural inputs to the brain typically originate from sensory systems. Neurons that carry impulses to the brain are called *afferent*, sensory or receptor nerves. Typically, these nerves originate in a sensory system such as the eyes or nose. Neurons that carry impulses away from the brain are called *efferent*, motor or effector nerves. Typically, they will innervate muscles or glands.

CHAPTER 2

Passive Membranes

Electrophysiology is the study of the electrical properties of biological materials from the molecules to the entire body. Although all materials in the body can be characterized by their electrical properties, the nervous and muscular system in particular use electrical impulses to communicate information between cells. In this chapter, we will lay down the basic principles of cellular electrophysiology which may be applied to any cell in the body. In Ch. 3, we will consider the more specialized electrical properties of neurons.

2.1 CELLULAR ELECTROPHYSIOLOGY

As we are primarily concerned with electrical properties, we must first define what voltages (difference in potential) and currents (flow of charged particles) mean in the context of a cell. All of the normal principles of electricity apply, however, most electrical texts consider the movement of negatively charged electrons. In biological systems, currents are a flow of ions (e.g., Na^+, Cl^-) and voltages are differences in potential created by different ionic concentrations.

2.1.1 Cellular Voltages

The *transmembrane voltage*, V_m, is defined as the difference in potential across the cell membrane (Fig. 2.1) and is typically measured in mV.

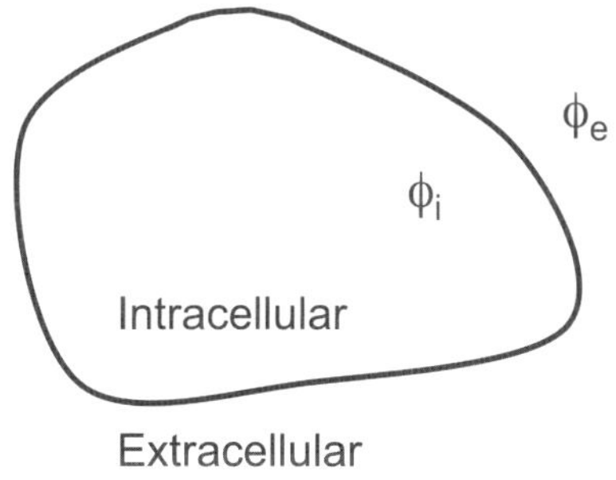

Figure 2.1: Definition of membrane voltage, V_m.

$$V_m = \phi_i - \phi_e = -\Delta\phi \tag{2.1}$$

where ϕ_i is the potential inside the cell and ϕ_e is the potential outside the cell. Surprisingly, the transmembrane voltage will reach a steady state called the *resting membrane voltage*, V_m^{rest}, that is not zero. In most neurons $V_m^{\text{rest}} \approx -60mV$, meaning that at rest the inside of the cell has a more negative potential than the outside. Any positive change in V_m is called a *depolarization* (Fig. 2.2) and may occur because the inside has become more positive or the outside of the cell has become more negative. Any negative change in V_m is called a *repolarization*. Any time the membrane voltage drops below V_m^{rest} is known as *hyperpolarization*.

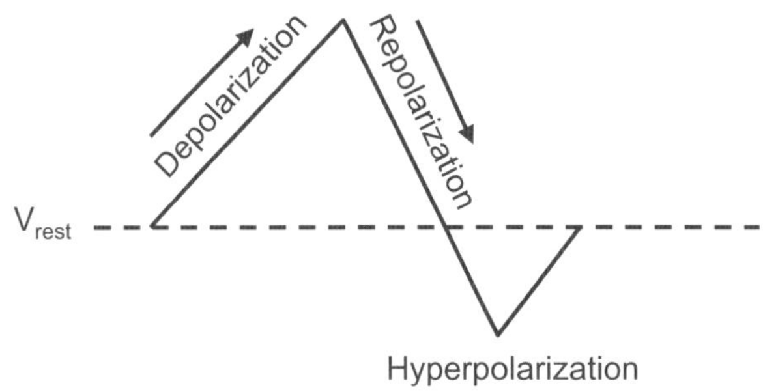

Figure 2.2: Depolarization, repolarization, and hyperpolarization of V_m.

2.1.2 Cellular Currents

The *transmembrane current*, I_m, is a measure of the movement of ions across the membrane and is typically measure in units of $\mu A/cm^2$. By definition, a positive I_m is positive charge leaving the cell. Therefore, a positive I_m will repolarize the cell membrane voltage. The transmembrane current is the sum of four currents

$$I_m = I_{cm} + I_{\text{ion}} + I_{\text{syn}} - I_{\text{stim}} \ . \tag{2.2}$$

The *capacitive current*, I_{cm}, is a result of the natural capacitance of the cell membrane. Recall that a capacitor is nothing more than an insulator sandwiched between two conductors. The cell membrane is composed of lipids which are a natural isolator. The intra and extracellular solutions are nothing more than salty water and therefore are good conductors. We will examine the role of the membrane capacitance, C_m, in more detail in Sec. 2.1.4.

The *ionic current*, I_{ion}, is the primary way for ions to cross the membrane. I_{ion} may be composed of many different currents (carrying different ions) summed together. We will consider I_{ion} in greater detail in Ch. 3.

The *synaptic current*, I_{syn}, is any one of hundreds of specialized currents that allow neurons to communicate chemically across a synapse. We will consider I_{syn} in Ch. 6.

The *stimulus current*, I_{stim}, is an externally applied current and will be considered in more detail in Sec. 2.2. Note that the sign of I_{stim} is opposite of the other currents.

2.1.3 Membrane Circuit Analog

It is customary to use circuit analogies to describe the relationship between voltages and currents. In Fig. 2.3, the top node represents the inside of the cell and the bottom node represents outside of the cell.

All currents are represented by a unique parallel pathway through which current may cross the membrane. In the literature, this circuit is therefore known as the *parallel conductance model*. By conservation of current, the total membrane current, I_m, must sum to zero, so

$$0 = I_{cm} + I_{\text{ion}} + I_{\text{syn}} - I_{\text{stim}} \,. \tag{2.3}$$

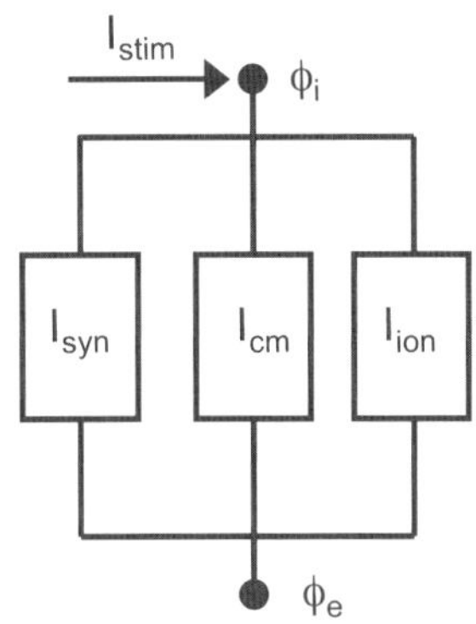

Figure 2.3: Parallel conductance model.

Substituting the know the relationship between voltage and current for a capacitor,

$$0 = C_m \frac{dV_m}{dt} + I_{\text{ion}} + I_{\text{syn}} - I_{\text{stim}} \tag{2.4}$$

$$\frac{dV_m}{dt} = \frac{1}{C_m}\left[-I_{\text{ion}} - I_{\text{syn}} + I_{\text{stim}}\right] \,. \tag{2.5}$$

Equation (2.5) is a differential equation that describes how V_m evolves over time based upon the currents that flow across the cell membrane.

2.1.4 The Membrane Capacitance

The value of the membrane capacitance can be directly computed by

$$C_m = \frac{k\epsilon_0}{d} \tag{2.6}$$

where k is the dielectric constant of the insulator, ϵ_0 is the permittivity of free space ($1 \times 10^{-9}/36\pi\, F/m^2$), and d is the membrane thickness. Using $k = 3$ (value for oil) and $d = 3nm$:

$$C_m = \frac{3 \times 10^{-9}}{36\pi(3 \times 10^{-9})} = 0.009\frac{F}{m^2} = 0.9\frac{\mu F}{cm^2} \,. \tag{2.7}$$

For simplicity, $1\mu F/cm^2$ is often used as an approximation for C_m. As our estimate for C_m is measured per unit area, it is independent of the size of the cell.

2.2 STIMULATING THE PASSIVE MEMBRANE

If an external current (I_{stim}) is applied to the membrane of a cell such that positive ions are forced into the cell, the membrane voltage will depolarize. For now we will assume that there are no synapses ($I_{\text{syn}} = 0$), so

$$\frac{dV_m}{dt} = \frac{1}{C_m}\left[-I_{\text{ion}} + I_{\text{stim}}\right] . \tag{2.8}$$

The membrane, although composed primarily of lipids, has "leaky" channels that will allow some current to pass. Experimentally, it has been observed that when V_m is close to V_m^{rest}, the leakage of ions is proportional to V_m. Therefore, we can approximate I_{ion} using Ohm's Law:

$$I_{\text{ion}} = \frac{\phi_i - \phi_e}{R_m} = \frac{V_m}{R_m} \tag{2.9}$$

where R_m is the *specific membrane resistivity* ($k\Omega cm^2$) to current flow and measures the "leakiness" of the membrane. Given our assumption of linear R_m,

$$\frac{dV_m}{dt} = \frac{1}{C_m}\left[-\frac{V_m}{R_m} + I_{\text{stim}}\right] . \tag{2.10}$$

Rearranging Eq. (2.10)

$$R_m C_m \frac{dV_m}{dt} + V_m = R_m I_{\text{stim}} . \tag{2.11}$$

It has become customary rewrite Eq. (2.11) as:

$$\tau_m \frac{dV_m}{dt} + V_m = V_\infty \tag{2.12}$$

where

$$\tau_m = R_m C_m \tag{2.13}$$

is the subthreshold *membrane time constant* measured in $msec$ and

$$V_\infty = R_m I_{\text{stim}} \tag{2.14}$$

is the steady-state voltage as time $\to \infty$.

When I_{stim} is applied, V_m will charge up the capacitor to V_∞ at a rate governed by τ_m. If the current remains on, the voltage will remain at V_∞. Solving Eq. (2.12) during stimulation,

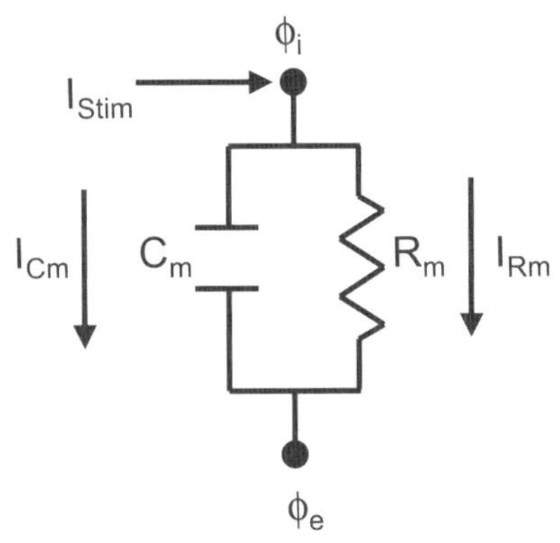

Figure 2.4: RC circuit analog of a passive cell membrane.

$$V_m(t) = V_\infty(1 - e^{-t/\tau_m}) \tag{2.15}$$

where t is the time after applying I_{stim}.

When I_{stim} is turned off, V_m will be at some initial voltage V_0 due to charging. From this initial value, V_m will return to V_m^{rest}, again at a rate governed by τ_m

$$V_m(t) = V_0 e^{-t/\tau_m} \ . \tag{2.16}$$

2.2.1 Finding Membrane Properties from $V_m(t)$

The left-hand side of Fig. 2.5 is a plot of $V_m(t)$ over time as a current pulse is applied and then removed. The rising phase is governed by Eq. (2.15) while the falling phase is governed by Eq. (2.16). The rising or falling phase can be used to determine the membrane properties, R_m, C_m, and τ_m. The right-hand side of Fig. 2.5 shows a stimulus that is applied only for a short time, so V_m does not have time to reach V_∞ before the stimulus is turned off.

2.2.2 The Passive Membrane

To perform the analysis above, we assumed that the current flowing through the membrane was linearly proportional to the membrane voltage. It turns out that this assumption is a good approximation as long as V_m is below some *threshold* voltage, V_m^{th}. When $V_m < V_m^{th}$, the membrane is called *passive*, R_m is a constant and the membrane can be represented as the RC circuit in Fig. 2.4. In a real neuron, V_m^{th} is typically between 5mV and 10mV higher than V_m^{rest}. When V_m depolarizes above the threshold, however, the membrane will become *active* and R_m will no longer be a constant. In Ch. 3 we will consider the nonlinear relationship between R_m and V_m above the threshold.

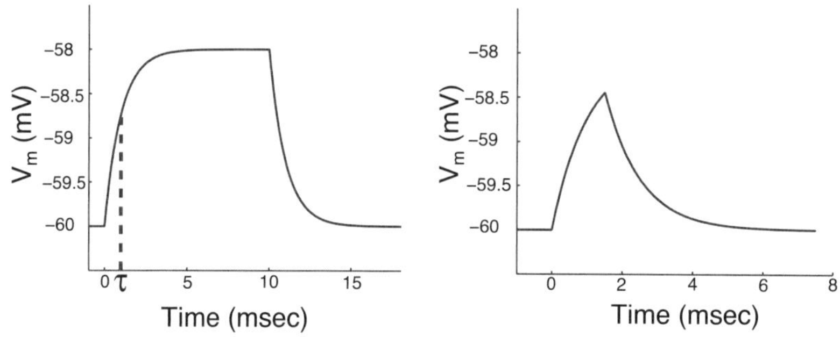

Figure 2.5: Charging and discharging of a passive membrane.

2.3 STRENGTH-DURATION RELATIONSHIP

To perform our analysis of the passive membrane we assumed that $V_m < V_m^{th}$. It would therefore be helpful to know in what situations our assumption is valid. Examining Eq. (2.15), $V_m(t)$ during the stimulus is dependent on $V_\infty = I_{\text{stim}} R_m$ and $\tau_m = R_m C_m$. Consider that I_{stim} may be small such that $V_\infty < V_m^{th}$. In this case, the stimulus may remain on forever and V_m will charge up to a steady state value below V_{th}. In other words, the membrane will remain passive. If I_{stim} is steadily increased, eventually V_∞ will be equal to V_m^{th}. In this case, the steady state will level out exactly at the threshold but it may take a very long time for V_m to reach threshold. If I_{stim} is increased further, V_∞ will clearly be greater than V_m^{th}. As the right side of Fig. 2.5 indicates, however, even if $V_\infty > V_m^{th}$ the stimulus may be on for such a short time that the membrane does not have time to charge to the threshold. It follows that there is an interplay between the strength of I_{stim} and the duration of I_{stim}, which can be represented graphically in a *strength-duration* plot. Given a constant value for V_m^{th}, Fig. 2.6 shows the combinations of strength and duration that exactly charge the membrane to the threshold. Mathematically, we can set $V_m = V_m^{th}$ and substitute into Eq. (2.15)

$$V_m^{th} = I_{\text{stim}} R_m (1 - e^{-D/\tau_m}) \tag{2.17}$$

where D is the duration of the stimulus. One interesting limit is to find the minimum current that can bring the membrane to V_m^{th}. If $D \to \infty$

$$V_m^{th} = I_{\text{stim}} R_m \tag{2.18}$$

$$I_{\text{stim}} = I_{\text{rhe}} = \frac{V_m^{th}}{R_m} \tag{2.19}$$

which explains the value of asymptote in Fig. 2.6. This value is called *rheobase*, I_{rhe}, and is a measure of current. Another important measure can be derived by assuming that we apply two times the rheobase current to find the corresponding time to charge the membrane to V_m^{th}.

$$V_m^{th} = 2 I_{\text{rhe}} R_m (1 - e^{-D/\tau_m}) \tag{2.20}$$

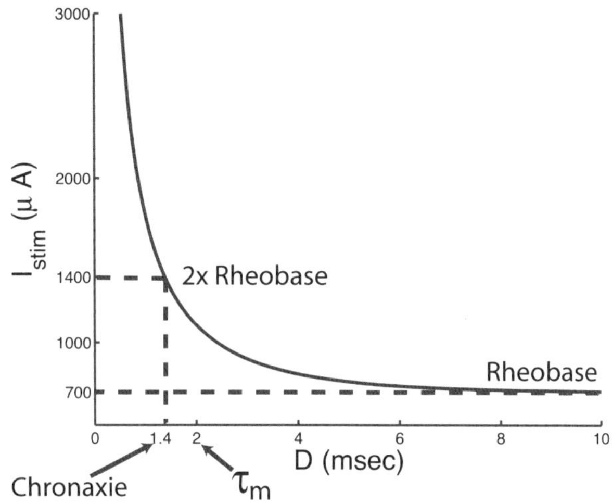

Figure 2.6: Strength-duration curve.

and solving for D

$$D = -\tau_m \cdot ln\left[1 - \frac{V_m^{th}}{2I_{\text{rhe}}R_m}\right] \tag{2.21}$$

but we know that $I_{\text{rhe}}R_m = V_m^{th}$ so

$$D = -\tau_m \cdot ln\left[1 - \frac{1}{2}\right] \tag{2.22}$$

$$T_c = D = 0.693\tau_m \ . \tag{2.23}$$

The time T_c is known as *chronaxie*. Therefore, given a strength duration curve, and Eqs. (2.19) and (2.23) one could find the passive membrane properties.

2.3.1 A Fluid Analogy

An alternative way to think about the passive membrane is as a cup with a hole in the bottom and a line painted half way up the side. In the analogy, the volume of the cup is similar to the capacitance in that it can store the quantity of interest (water instead of charge). The size of the hole is similar to the resistance in that it tends to leak out the quantity of interest (water instead of charge). The line on the cup represents the threshold voltage. As long as the water level remains below the threshold, the water will leak from the hole at a constant rate. A stimulus (a flow rate of charges) in our analogy would be the steady poring of water into the cup.

Using our analogy we can gain some intuition about charging and discharging in response to a stimulus. For example, if water is poured in slower than the rate at which it leaks out of the hole, then the threshold line will never be reached. This is analogous to being below I_{rhe}. You could imagine, however, that more aggressive pouring would eventually result in the threshold being reached but only very slowly. It may also be possible to pour water in only for some short period of time. If this is the case, then the rate at which you pour and the duration of your pour will determine if the water will reach the threshold. You could in fact create a strength-duration plot for this cup-water system. Furthermore, if you assume that the cup already has water in it, you may ask how long it will take for the cup to drain. The two parameters of interest will be the size of the cup (capacity) and the size of the hole (resistance). The combination of the two, as in the circuit analogy, can be used to define the time constant of the filling or draining of the cup.

2.4 THE MEMBRANE AT REST

A careful examination of Fig. 2.4 will reveal that V_m^{rest} must be $0mV$, which biologically is not true. In this section we will examine how a cell can maintain a nonzero resting potential.

2.4.1 The Forces Driving Ion Movement

Consider Fig. 2.7 where circles and dots represent two types of positively charged ions but the membrane will only allow dots to pass through. The special property of the membrane to easily pass some ions but not others is called *selective permeability* and results in a nonzero resting potential.

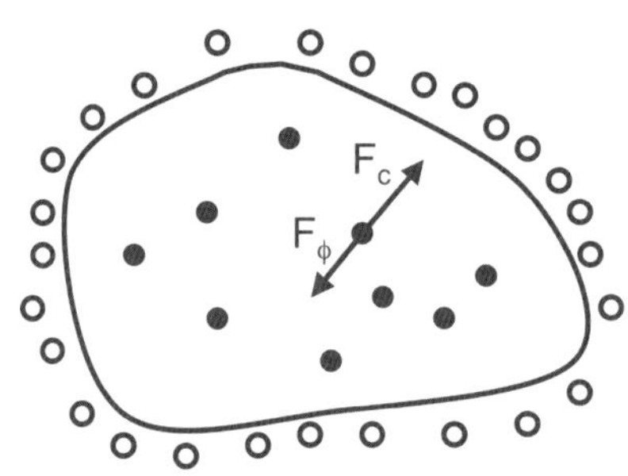

Figure 2.7: Concentrations and electrical potentials in a cell.

There are two forces driving the motion of a ion. The electric force, F_ϕ, is due to any difference in potential ($\Delta\phi$) between the inside and outside of the cell. Remember that potential differences arise because there are different amounts of charge on either side of the membrane. The potential is therefore due to *all* of the charges, whether they are able to cross the membrane or not. The chemical force, F_C, is due differences in *specific* ionic concentrations (ΔC) across the membrane and will act only on a single type of ion (e.g., dot or circle in Fig. 2.7). It is because F_ϕ is a function of all charges and F_C is a function of only a specific ion that the membrane can support a nonzero rest potential.

2.4.2 A Helium Balloon Analogy

An alternative way to think of the balance of two forces is using an analogy to a Helium balloon. There are always two forces acting on the balloon: gravity pulling the balloon downward and, because Helium gas is less dense than air, an upward buoyant force. When the balloon is first bought it is full of Helium gas that will not easily pass through the balloon and the buoyant force is much stronger than the pull of gravity. Therefore, the balloon will tend to sail into the air. As the balloon sails higher, however, the air becomes less dense, and so the buoyant force begins to decrease. At some elevation, the buoyant force and gravitational forces will exactly cancel out and the balloon will no long move up or down. The height of the balloon is the point at which two forces balance and is similar to the resting membrane voltage. As the balloon slowly leaks Helium, however, the buoyant force will decrease. The partially deflated balloon will sink to reestablish the balance of buoyant gravitational forces. Eventually, the balloon will have leaked so much Helium that it can not overcome the force of gravity.

2.4.3 Definition of Resting Membrane Voltage

Although we have an intuitive feel for V_m^{rest}, we can now derive a more formal mathematical definition. At rest the membrane voltage is not changing so $\frac{dV_m}{dt} = 0$. If no stimulus is being applied, then Eq. (2.8) reveals that I_{ion} must equal zero. The meaning is that no *net* current is crossing the membrane. We can only say net current because in general, I_{ion} may be composed of many currents which may balance one another.

As F_ϕ and F_C are the driving forces behind ion movement, if $F_C \neq F_\phi$ then charged particles will cross the membrane, i.e., $I_{\text{ion}} \neq 0$). For $I_{\text{ion}} = 0$ this means that $F_C = F_\phi$, or alternatively, the current due to the potential gradient, I_ϕ, is equal to the current due to the concentration gradient, I_c. In the next sections we will use Fick's Law and Ohm's law to define I_ϕ and I_c.

2.4.4 Fick's Law and Chemical Gradients

Fick's Law describes the *flux* $(\frac{\text{mol}}{m^2 \cdot s})$ of ions through an area of membrane with a thickness of dx due to a concentration gradient.

$$I_c = -D\frac{dC}{dx} \tag{2.24}$$

where the *diffusion coefficient*, D $(\frac{m^2}{s})$, is a material property of the membrane and is a measure of how easily ions can pass. C is a concentration in $\frac{\text{mol}}{m^3}$.

2.4.5 Ohm's Law and Electrical Gradients

As stated earlier, Ohm's Law for a passive membrane is

$$V_m = I_\phi R_m \tag{2.25}$$

$$I_\phi = \frac{V_m}{R_m} = \frac{\phi_i - \phi_e}{R_m}\,. \tag{2.26}$$

The specific membrane resistivity, R_m, may be thought of as the ability of some ion to pass across the thickness (dx) of the membrane

$$R_m = \frac{dx}{\mu_p C} \cdot \frac{|Z|}{Z} \tag{2.27}$$

where μ_p is the ion mobility. Z is the ion valence, so $Z/|Z|$ is the charge sign (+ or -). C is the ionic concentration.

In 1905 Einstein recognized that mobility (μ_p) and the diffusion coefficient (D) are related by

$$\mu_p = \frac{D|Z|F}{RT} \tag{2.28}$$

where R is the ideal gas constant, T is the temperature, and F is Faraday's constant. We can therefore, rewrite the equation for R_m as

$$R_m = \frac{dx \cdot RT}{DCFZ} \,. \tag{2.29}$$

Substitution back into Eq. (2.26) yields

$$I_\phi = \frac{[\phi_i - \phi_e]\, DCFZ}{dx \cdot RT} \tag{2.30}$$

$$I_\phi = \frac{DCFZ}{RT}\frac{d\phi}{dx} \,. \tag{2.31}$$

2.4.6 The Nernst Equation

Using Eqs. (2.24) and (2.31)

$$I_{\text{ion}} = I_c + I_\phi \tag{2.32}$$

$$I_{\text{ion}} = -D\left[\frac{dC}{dx} + \frac{CFZ}{RT}\frac{d\phi}{dx}\right] \,. \tag{2.33}$$

At rest we know that $I_{\text{ion}} = 0$

$$0 = -D\left[\frac{dC}{dx} + \frac{ZCF}{RT}\frac{d\phi}{dx}\right] \tag{2.34}$$

and after some rearranging

$$V_m^{\text{rest}} = E_{\text{rest}} = \frac{RT}{ZF} ln\left[\frac{C_e}{C_i}\right] \,. \tag{2.35}$$

Equation (2.35) is the *Nernst* equation that relates V_m^{rest} to the difference in intracellular and extracellular concentrations. In our derivation we assumed that C was a positive ion. The only different in Eq. (2.35) for negative ions is that C_e and C_i are switched in the numerator and denominator. Figure 2.8 is the circuit analog for the passive membrane where the Nernst potential is represented by a battery, E_{rest}.

The ionic current that flows through the resistor is described by

$$I_{\text{ion}} = \frac{1}{R_m}[V_m - E_{\text{rest}}] \,. \tag{2.36}$$

It can now be understood why the membrane settles to $V_m^{\text{rest}} = E_{\text{rest}}$. If $V_m > V_m^{\text{rest}}$, then I_{ion} will be positive. If $V_m < V_m^{\text{rest}}$ then I_{ion} is negative. Therefore, E_{rest} is sometimes also referred to as the *reversal potential.*

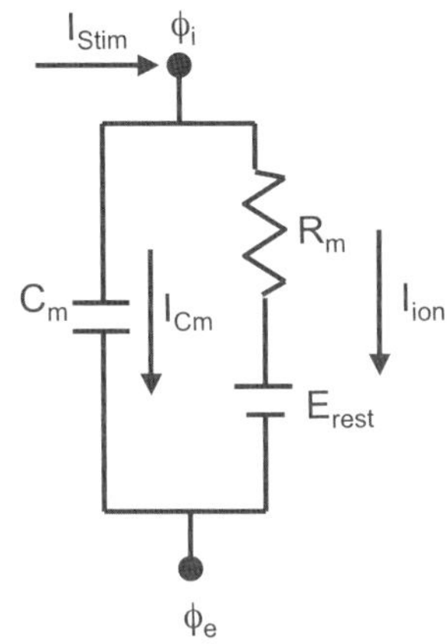

Figure 2.8: Passive membrane circuit analog with a resting potential.

2.4.7 The Goldman-Hodgkin-Katz Equation

In reality, there are many ions with different abilities to cross the membrane. There are also some large charged particles (e.g., proteins) that cannot cross the membrane. The result is that achieving $I_{\text{ion}} = 0$ to calculate V_m^{rest} is somewhat more difficult. We will not provide a derivation here but as the major ions involved are Sodium, Potassium, and Chloride, V_m^{rest} can be approximated by the Goldman-Hodgkin-Katz equation:

$$E_m = \frac{RT}{F} ln \left[\frac{P_K[K^+]_e + P_{Na}[Na^+]_e + P_{Cl}[Cl^-]_i}{P_K[K^+]_i + P_{Na}[Na^+]_i + P_{Cl}[Cl^-]_e} \right] \tag{2.37}$$

where the P-terms are the relative *permeabilities* (unitless) of the respective ions.

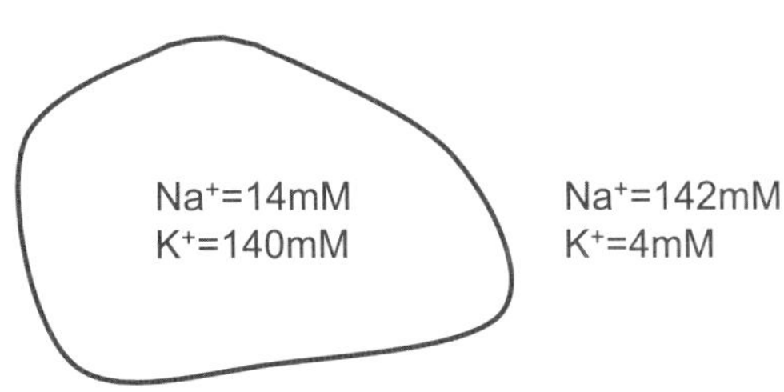

Figure 2.9: Multiple ions species and the Goldman-Hodgkin-Katz equation.

Figure 2.9 shows a simple example for only Na^+ and K^+ ions when the cell is at rest. Given that $P_K = 1$, $P_{Na} = 0.002$, $R = 8.316\frac{J}{K\text{mol}}$, $T = 300K$, and $F = 96487\frac{C}{\text{mol}}$,

$$\frac{RT}{F} = 0.0258\frac{J}{C} = 25.8mV \tag{2.38}$$

and

$$25.8 \cdot ln\left[\frac{1 \times 4mM + 0.002 \times 142mM}{1 \times 140mM + 0.002 \times 14mM}\right] = -90mV \ . \tag{2.39}$$

The individual Nernst potentials for K and Na may be found using Eq. (2.35). Perhaps surprisingly the computations show that $E_{Na} = 59.8mV$ and $E_K = -91.7mV$, neither of which are exactly at -90mV. It is worth noting that V_{rest} is much closer to E_K than E_{Na} at rest. It is also important to consider that Na^+ and K^+ may still be able to cross the membrane as long as the total current, I_{ion}, sums to zero. The direction of current flow may also be found by comparing V_m^{rest} and the reversal potential for an ion. In general, a particular ion will flow in the direction that will send V_m closer to its own Nernst potential. Remember that the definition of positive current is positive charge leaving the cell.

2.5 NUMERICAL METHODS: THE EULER METHOD

In this chapter we have defined a differential equation to relate V_m to membrane currents. Analytical solutions were possible because we assumed that all terms were linear. Once terms such as I_{ion} become nonlinear (as they will in Ch. 3), analytical solutions are no longer possible. In these situations we can turn to numerically simulations using a computer. Perhaps the simplest and most used method of finding $V_m(t)$ given $\frac{dV_m}{dt}$ is the *Euler* method. The general strategy is to use the current value, $V_m(t)$, and the current slope, $\frac{dV_m}{dt}$ to predict what the new membrane potential $(V_m(t + \Delta t)$ will be Δt forward in time (see Fig. 2.10). Mathematically,

$$\frac{dV_m}{dt} = \frac{1}{C_m}[-I_{\text{ion}} - I_{\text{syn}} + I_{\text{stim}}] \tag{2.40}$$

$$\frac{\Delta V_m}{\Delta t} = \frac{1}{C_m}[-I_{\text{ion}} - I_{\text{syn}} + I_{\text{stim}}] \tag{2.41}$$

$$\Delta V_m = \frac{\Delta t}{C_m}[-I_{\text{ion}} - I_{\text{syn}} + I_{\text{stim}}] \tag{2.42}$$

$$V_m^{\text{new}} = \Delta V_m + V_m^{\text{old}} \ . \tag{2.43}$$

Therefore, to solve a coupled set of nonlinear differential equations, we can apply Eq. (2.40) over and over again to march forward in time. One word of warning is that if Δt is large, you may overshoot the real solution.

Homework Problems

(1) Assuming that $V_m^{\text{rest}} = -70mV$ and the initial voltage is $V_m = -60mV$, draw a V_m trace that repolarizes from $t = 0msec$ to $t = 30msec$, then hyperpolarizes from $t = 30msec$ to t=35msec, depolarizes from $t = 35msec$ to $t = 50msec$, and then repolarizes to rest from $t = 50msec$ to $t = 70msec$. The exact values of the peaks and valleys are not important.

(2) A negative stimulus of $-15.7pA$ was applied to a membrane at $t = 1msec$ and produced the following trace. Note that hyperpolarizing pulses are always considered to be subthreshold.

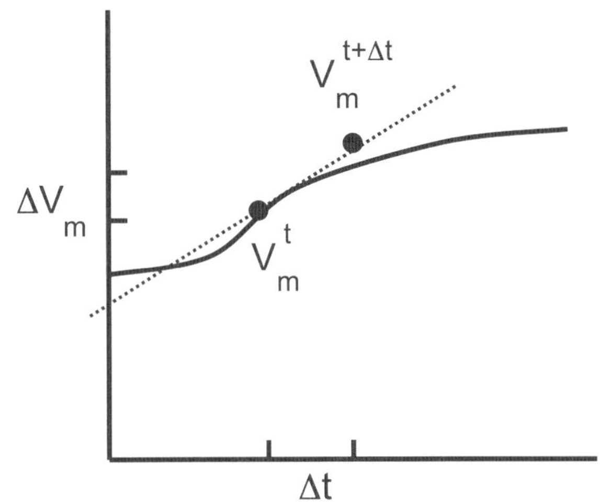

Figure 2.10: Euler method for time integration of a differential equation.

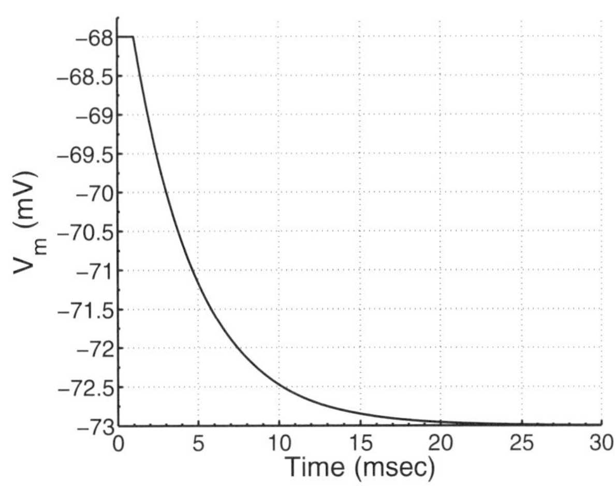

a) What is the resting membrane potential?
b) Compute the total R_m in $M\Omega$. Explain your reasoning.
c) Compute τ_m in $msec$. Explain your reasoning.
d) Compute the total cellular C_m in pF.
e) Given that the specific C_m is $1\mu F/cm^2$, find the cell surface area.

(3) The following data were obtained from a neuron at 300K (see Table 2.1).

Table 2.1:

Ion	Intracellular (mM)	Extracellular (mM)	Permeability
$[K^+]$	280	10	0.001
$[Na^+]$	51	485	1
$[Cl^-]$	46	340	0.5

a) Report the Nernst potential for each ion.
b) Compute the resting potential.
c) Report the direction that each ionic current moves at rest.

(4) Perform the derivation from Eqs. (2.34) and (2.35).

(5) Below is a strength-duration curve for a membrane with a threshold $7mV$ above rest.

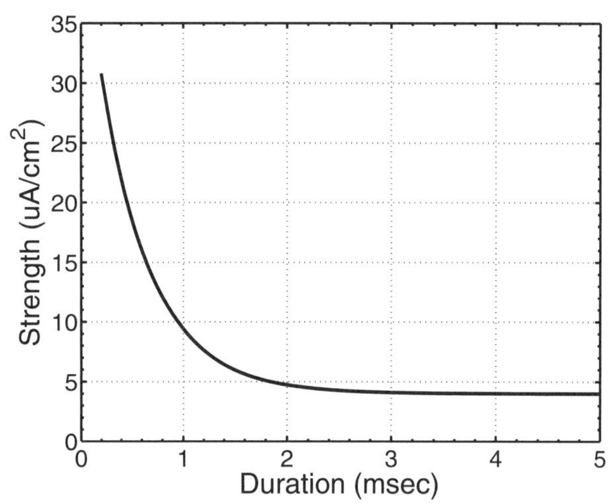

Given the figure, answer the following questions.
a) If a stimulus is above the curve, will the membrane behave linearly or nonlinearly?
b) What is I_{rhe}?
c) What is R_m in $k\Omega cm^2$?
d) What is T_{ch} in msec?
e) What is τ_m in msec?
f) What is C_m in $\mu F/cm^2$?

(6) Draw Strength Duration curve given $C_m = 1\mu F/cm^2$, $R_m = 3k\Omega cm^2$, $V_m^{\text{rest}} = -60mV$ and $V^{th} = -50mV$. Be sure to label I_{rhe} and T_c and report numerical values for each.

(7) Given that a negative stimulus of -15.7pA applied at t=1msec produced the following:

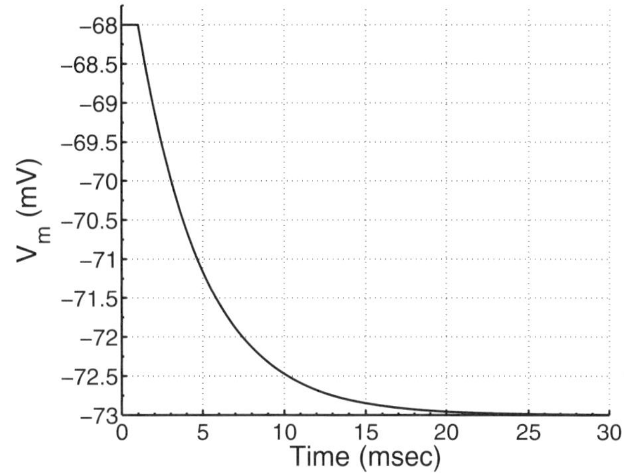

a) What is the resting membrane potential?
b) Compute total cellular R_m in $M\Omega$. Explain your reasoning.
c) Compute τ_m in msec. Explain your reasoning.
d) Compute the total cellular C_m in pF. Explain your reasoning.

(8) Demonstrate that the units in Eq. (2.11) are the same on the left- and right-hand side of the equation.

(9) Cellular currents are often measured in units of $\mu A/pF$. Explain why this may be a good measure to use and explain how you would convert Eq. (2.11) to the proper form.

Simulation Problems

(1) Program a passive membrane with input variables of Rm, Cm, dt, endtime, Stim start, Stim End, and Stim Strength.

(2) Program a passive membrane with input variables of Rm, Cm, dt, endtime, Stim start, Stim End, and Stim Strength.

CHAPTER 3

Active Membranes

In the previous chapter we considered the response of a membrane to a small stimulus. In these situations the resistance of the membrane was linear and I_{ion} was modeled as a simple resistor and battery in series. In this chapter, we consider what happens when a stimulus causes V_m to reach threshold. The result is that R_m no longer behaves linearly a property that may be represented in a circuit model as the *variable resistance* in Fig. 3.1.

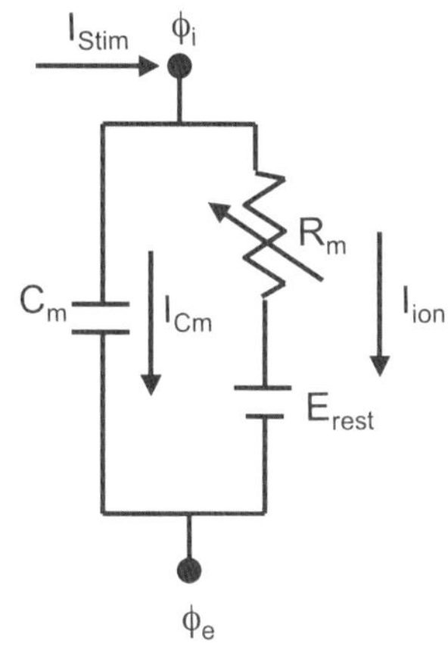

Figure 3.1: A nonlinear resistive membrane.

3.1 THE HODGKIN-HUXLEY MODEL

The first physiologically accurate nonlinear model of I_{ion} was published in 1952 by Hodgkin and Huxley. To create the model, Hodgkin and Huxley combined a brilliant experiment a number of assumptions were about the biophysics of ion channels. Despite its relative simplicity, it remains the gold-standard of ionic membrane models.

3.1.1 The Parallel Conductance Model

The first assumption was that I_{ion} was composed of three currents that acted independently of one another. These currents were Sodium (I_{Na}), Potassium (I_K), and a generic *leakage* current (I_L). Mathematically,

$$I_m = C_m \frac{dV_m}{dt} + I_{Na} + I_K + I_L \ . \tag{3.1}$$

Experimental data showed that I_L was a linear current, while I_{Na} and I_K were nonlinear. Therefore, the circuit analog, taking into account the independence assumption and nonlinear ionic currents, can be represented as in Fig. 3.2. The resistors have been replace by G-terms that represent *conductances*. Conductance is simply the inverse of resistance ($G = \frac{1}{R}$) and is measured in units of *siemens*, abbreviated with a capital 'S'. Therefore, as resistance is increased, conductance is decreased. For biological membranes the most commonly used unit for conductance is mS/cm^2.

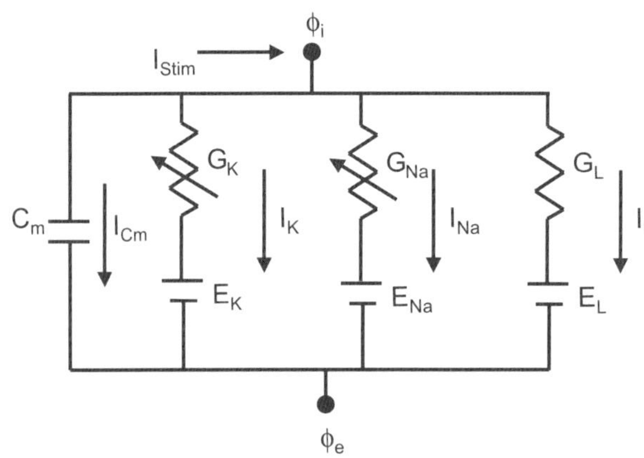

Figure 3.2: Hodgkin-Huxley circuit analog.

3.1.2 The Leakage Current

The leakage current is linear and so can be formulated in the same way as the linear current in Eq. (2.36).

$$I_L = G_L[V_m - E_L] \tag{3.2}$$

where G_L is the *leakage conductance* and E_L is the leakage Nernst potential. The nature of the leakage current was not fully known to Hodgkin and Huxley but they guessed (again correctly) that it was some combination of other ionic currents.

3.1.3 Nonlinear Currents and the Voltage Clamp

Characterization of the two nonlinear currents (I_{Na} and I_K) is nontrivial because there is a natural feedback loop between I_{ion} and V_m. Consider the following generic nonlinear current

$$I_{nl} = G_{nl}(V_m)[V_m - E_{nl}] \, . \tag{3.3}$$

Notice that the conductance, G_{nl}, is a function of V_m. Therefore, a change in V_m causes a change in G_{nl} which has an impact on I_{nl}. As I_{nl} is a component of I_{ion}, any nonzero value for I_{nl} may in fact lead to a further change in V_m. Untangling this interdependence in an experiment was a monumental step first taken by Cole and later used by Hodgkin and Huxley.

The *voltage clamp* is a method of forcing V_m to be constant at any desired holding voltage, V_h, using an external circuit (Fig. 3.3). In this way, the feedback loop between V_m and I_m is effectively cut.

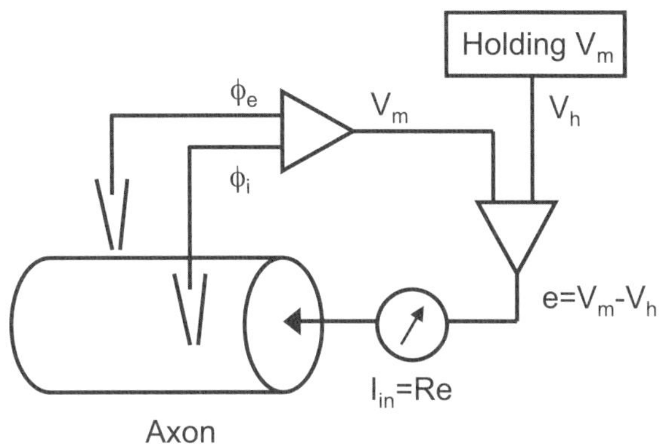

Figure 3.3: Voltage clamp circuit.

If the circuit can react faster than the membrane, an exactly counterbalancing current, I_{in}, may be sent into the cell to maintain a specific voltage level. The transient of this counterbalancing current can be used to determine how fast the membrane can react to changes when $V_m = V_h$, while the steady-state current can be used to determine the maximum current at the clamped V_m. The combination of these two parameters, reaction time and maximum current, as a function of V_h, enabled Hodgkin and Huxley to develop the first mathematical model of an excitable cell.

It is clear from Fig. 3.3 that at least one electrode must have access to the inside of a cell. At the time of Hodgkin and Huxley, electrodes were relatively large compared to mammalian neurons. To create their model, Hodgkin and Huxley chose to study the large axon (up to $1mm$ in diameter) of the giant squid. Since 1952, the size of electrodes has drastically shrunk, allowing for many types of neurons to be studies. The electronic feedback loop of the voltage clamp, however, has remained largely unchanged.

3.1.4 The Sodium Current

The Sodium current is formulated in the same way as the leakage current.

$$I_{Na} = G_{Na}(V_m)[V_m - E_{Na}] \tag{3.4}$$

but now the leakage term G_{Na} is a nonlinear function of V_m. To begin, Hodgkin and Huxley assumed (again correctly) that the nonlinear nature of the flow of Sodium ions was due to the opening and closing of proteins that spanned the membrane. These proteins could somehow change their shape to be either open or closed in response to pH, intra and extracellular concentrations of ions and molecules, temperature, or even V_m. Hodgkin and Huxley focused on the impact of changes in the membrane voltage. Yet another assumption was that each ion channel would be either entirely open or entirely closed. In other words, there would be no in between states. To create their model, however, they assumed that there were thousands or millions of Sodium channels embedded in a small patch of membrane. Therefore, the model would be based on the *probability* that any individual channel would be open. So, although each ion channel would be either open or closed, 30% of the channels may be open and 70% are closed. To capture the aggregate behavior, Hodgkin and Huxley defined a variable, O, that would be 0.3, or the

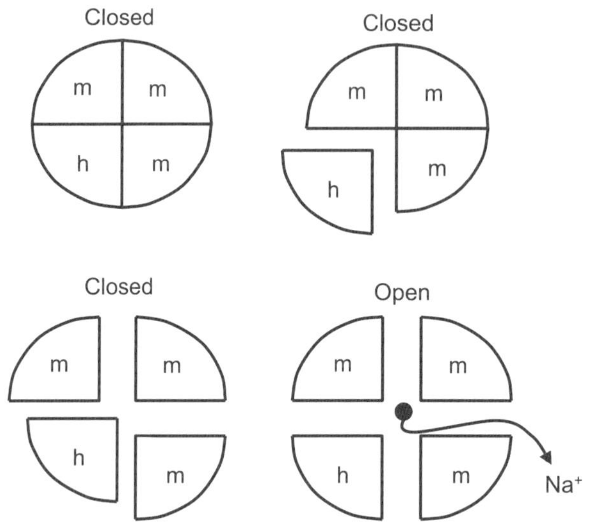

Figure 3.4: Schematic of Sodium channel dynamics.

probability of randomly picking an open channel. Next they next assumed that the rate at which channels opened (on average) was not necessarily the same as the rate at which channels closed. This situation can be represented kinematically by

$$O \leftrightharpoons \alpha(V_m)\beta(V_m)C$$

where rate constants, $\alpha(V_m)$ and $\beta(V_m)$, would be functions of V_m. Using this formulation they could write down a differential equation to describe the dynamics of the O variable.

$$\frac{dO}{dt} = \alpha(V_m)(1 - O) - \beta O(V_m)\ . \tag{3.5}$$

Given the variable O, they defined the Sodium current as:

$$I_{Na} = g_{Na} \times O(V_m)[V_m - E_{Na}] \tag{3.6}$$

where g_{na} is the *maximum conductance*. The meaning of g_{na} is that if $O = 1$, i.e., every sodium channel is open), the maximum possible current will flow. O, on the other hand, can vary between a probability of 0 and 1. When Hodgkin and Huxley fit parameters for α and β using the voltage clamp, they found that the channel was more complicated than the simple O variable could capture. They assumed that the Sodium channel was composed of four parts (Fig. 3.4) and for the channel to be fully open, all four parts needed to be in the right configuration. They therefore assumed I_{Na} to take the form of:

$$I_{Na} = g_{Na}O_{1(V_m)}O_2(V_m)O_3(V_m)O_4(V_m)[V_m - E_{Na}] \tag{3.7}$$

where $O_{1,2,3,4}$ were the four parts. Further analysis showed that three of the parts functioned in the same way, i.e., α and β functions were identical). These channel parts were each called m and the remaining part was called h. Their model of the Sodium current was described by

$$I_{Na} = g_{Na} m^3 h(V_m)[V_m - E_{Na}] \tag{3.8}$$
$$\frac{dm}{dt} = \alpha_m(1-m) - \beta_m m \tag{3.9}$$
$$\frac{dh}{dt} = \alpha_h(1-h) - \beta_h h \tag{3.10}$$

where m and h are known as *gating* variable because they control how ions are gated through the channel. The only difference between the h and m variables is that the functions of α and β are different.

Using their experimental data and some curve fitting, Hodgkin Huxley found the following fits for the α's an β's:

$$\alpha_m = 0.1 \frac{25 - v_m}{e^{(25-v_m)/10} - 1} \tag{3.11}$$
$$\beta m = 4e^{-v_m/18} \tag{3.12}$$
$$\alpha_h = 0.07e^{v_m/20} \tag{3.13}$$
$$\beta_h = \frac{1}{e^{(30-v_m)/10} + 1} \tag{3.14}$$

where v_m is scaled by the resting potential such that $v_m = V_m - V_m^{rest}$.

3.1.5 The Potassium Current

The Potassium current may be assumed to be of the nonlinear form

$$I_K = g_K O(V_m)[V_m - E_K]\,. \tag{3.15}$$

Recall, however, that the first assumption was that each current was independent of the other currents present. So, simply performing the voltage clamp as described above would yield data on $I_K + I_{Na}$. To separate the two currents, Hodgkin and Huxley used tetrodotoxin (TTX), a Sodium channel blocker, to isolate I_K. They then performed the same experiment without TTX to yield $I_K + I_{Na}$ and by simple subtraction they isolated I_{Na}. Using this two-step procedure, they found that $O(V_m)$ for I_K was the product of four gating variables that were all identical:

$$I_K = g_K n^4[V_m - E_K] \tag{3.16}$$
$$\frac{dn}{dt} = \alpha_n(1-n) - \beta_n n\,. \tag{3.17}$$

Experimental data for the n gating variable was fit to the following α and β functions:

$$\alpha_n = 0.01 \frac{10 - v_m}{e^{(10-v_m)/10} - 1} \tag{3.18}$$
$$\beta_n = 0.125e^{-v_m/80}\,. \tag{3.19}$$

3.1.6 Steady-State and Time Constants

An alternative, and possibly more intuitive, way of writing the gating differential equations is

$$\frac{dm}{dt} = \frac{m_\infty - m}{\tau_m} \tag{3.20}$$

$$m_\infty = \frac{\alpha_m}{\alpha_m + \beta_m} \tag{3.21}$$

$$\tau_m = \frac{1}{\alpha_m + \beta_m} \tag{3.22}$$

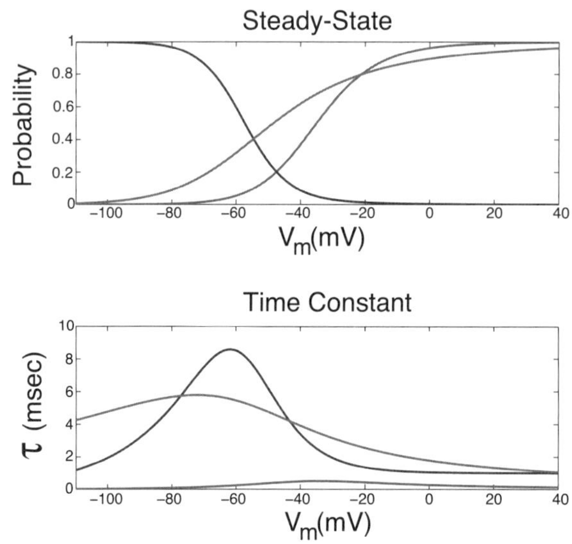

Figure 3.5: Steady-state and time constant for Hodgkin-Huxley gates.

with similar equations to describe the h and n variables. Plots of the steady-state and time constant curves for m (red), h (blue), and n (green) are shown in Fig. 3.5. The reason for writing the equations this way is that the steady state (∞-terms) and time constants (τ-terms) have a physical interpretation. The solution to Eq. (3.20) is

$$m(t) = m_\infty - (m_\infty - m_0)e^{-t/\tau_m} \tag{3.23}$$

where m_0 is the initial value of m. In the context of the voltage clamp, consider that V_m is equal to a value A and has been there for a long time, i.e., $t \to \infty$). Therefore, $m \to m_\infty(A)$ which could be read directly from Fig. 3.5. Next, consider that the membrane potential is suddenly changed to $V_m = B$. Since the gating variable, m, cannot change instantaneously, the initial condition m_0 is equal to $m_\infty(A)$. The steady-state and time constants, however, do change instantaneously. Therefore, to compute $m(t)$ at any point after V_m is clamped to B

$$m(t) = m_\infty(B) - (m_\infty(B) - m_0)e^{-t/\tau_m(B)} \quad (3.24)$$
$$m(t) = m_\infty(B) - (m_\infty(B) - m_\infty(A))e^{-t/\tau_m(B)} \,. \quad (3.25)$$

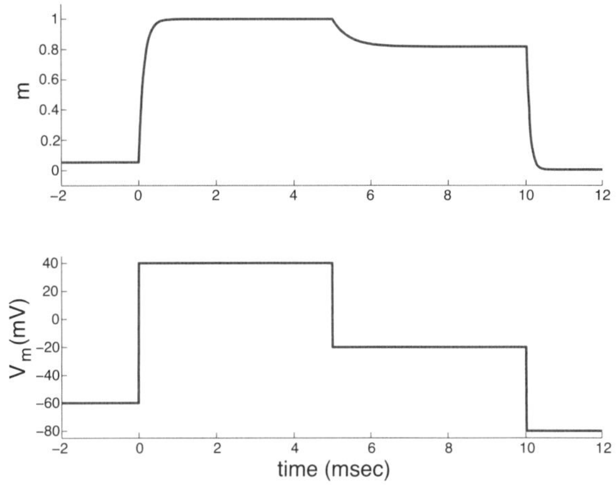

Figure 3.6: m-gate changes over time for changes in holding potential.

The intuitive interpretation is that m_∞ is the target for m and τ_m is the rate at which this new target is reached. To demonstrate this idea, Fig. 3.6 shows three changes in V_h with the corresponding change in m. It may be helpful to think about how Figs. 3.5 and 3.6 are related.

3.1.7 A Game of Tag

An analogy to the steady-states and time constants may be thought of as a game on an obstacle course. You and 100 of your friends all have the goal of reaching a diamond that will be set somewhere on the course. You can imagine that if the diamond stays in one location, you would simply navigate through water, hoops, sand, and gravel to get to it. Although this may take time, eventually you would all reach the diamond. The location of the diamond therefore is analogous to the steady state. If the location of the diamond, however, suddenly changed, you and your friends would again set out across the course to reach it. The type of obstacles in your way (not necessarily the distance) would determine how long it takes to reach the diamond. The way that your time to reach the diamond is dependent upon the obstacles in your way is analogous to the time constant. Now imagine that the diamond is moving around the course, suddenly appearing in one location and then quickly moving to a new location, faster than you and your friends can keep up.

Next imagine that at the same time that you and your friends are chasing a diamond, another group of 101 people are also chasing a ruby, and a third set of 101 people are chasing an emerald. Here, the analogy is at the diamond is the m-gate, the ruby is the h-gate, and the emerald is the n-gate. Although our game is not a perfect analogy to the dynamics of gates, it may help understand the next section where the Hodgkin-Huxley action potential is examined.

3.2 THE HODGKIN-HUXLEY ACTION POTENTIAL

When stimulated, the Hodgkin-Huxley model generates a rapid depolarization followed by a slower depolarization back to rest. This cycle is known as an *action potential* and is the basic functional measure of all excitable cells. Below we discuss the four phases of the action potential as summarized in Fig. 3.7.

3.2.1 Phase 0 - Rest

When V_m is at rest, $m = m_\infty \approx 0.05$ and $h = h_\infty \approx 0.6$ and therefore $g_{na}m^3h$ is small. On the other hand, V_m is close to E_K, so the driving force for I_K is small. For these different reasons, both I_{Na} and I_K are small. The result is that at rest the linear leakage current, I_L, will dominate and the membrane behaves linearly.

3.2.2 Phase 1 - Activation

If a stimulus is applied, V_m will depolarize. As V_m changes, the α, β, time constants and steady-state values will also change. Consider the following:

1. The driving force for I_{Na} (e.g., $V_m - E_{Na}$) at rest is large and negative because V_m is much smaller than E_{Na}. If given the chance, Na^+ ions would therefore move *into* the cell.

2. Na^+ ions are prevented from crossing the membrane because m is small.

3. Although the m_∞ curve is flat at V_m^{rest}, if the membrane is depolarized slightly, m_∞ will become relatively large. This is because of the steep slope of the m_∞ curve.

4. τ_m is small so any change to m_∞ will result in a fast change in m.

5. The driving force for I_K at rest is small because V_m is close to E_K.

Given these properties, consider that a small depolarizing change in V_m would cause m_∞, and therefore m, to increase quickly. The result is that $g_{Na}m^3h$ and thus I_{Na} increase and Na^+ will rush into the cell. As Na^+ rushes into the small volume of the cell, the intracellular potential, ϕ_i, will increase. The extracellular space, however, is large and so ϕ_e will change only slightly. The result is that the cell membrane will become more depolarized.

The fast rise in the membrane potential is called *activation* or *upstroke* and the m gate is therefore called the *activation* gate. The origin of a threshold voltage can now be understood. Below V_m^{th}, m_∞ and m are not large enough to create a large I_{Na}. But, when V_m reaches V_m^{th}, m is on the steep part the m_∞ curve. So, a small depolarizing I_{Na} current pushes V_m and m to larger and larger values. This runaway effect is counterbalanced by *repolarization*.

3.2.3 Phase 2 - Repolarization

As the I_{Na} current attempts to drive $V_m \rightarrow E_{Na}$, two other players become important. First, during the upstroke h_∞ changes from ≈ 0.6 to ≈ 0, but does so slowly, i.e., τ_h is relatively large compared to τ_m). Over time, however, h becomes small enough that the m^3h term, and therefore I_{Na}, is reduced. The h gating variable is therefore called the *inactivation* gate because it turns the Sodium current off after activation. Second, after depolarization, V_m is no longer close to E_K and the driving force for I_K is increased. Due to the sign of E_K, I_K will be a positive current, i.e., K^+ flows *out* of the cell). Positive charge exiting the cell causes V_m to decrease or *repolarize*.

3.2.4 Phase 3 - Hyperpolarization

In the Hodgkin-Huxley model, the n-gate does not return the membrane directly to rest. Instead, V_m drops slightly below V_m^{rest}, a *hyperpolarization*, and then depolarizes back to rest.

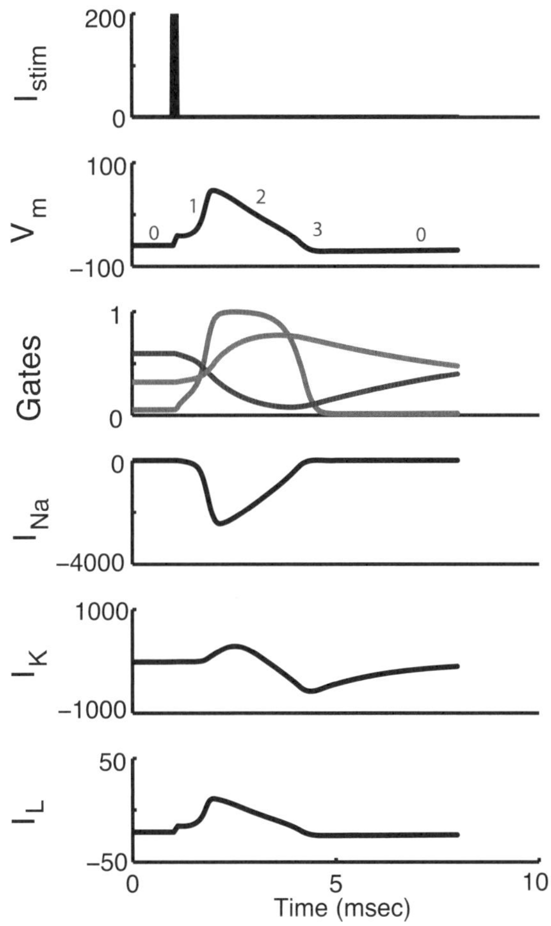

Figure 3.7: Hodgkin-Huxley action potential and currents.

3.3 PROPERTIES OF NEURONAL ACTION POTENTIALS

The Hodgkin-Huxely model was successful largely because it explained the origin of four of the most important features of an action potential. We review these properties below and related them to the relevant parts of the model.

3.3.1 All Or None

When a stimulus of a small amplitude is applied to an active membrane, the membrane will act in the same way as a passive membrane. We argued in Ch. 2 that at some combination of strength and duration (recall Fig. 2.6) the membrane will reach a threshold, V^{th}, after which the membrane becomes nonlinear. Furthermore, the Hodgkin-Huxely model correctly predicts that once the membrane is above threshold, the Sodium current begins the positive feedback cycle that results in the upstroke of the action potential. Therefore, once the Sodium current is activated, an action potential is inevitable. This abrupt *all or none* response of a neuron also highlights the importance of the strength-duration curve because any stimulus above the curve will result in an action potential.

3.3.2 Refractory Periods

After I_{Na} has caused the upstroke of the action potential and repolarization has begun, m will remain relatively high. The h-gate, however, will slowly drop to a value forcing I_{Na} to become small. But remember that τ_h is large so any changes in h occur slowly. Therefore, h will remain small for some time while the repolarization of the action potential is occurring. During this time, no additional change in m will cause a second large influx of I_{Na}. The time when no second action potential is possible is called the *absolute refractory period* (see Fig. 3.8). Following the absolute refractory period is the *relative refractory period.* During the relative refractory period a second action potential is possible, but requires a greater depolarization (e.g., larger I_{stim}) than the first action potential.

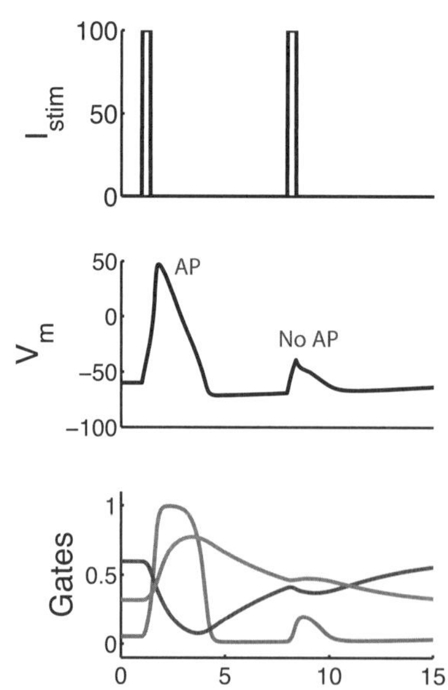

Figure 3.8: Demonstration of action potential refractory period.

3.3.3 Anode Break

An interesting phenomenon can occur if a hyperpolarizing current is applied to the membrane for a long time. In this case, m is forced low and because the current is on for a long time, h is forced high. When the stimulus is turned off, m will respond quickly and increase. h, on the other hand, will respond slowly and stay at approximately 1. The combination of these two factors can cause $g_{Na}m^3h$ to reach a level high enough for I_{Na} to increase in magnitude. As in a normal activation, the increase in I_{Na} can lead to the run-away process that causes the upstroke. The firing of an action potential using a long duration hyperpolarizing current is known as *anode break* and is shown in Fig. 3.9.

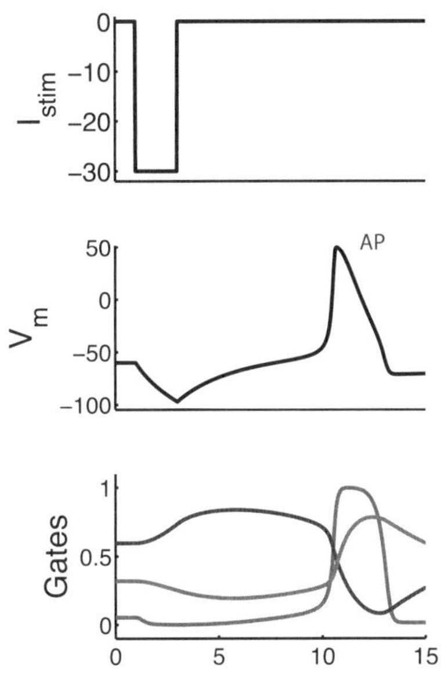

Figure 3.9: Demonstration of an action potential generated by anode break.

3.3.4 Accommodation

Another interesting phenomenon can be observed if the membrane is very slowly depolarized. A slow depolarization can be accomplished by making I_{stim} a ramp with a small slope as in Fig. 3.10. If the depolarization is slow enough, the h gate will have time to reach h_∞ (unlike in a normal depolarization). Therefore, even though m increases, h will decrease such that $g_{Na}m^3h$ will not be a large enough to initiate the run-away I_{Na} current. The effect is that V_m can become higher than V_m^{th} without causing an action potential to fire. This phenomenon is called *accommodation* because the h gate is impacting the value of the threshold. The meaning of both anode break and accommodation is that what we typically think of as the threshold voltage is not an exact number and depends on the path taken to cause the run-away I_{Na}. Both are related to the time constant of h.

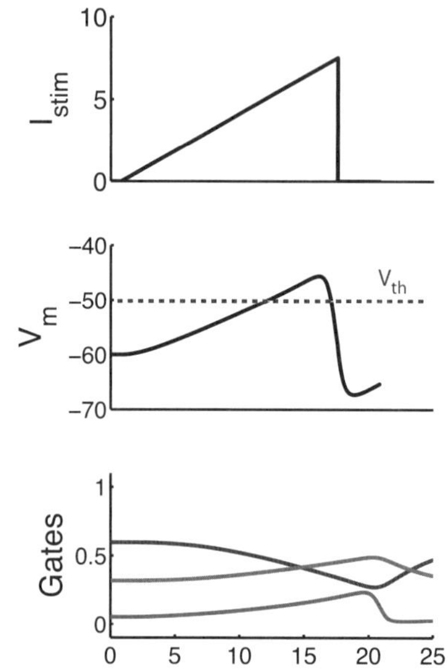

Figure 3.10: Demonstration of accommodation of threshold voltage.

3.4 COMPLEX IONIC MODELS

While Hodgkin and Huxley created the first physiologically based mathematic model of a neuron, there were many neuronal phenomenon that it could not reproduce. Below we review some of the ways researchers have extended the Hodgkin-Huxley model.

3.4.1 More Currents

As the electrodes used for the voltage clamp method have become smaller and smaller, and new channel blocking drugs were found, it became possible to isolate more than the three simple currents of the Hodgkin-Huxely model. The result has been that more recent models have many more currents included in I_{ion}. For example, the unspecified leakage current was found to be due to other ions such as Ca^{2+}, Mg^{2+}, and Cl^{-}. To include these currents we would simply add them to the parallel conductance I_{ion} term.

$$I_{\text{ion}} = I_{Na} + I_K + I_{Ca} + I_{Mg} + I_{Cl} \, . \tag{3.26}$$

Here the leakage current has been replaced by $I_{Ca} + I_{Mg} + I_{Cl}$. It was also found that I_K may in fact be the summation of many different types of Potassium ion channels.

$$I_{\text{ion}} = I_{Na} + I_{K1} + I_{K2} + I_{K3} + I_{Ca} + I_{Mg} + I_{Cl} \tag{3.27}$$

where I_K has been split into three different currents. Similarly, I_{Ca} may be decomposed into many currents. For each current added, there will be new gating variables and new differential equations.

It may be questioned why the addition of these currents is necessary. After all, the Hodgkin-Huxely model successfully reproduced the properties of the action potential. There are at least three reasons why the extra currents, and complexity, are important. First, the giant squid axon does not display all of the behavior of mammalian neurons. Second, is often the case that a genetic mutation, drug, or hormone will only effect one type of ion channel. For example, Nifedipine specifically blocks a certain type of Calcium ion channel, while allowing other Calcium channels to remain functional. Third, when a single action potential fires, Na^+, K^+, and other ions cross the membrane. Surprisingly, it does not require many ions to cross the membrane to achieve the $\approx 100mV$ change in V_m during the action potential. Therefore, the intracellular and extracellular concentrations do not change significantly during a single action potential. If many action potentials fire in rapid succession, however, the concentrations will eventually begin to change. Cells therefore have a variety of *pumps* and *exchangers* that slowly act in the background, transporting ions back across the membrane. As these pumps are typically working against a concentration gradient, they require ATP. Similar to other currents, they may be added to I_{ion}. We will consider how to incorporate changing concentrations in Sec. 3.4.5.

3.4.2 The Traub Model of the Pyramidal Neuron

A model with additional currents is the Traub model. We present the model here as an example of how extra currents lead to more realistic behavior and because it will be encountered again in Ch. 7.

$$
\begin{aligned}
I_{\text{ion}} &= I_{Na} + I_K + I_{Ca} + I_{kCa} \\
I_{Na} &= g_{na}m^3h[V_m - E_{Na}] \\
I_K &= g_K n^4 y[V_m - E_K] \\
I_{Ca} &= g_{Ca}s^5 r[V_m - E_{Ca}] \\
I_{kCa} &= g_{kCa}q[V_m - V_k]
\end{aligned}
$$

where the gates are defined in Table 3.1.

Table 3.1:

gate	α	β
s	$\frac{0.04(60-V_m)}{\exp\left(\frac{60-V_m}{10}\right)-1}$	$\frac{0.005(V_m-45)}{\exp\left(\frac{V_m-45}{10}\right)-1}$
m	$\frac{0.32(12-V_m)}{\exp\left(\frac{12-V_m}{4}\right)-1}$	$\frac{0.28(V_m-40)}{\exp\left(\frac{V_m-40}{5}\right)-1}$
h	$0.128\exp\left(\frac{17-V_m}{18}\right)$	$\frac{4}{\exp\left(\frac{20-V_m}{5}\right)+1}$
n	$\frac{0.032(15-V_m)}{\exp\left(\frac{15-V_m}{5}\right)-1}$	$0.5\exp\left(\frac{10-V_m}{40}\right)$
y	$0.028\exp\left(\frac{15-V_m}{15}\right)+\frac{2}{\exp\left(\frac{85-V_m}{10}\right)+1}$	$\frac{0.4}{\exp\left(\frac{40-V_m}{10}\right)+1}$
r	0.005	$\frac{0.025(200-x)}{\exp\left(\frac{200-x}{20}\right)-1}$
q	$\frac{0.005(200-x)\exp\left(\frac{V_m}{20}\right)}{\exp\left(\frac{200-x}{20}\right)-1}$	0.002

and the differential equations for each gate are of the form shown in Eq. (3.5). One additional x-gate is defined by

$$\frac{dx}{dt} - \frac{c \cdot I_{Ca}}{A \cdot d} - \beta_x \cdot x \ .$$

The constants for the Traub model appear in Table 3.2.

Table 3.2:

$\beta_x = 0.1$	
$A = 3320$	$c = 5.2$
$d = 0.0005$	$E_{Ca} = 140mV$
$E_{Na} = 115mV$	$E_k = -15mV$
$g_{Na} = 3.32\mu S$	$g_{kCa} = 0.1\mu S$
$g_{Ca} = 6.64\mu S$	$g_K = 3.98\mu S$

The additional phenomena the Traub model enables is a repeating cycle of depolarization and repolarization called *bursting*. Bursting is often modulated by the strength of a sustained input current and is shown in Fig. 3.11.

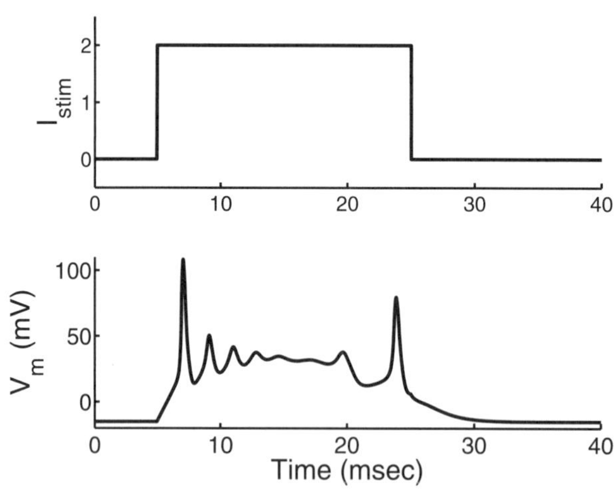

Figure 3.11: Bursting in the Traub membrane model.

3.4.3 Complex Gating

Hodgkin and Huxley assumed that ion gates could be either open or closed. But, the h gate was a clue to the more complex nature of channels. In reality, some channels may be open in more than one way, closed in more than one way, and even inactivated in more than one way. The method of modeling the more complex dynamics of an ion channel is through a *Markov State* model. Figure 3.12 is an example of the more complex dynamics of a Sodium channel. Each circle is a state that the channel may be in. The α and β terms are the rates of transition from one state to another and typically do not depend upon V_m. To derive the differential equations for a state, we subtract the product of the state and all outflowing rates (exiting arrows) from the product of the entering states and inflow rates (entering arrows). For example, the equations for the model in Fig. 3.12 are:

$$\frac{dC3}{dt} = \beta_{23}C2 - \alpha_{32}C3 \tag{3.28}$$
$$\frac{dC2}{dt} = \beta_{12}C1 + \alpha_{32}C3 - (\alpha_{21} + \beta_{23})C2 \tag{3.29}$$
$$\frac{dC1}{dt} = \beta_{01}O + \alpha_{21}C2 - (\alpha_{10} + \beta_{12})C1 \tag{3.30}$$
$$\frac{dIF}{dt} = \beta_{45}IS + \alpha_{40}O - (\alpha_{54} + \beta_{40})IF \tag{3.31}$$
$$\frac{dIS}{dt} = \alpha_{54}IF - \beta_{45}IS \tag{3.32}$$
$$\frac{dO}{dt} = \beta_{40}IF + \alpha_{10}C1 - (\alpha_{40} + \beta_{10})O\ . \tag{3.33}$$

The advantage of this type of model is that it is a more direct link between the genetic code, expressed protein, structure of the channel, and electrophysiologic function. It is hoped that someday we will be able to engineer sections of genetic code and simulate their function before they are made. In this way, diseased channels could be repaired by making only the correction that is needed. Such a simulation of the function will require a Markov model.

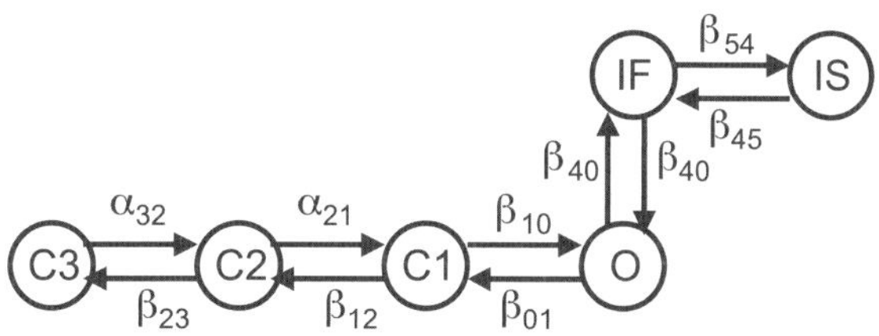

Figure 3.12: Markov model of Sodium channel.

3.4.4 Gating Dependence on pH, Temperature and Concentrations

It was mentioned in Sec. 3.1.4 that ion channels are often gated by V_m but could also be gated by pH, ionic or molecular concentrations, temperature, or some combination of all of these gating mechanisms. For example, the functions for α and β in the Hodgkin-Huxely model were for a particular temperature. In general, these function may be much more complex, for example $\alpha(V_m, pH, [Mg^{2+}]_e, [Atropine], T)$. Although this may seem to be a daunting modeling task, there are some very important ion channels that are not gated by V_m and are only sensitive to the concentration of some ion or molecule. These channels are said to be *Ligand*-gated and typically play an important role in mutations, the impact of drugs and the onset of diseases. In Ch. 6, the dynamics of the synapse will be modeled as gating by concentration of neurotransmitter.

3.4.5 Changes in Nernst Potentials

Earlier it was stated that a bursting neuron may begin to change the intra and/or extracellular concentrations. Even with pumps and exchangers, the cell may build up concentrations of some ions. Accompanying these concentration changes, will be a change in the Nernst potentials, and thus the driving forces behind the action potential. It is therefore important for some models to incorporate a way for concentrations to change. The most common technique is to write differential equations for the intra and extracellular ionic concentrations. For example, consider that there may be several I_K outward currents that could potentially deplete the intracellular Potassium concentration, $[K^+]_i$. This change in $[K^+]_i$ would impact E_K as well as V_m^{rest} through the Goldman-Hodgkin-Katz equation. A common formulation would be:

$$\frac{d[K^+]_i}{dt} = -\frac{I_{K1} + I_{K2} + I_{K3}}{FV_i} \tag{3.34}$$

$$\frac{d[K^+]_e}{dt} = \frac{I_{K1} + I_{K2} + I_{K3}}{FV_s} \tag{3.35}$$

where F is Faraday's constant, V_i is the volume of the intracellular space, and V_s is the small *shell* of extracellular space surrounding the outside of the cell. In this formulation, any K^+ ions leaving the intracellular space must enter the extracellular space.

A situation where ionic concentrations may drastically change is during a disease. For example, during ischemia (e.g., lack of O_2) the concentration of ATP drops and the pumps that restore concentration gradients become less efficient and may even fail. The result is a buildup of ions. If this buildup continues, cells may die and lyse (e.g., pop) spilling their ions into the extracellular space and lead to a drastic change in concentrations.

3.4.6 Intracellular Compartments and Buffers

The intracellular space in a cell is populated by many smaller organelles. Some of these organelles have their own membranes, with their own ion channels, and are capable of transporting ions to and from the intracellular space to the intra-organelle space. When this transport occurs, the concentration of that ion will change in the intracellular space. Functionally, these organelles behave as a *buffer*. As in the previous section, these changes in concentration can impact the Nernst potential of that ion. Figure 3.13 is a schematic of a Ca^{2+} buffer.

3.5 PHENOMENOLOGICAL MODELS

Although the Hodgkin-Huxley mathematical model is simple to solve using todays computers, Hodgkin and Huxley performed all of their calculations using calculators. In particular, computing exponentials was difficult, so a number of simplified models were developed that captured the basic features of neuronal action potentials.

3.5.1 Fitzhugh-Naghumo Model

In 1961, Fitzhugh and Nagumo independently developed a model based upon simple polynomials. In the Fitzhugh-Naghumo model, the m, h, n, and V_m variables were reduced to only two differential equations.

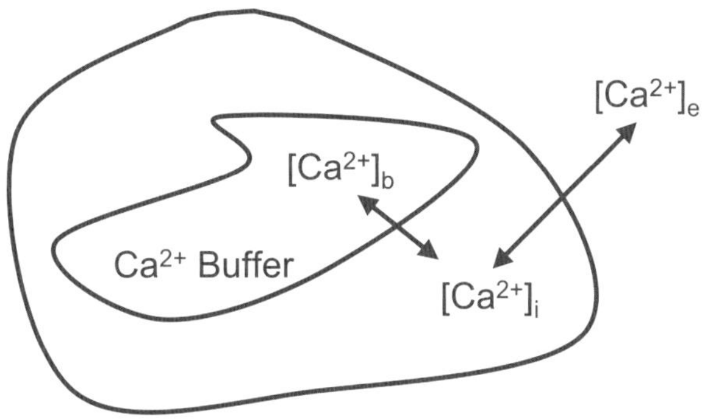

Figure 3.13: Ionic buffering by intracellular compartments.

$$\frac{dV_m}{dt} = V_m - \frac{V_m^3}{3} - W + I_{\text{stim}} \tag{3.36}$$
$$\frac{dW}{dt} = a\left[V_m + b - cW\right] . \tag{3.37}$$

The model has a steady-state resting potential, fast all-or-none upstroke, and slower repolarization. Furthermore, the parameters a, b, and c can be tuned to generate action potentials of different shapes and durations.

3.5.2 Hindmarsh-Rose Model

The Fitzhugh-Naghumo model does not generate bursting behavior. In 1984, Hindmarsh and Rose developed a set of three differential equations that would allow for the phenomenon of bursting:

$$\frac{dV_m}{dt} = -aV_m^3 + bV_m^2 + y - z + I \tag{3.38}$$
$$\frac{dy}{dt} = -dV_m^2 - y + c \tag{3.39}$$
$$\frac{dz}{dt} = rsV_m - rz - rsV_{\text{rest}} \tag{3.40}$$

where a, b, c, d, r, s, and V_{rest} are constants and may be tuned to producing different bursting behavior.

3.5.3 Integrate and Fire Model

In 1907, long before Hodgkin and Huxley, Lapicque proposed that the firing of an action potential could be modeled simply as a spike in voltage. As I_{stim} is applied, V_m will depolarize according to the familiar passive model

$$\frac{dV_m}{dt} = \frac{1}{C_m}\left[-G_L(V_m - E_L) + I_{\text{stim}}\right] . \tag{3.41}$$

When V_m reaches some predefined V_m^{th}, however, all solving of differential equations is suspended. The simulation then abruptly jumps V_m to some value, V_m^{peak}, to simulate the upstroke of the action potential. In some implementations, V_m is clamped to V_m^{peak} for some short time and then *reset* back to the resting value as in Fig. 3.14. In other implementations, the reset is not exactly back to rest but below V_m^{th}. This simple yet elegant model is called the *integrate and fire* model. It is still used in neural modeling because it requires no updating of gating variables.

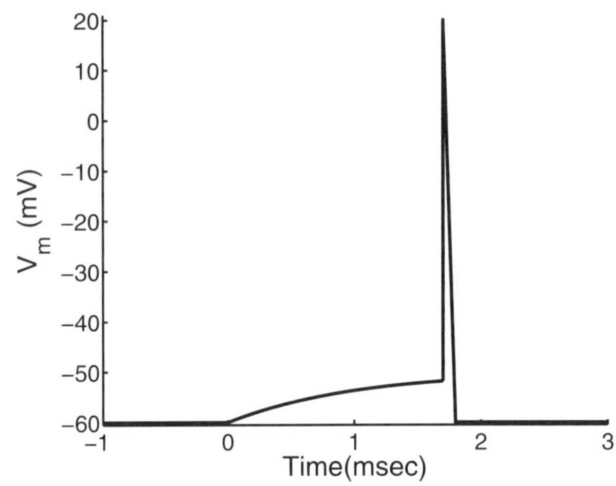

Figure 3.14: Integrate and fire action potential.

3.6 NUMERICAL METHODS: TEMPLATE FOR AN ACTIVE MEMBRANE

In Sec. 2.5, a way of numerically solving a differential equation was outlined. In the active membrane models, we need to keep track of several differential equations as well as compute rate constants, steady-state values and currents. Below is a template for how to write a program to solve the active equations.

```
Define constants (e.g., G_L, g_Na, g_K, C_m, dt)
Compute initial αs and βs
Compute initial conditions for state variables
(e.g., V_m^rest, m, h, n)

for (time=0 to time=end in increments of dt)

        Compute Currents (e.g., I_L, I_Na, I_K)
        Compute αs and βs
        Update Differential Equations (V_m, m, n, h)
        Save values of interest into an array (e.g., V_m)
```

```
end

Store values of interest to a file
```

Homework Problems

(1) If a stimulus is above a strength duration curve and the membrane is active, will and action potential fire? Explain.

(2) The following data were recorded at the peak of a Hodgkin-Huxely action potential:

$$I_K = 353.574 \frac{\mu A}{cm^2}$$

$I_{Na} = -394.59 \frac{\mu A}{cm^2}$
$g_K = 3.36737 \frac{mS}{cm^2}$
$g_{Na} = 130.7429 \frac{mS}{cm^2}$
$m = 0.889$
$h = 0.288$
$E_K = -75.0mV$
$E_{Na} = 50mV$.

a) What is the peak magnitude of the action potential in mV?
b) How does this compare to E_{Na}? What does this mean?
c) Compute the fraction of K+ channels open.
d) What is the value of I_{leak}?

(2) In a voltage-clamp experiment, the transmembrane potential (V_m) was changed from rest (-60mV) to 0mV, kept at 0mV for 100ms, and then changed to -50mV. Relevant parameters are given in the table below (time constants in msec).

Table 3.3:

V_m	α_m	τ_m	α_h	τ_h
-60mV	0.225	0.238	0.0697	8.51
-50mV	0.433	0.368	0.0423	6.16
0mV	3.62	0.266	0.00347	1.05

a) Sketch the holding potential and label the times and voltage levels.
b) Compute the fraction of h gates open at $t = 100msec$.
c) Compute the fraction of h gates that are open at $t = 104msec$.
d) Compute the probability of an Na+ channel being open at $t = 104msec$.

(3) Extend the Hodgkin-Huxely model to include Potassium buffering by an intracellular organelle. Be sure to write down differential equations for the concentrations of K^+ in the organelle, cell,

and extracellular space. Include two additional currents, one between intra and extracellular space and one between intracellular and organelle space.

(4) Write down the differential equations to describe the Markov Model of the K^+ channel shown below.

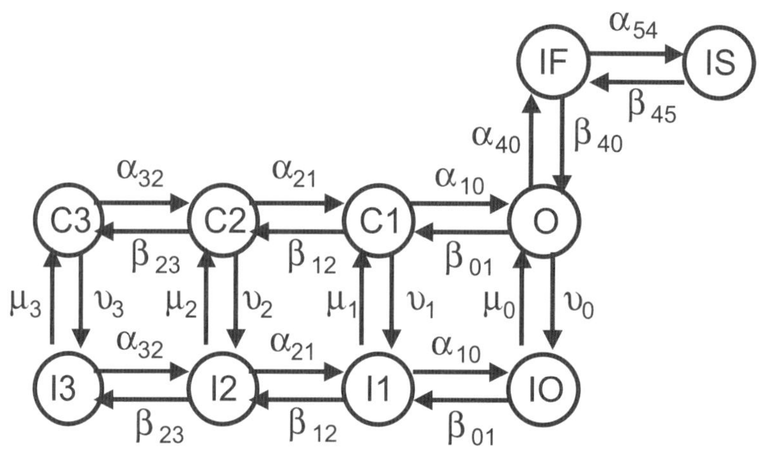

Simulation Problems

(1) Program the Hodgkin-Huxely membrane model and reproduce one of the properties in Sec. 3.3. Assume $G_L = 0.3mS/cm^2$, $g_{na} = 120.0mS/cm^2$, $g_K = 36.0mS/cm^2$, $E_L = -50mV$, $E_{Na} = 55mV$, $E_K = -72.0mV$ $C_m = 1\mu F/cm^2$.

(2) Create a strength-duration curve for anode break of the Hodgkin-Huxely model.

(3) Program the Integrate and Fire membrane model with $R_m = 2k\Omega cm^2, C_m = 1.0\mu F/cm^2, V_m^{\text{rest}} = -70mV$, $V_m^{th} = -60mV$, and $V_m^{\text{peak}} = 20mV$. Assume that the voltage is reset to rest after firing.

(4) Find I_{rhe} for the Integrate and Fire model in problem 3. Then, proceed to problem 5.

(5) Program the Hindmarsh-Rose model with the parameters $a = 0.4$, $b = 2.4$, $c = 1.5$, $d = 1.0$, $s = 1.0, r = 0.001$. Show that by adding a short-duration I_{stim} to the dx/dt term, these parameters will cause periodic bursting.

(6) For the Hindmarsh-Rose model in problem 5, find a value for a continuous I_{stim} that will turn the bursting into a model that fires periodically.

(7) Program the Traub Neuron model and determine if the properties in Sec. 3.3 apply.

(8) For the Traub model, determine if a continuously applied current can induce repeated bursting. If stimulus strength is changed how does the behavior of the model change?

CHAPTER 4

Propagation

Experimental studies have shown that a neuron does not fire all at once. Rather, there is a wave of electrical activity that is passed from one small patch of membrane to the next, and so on, around the cell. Likewise, electrical impulses spread down dendrites to the soma by moving from one patch to the next. A similar process is involved in the spread of electrical impulses from the axon hillock down the axon. These waves of electrical activity are called *propagation*. Not all propagation, however, is the same. As the dendrites are composed of passive patches of membrane, they will leak charge out of each patch as propagation moves forward. Therefore, the strength of the impulse will be decreased or *attenuated* down the dendrite. Attenuated propagation is often called *passive propagation*. The axon, on the other hand, is composed of active patches of membrane that can generate their own currents (e.g., an action potential). Therefore, an impulse that enters the axon hillock will propagate *unattenuated* to the end of the axon. Furthermore, because of the refractory period, *propagation* can proceed in only one direction. Unattenuated propagation is often called *active propagation*.

In this chapter, we will develop models to describe propagation. As passive propagation is considerable more simple than active propagation, we will begin by considering propagation in the dendrites. Active propagation in an axon will be explained as a special case of passive propagation.

4.1 PASSIVE PROPAGATION IN DENDRITES

4.1.1 The Core Conductor Model

The dendrites of a neuron may be thought of as being composed of many small cylindrical patches of passive membrane connected together into a thin one-dimensional *cable* (Fig. 4.1). In fact, these assumptions are the same as those used by the pioneers of cable theory to describe propagation of electricity down a wire. Since the theory of propagation down a wire was developed before propagation in neurons was considered, the terms *cable theory* and *core conductor theory* have been adopted by electrophysiologists.

Consider the *discrete* cable in Fig. 4.2 where the passive elements are separated by dx and connected together in the intracellular and extracellular space by resistance per unit length, r_i and r_e (Ω/cm). From Fig. 4.2, we can choose a node in the center of this cable in Fig. 4.3 and write down intra and extracellular currents at nodes 1, 2, and 3 using Kirchhoff's Current Law.

At Node 2,

$$\frac{\phi_i^1 - \phi_i^2}{r_i \cdot dx} - \frac{\phi_i^2 - \phi_i^3}{r_i \cdot dx} - dx \cdot i_m = 0 \tag{4.1}$$

$$\frac{\phi_e^1 - \phi_e^2}{r_e \cdot dx} - \frac{\phi_e^2 - \phi_e^3}{r_e \cdot dx} + dx \cdot i_m = 0 \tag{4.2}$$

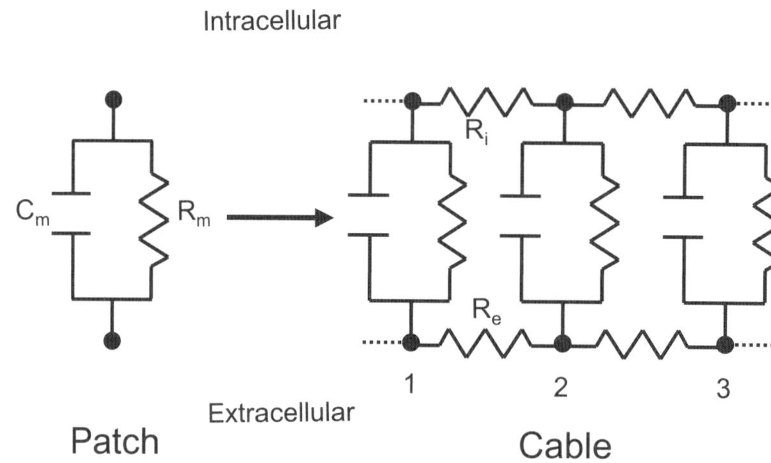

Figure 4.1: Construction of a cable from patches of membrane.

or more compactly,

$$\frac{\phi_i^1 - 2\phi_i^2 + \phi_i^3}{dx^2 \cdot r_i} = i_m \tag{4.3}$$

$$\frac{\phi_e^1 - 2\phi_e^2 + \phi_e^3}{dx^2 \cdot r_e} = -i_m \tag{4.4}$$

where i_m is defined as the passive circuit:

$$i_m = c_m \frac{dV_m}{dt} - \frac{V_m}{r_m} \tag{4.5}$$

rearranging Eqs. (4.3) and (4.4)

$$\frac{\phi_i^1 - 2\phi_i^2 + \phi_i^3}{dx^2} = r_i \cdot i_m \tag{4.6}$$

$$\frac{\phi_e^1 - 2\phi_e^2 + \phi_e^3}{dx^2} = -r_e \cdot i_m \tag{4.7}$$

and then subtracting Eq. (4.7) from Eq. (4.6)

$$\frac{(\phi_i^1 - \phi_e^1) - 2(\phi_i^2 - \phi_e^2) + (\phi_i^3 - \phi_3^3)}{dx^2} = (r_i - r_e)i_m \tag{4.8}$$

$$\frac{V_m^1 - 2V_m^2 + V_m^3}{dx^2} = (r_i - r_e)i_m \tag{4.9}$$

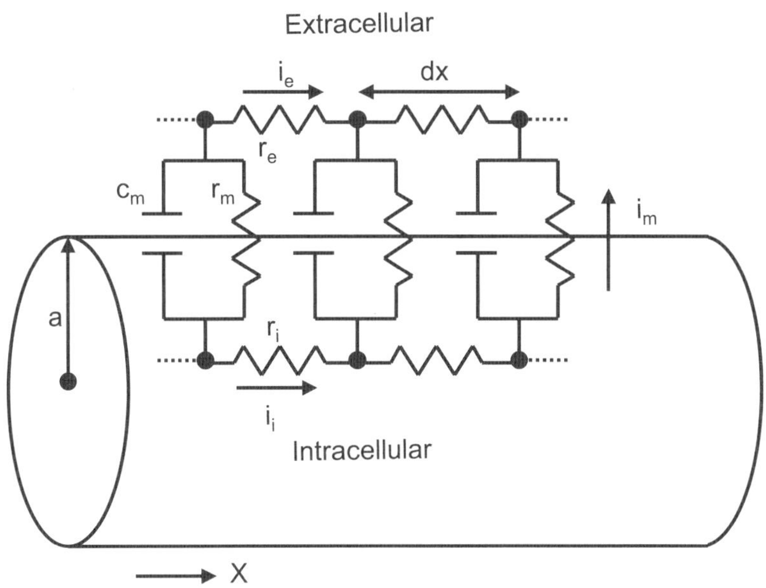

Figure 4.2: Circuit analog of a discrete cable.

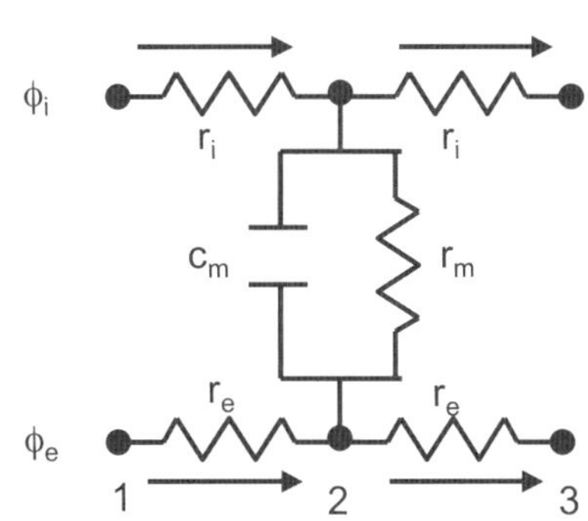

Figure 4.3: Example node in a discrete cable.

substitution for i_m yields

$$\frac{V_m^1 - 2V_m^2 + V_m^3}{dx^2} = (r_i - r_e)\left[c_m \frac{dV_m}{dt} + \frac{V_m}{r_m}\right]. \tag{4.10}$$

If we let $dx \to 0$, Eq. (4.10) becomes the continuous cable equation:

$$\frac{\partial^2 V_m}{\partial x^2} = (r_i - r_e)\left[c_m \frac{\partial V_m}{\partial t} + \frac{V_m}{r_m}\right] \tag{4.11}$$

where we have replaced the d/dt terms by $\partial/\partial t$ to indicate that this is a partial differential equation. A bit of algebra yields the following:

$$\frac{1}{r_i - r_e}\frac{\partial^2 V_m}{\partial x^2} = c_m \frac{\partial V_m}{\partial t} + \frac{V_m}{r_m} \tag{4.12}$$

$$\frac{r_m}{r_i - r_e}\frac{\partial^2 V_m}{\partial x^2} = r_m c_m \frac{\partial V_m}{\partial t} + V_m \tag{4.13}$$

or in the typical core-conductor form

$$\lambda^2 \frac{\partial^2 V_m}{\partial x^2} = \tau_m \frac{\partial V_m}{\partial t} + V_m \tag{4.14}$$

where $\tau_m = r_m c_m$ is the *membrane time constant* in $msec$, as defined in Sec. 2.2 and

$$\lambda = \sqrt{\frac{r_m}{r_i + r_e}} \tag{4.15}$$

is the *cable space constant* in units of cm.

4.1.2 A Simplification

One very common simplification to the cable equation is achieved by assuming that the extracellular bath is much more conductive than the intracellular space. As a result, we can assume that $r_e \approx 0$. As a result, all of the ϕ_e potentials are equal and the extracellular bath acts as a ground. Therefore, $V_m = \phi_i$ and $\lambda = \sqrt{\frac{r_m}{r_i}}$.

4.1.3 Units and Relationships

The units of the variables in Eqs. (4.1)–(4.15) can be confusing because they are different than the units used in Ch. 2. The tables show the units used in parameters of the core conductor model and their relationship to the membrane parameters.
Given these relationships we can rewrite Eq. (4.14) as:

$$\frac{1}{(R_i + R_e)}\frac{\partial^2 V_m}{\partial x^2} = \beta\left[C_m \frac{\partial V_m}{\partial t} + \frac{V_m}{R_m}\right] \tag{4.16}$$

where $\beta = 2/a$.

Table 4.1:

Parameter	Name	Unit
r_m	Membrane Resistance	Ωcm
r_i	Axial Intracellular Resistance	Ω / cm
r_e	Axial Extracellular Resistance	Ω / cm
c_m	Axial Membrane Capacitance	$\mu F / cm$
i_m	Axial Membrane Current	$\mu A / cm$
dx	Spatial Step Size	cm
a	Cable radius	cm

Table 4.2:

Variable	Name	Unit	Equivalence
R_i	Specific Intracellular Resistivity	Ωcm	$\pi a^2 r_i$
R_e	Specific Extracellular Resistivity	Ωcm	$\pi a^2 r_e$
R_m	Specific Membrane Resistivity	Ωcm^2	$2\pi a r_m$
C_m	Specific Membrane Capacitance	$\mu F / cm^2$	$c_m / (2\pi a)$
I_m	Specific Membrane Current	$\mu A / cm^2$	$i_m / (2\pi a)$

4.1.4 An Applied Stimulus

A stimulus, i_{stim}, may be applied at a particular location on the cable for some particular duration. Mathematically, a stimulus applied to a point can be represented using the Dirac delta function, $\delta(s)$. If the stimulus at this point is applied at $t = 0$ and remains on, the unit step function, $u(t)$, may be used. For the simple case where a stimulus is applied to a point in the middle of an infinitely long cable beginning at $t = 0$,

$$\lambda^2 \frac{\partial^2 V_m}{\partial x^2} = \tau_m \frac{\partial V_m}{\partial t} + V_m \pm r_m i_{\text{stim}} \delta(x) u(t) \tag{4.17}$$

where the $\pm$ is to take into account either an intracellular or extracellular stimulus.

4.1.5 Steady-State Solution

We can consider the steady state solution to Eq. (4.17) by assuming $\frac{\partial V_m}{\partial t} = 0$:

$$\lambda^2 \frac{d^2 V_m}{dx^2} - V_m = -r_m i_{\text{stim}} \delta(x) \tag{4.18}$$

where we have assumed the stimulus will depolarize the membrane. Equation (4.18) is an ordinary differential equation that no longer depends upon time. We can find the homogeneous solution by finding the solution to:

$$\lambda^2 \frac{d^2 V_m}{dx^2} - V_m = 0 \,. \tag{4.19}$$

The solution to Eq. (4.19) is the Helmholtz equation:

$$V_m(x) = Ae^{-|x|/\lambda} + Be^{|x|/\lambda} \,. \tag{4.20}$$

The second term does not make physical sense so

$$V_m(x) = Ae^{-|x|/\lambda} \,. \tag{4.21}$$

To find A we will integrate Eq. (4.18) around the stimulus ($x = 0$):

$$\lambda^2 \int_{0-}^{0+} \frac{d^2 V_m}{dx^2} \, dx - \int_{0-}^{0+} V_m \, dx = -r_m i_{\text{stim}} \int_{0-}^{0+} \delta(x) \, dx \,. \tag{4.22}$$

Evaluation shows that $A = \frac{r_m \cdot i_{\text{stim}}}{2\lambda}$ so the steady-state solution is

$$V_m(x) = \frac{r_m i_{\text{stim}}}{2\lambda} e^{-|x|/\lambda} \,. \tag{4.23}$$

Note that the value of the constant A will be dependent upon the nature of the stimulus. For example, the solution would change if we applied the stimulus to one end of the cable rather than at the middle.

4.1.6 Finding The Length Constant

If a cable with unknown membrane properties is encountered, Eq. (4.23) is a way to find λ and r_m. If a known stimulus is applied for a long time, the membrane will eventually reach steady-state. At the point of the stimulus, $x = 0$ so the exponential term becomes 1. Therefore, the voltage at the stimulus is $\frac{r_m \cdot i_{\text{stim}}}{2\lambda}$. Furthermore, this voltage level will fall off in space (in both directions because of $|x|$) at an exponential rate governed by λ. Therefore, λ can be found as the rate of fall off in a similar way to finding τ_m in Ch. 2. Once λ is known, r_m can be found from $\frac{r_m \cdot i_{\text{stim}}}{2\lambda}$.

4.1.7 Time and Space Dependent Solution

Although we will not show the solution here, it is possible to solve Eq. (4.17) for V_m as a function of both time and space:

$$\begin{aligned} V_m(x,t) &= \frac{r_m \cdot i_{\text{stim}}}{4\lambda} \left[e^{-|x|/\lambda} erfc\left(\frac{|x|}{2\lambda}\sqrt{\frac{\tau_m}{t}} - \sqrt{\frac{t}{\tau_m}} \right) \right. \\ &\quad \left. - \, e^{|x|/\lambda} erfc\left(\frac{|x|}{2\lambda}\sqrt{\frac{\tau_m}{t}} + \sqrt{\frac{t}{\tau_m}} \right) \right] \end{aligned} \tag{4.24}$$

where *erfc* is the complimentary error function defined by $erfc(y) = 1 - erf(y)$ and

$$erf(y) = \frac{2}{\pi} \int_y^0 e^{-z^2}\, dz \tag{4.25}$$

is defined as the error function (see Fig. 4.4). It is important to note that Eq. (4.24) is for an infinite cable with a negative i_{stim}. The leading terms in Eqs. (4.23) and (4.24) will change if the nature of the stimulus is changed or the cable is not infinitely long.

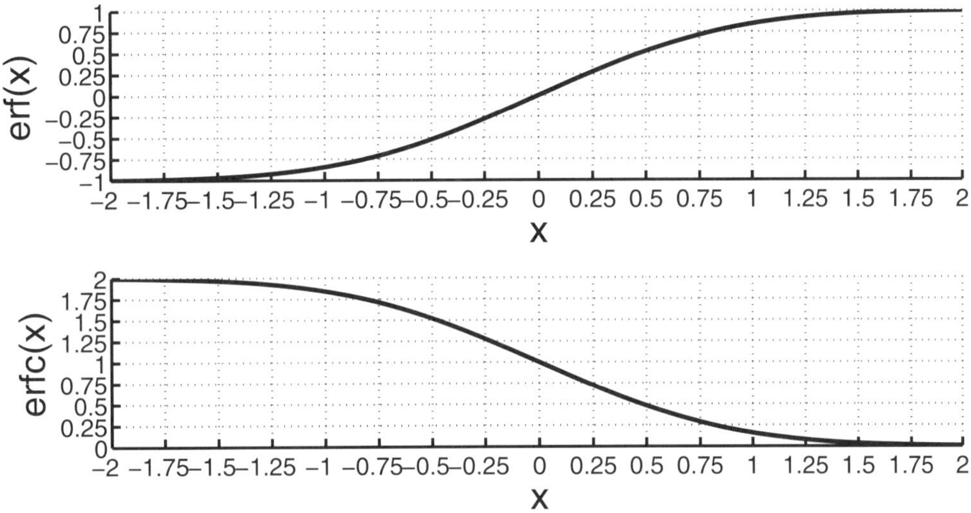

Figure 4.4: Error and complimentary error functions.

Figure 4.5 shows the time and space solutions to Eq. (4.24) for different times (left panel) and different locations in space (right panel). Because the specific numbers may vary for any given cable, the plot is shown in terms of the general passive properties of the cable, i.e., λ and τ_m). It is important to note that the fall off in space is exponential only for the steady-state ($t \to \infty$). Likewise, the rise in time is exponential (as in an RC circuit) only at $x = \lambda$.

4.2 ACTIVE PROPAGATION IN AXONS

In Sec. 4, we considered passive propagation down a cable as a model of dendritic propagation. The membrane of the axon, however, has many nonlinear ion channels that are capable of generating an action potential. The only modification needed to create an *active cable* is to replace the passive $I_{ion} = V_m/r_m$ with a more complex model as outlined in Ch. 3. Whereas a stimulus applied to a passive cable will be *attenuated*, in an active cable the action potential will propagate unattenuated. In this way, a signal that reaches the axon hillock will be propagated unattenuated to the end of the axon. In all but the simplest cases, it is not possible to derive an analytic solution when propagation is active.

To demonstrate the concept of active propagation, Fig. 4.6 shows propagation down an active $3cm$ long axon. The superthreshold stimulus was delivered to the left end of the axon at $t = 0msec$. In

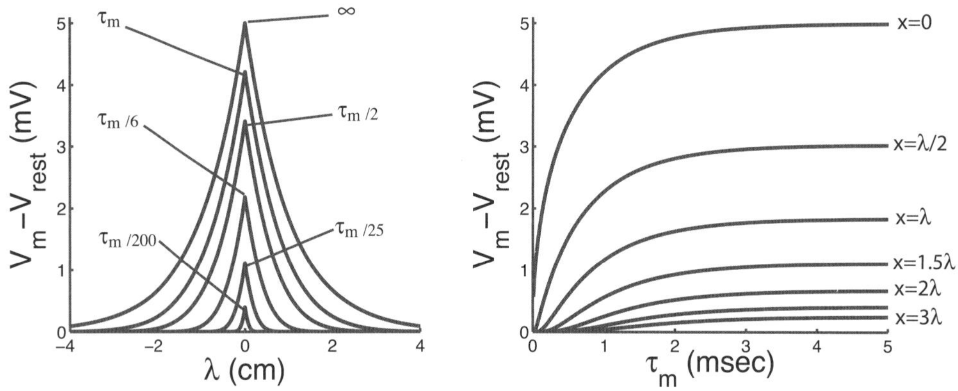

Figure 4.5: Time and space solution in a passive cable.

the left panel, action potentials shown at $x = 1cm$ (black) and $x = 2cm$ (red). Notice that although the action potential shape does not change, there is a delay in timing of the upstroke. The right panel shows a snapshot of the voltage along the cable at $t = 40msec$. Notice that the when propagation is from left to right, the spatial plot has the shape of a reversed action potential. To understand why, consider the location at $x = 2cm$. In the right panel, at $40msec$, the point is at rest. The shape in the left panel, however, is moving the right so eventually the sharp spike will reach $x = 2cm$. From the right panel we know that this occurs at approximately $t = 55msec$. In space, the reversed action potential shape will continue to move the right causing the location at $x = 2cm$ to undergo all of the phases of an action potential.

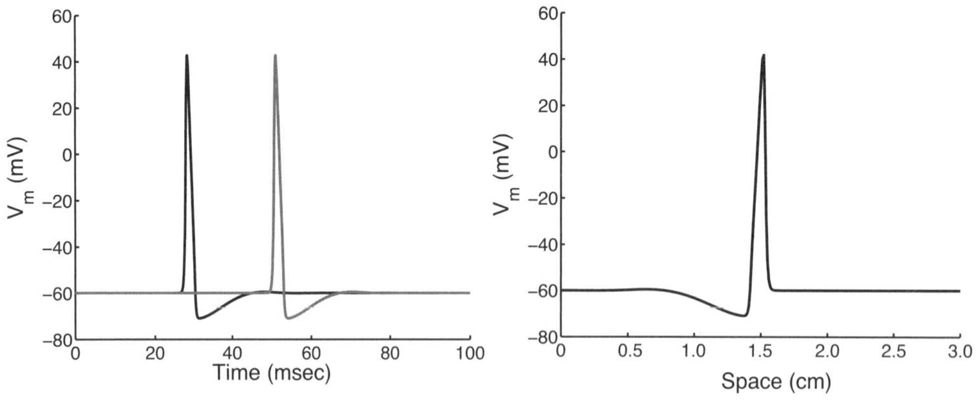

Figure 4.6: Propagation down an active cable.

For uniform propagation down a cable and uniform I_{ion} everywhere in the cable, we say that

$$V_m(x,t) = V_m(x - \theta\ (t) \qquad (4.26)$$

where θ is the *propagation velocity* and is a measure of the speed at which the action potential moves down the cable. The reason Eq. (4.26) works is because the shape of the action potential is the same at every point on the cable. The only parameter that changes is *when* the action potential occurs in time. Using the chain rule twice

$$\frac{\partial V_m}{\partial x} = \frac{1}{\theta}\frac{\partial V_m}{\partial t} \tag{4.27}$$

$$\frac{\partial^2 V_m}{\partial x} = \frac{1}{\theta^2}\frac{\partial^2 V_m}{\partial t}. \tag{4.28}$$

Substitution into Eq. (4.16) yields

$$I_m = \frac{a}{2R_i\theta^2}\frac{d^2 V_m}{dt}. \tag{4.29}$$

Upon careful inspection, Eq. (4.29) has a deeper meaning. As long as I_{ion} does not change throughout the cable, a family of solution exists with the only requirement being:

$$\frac{a}{2R_i\theta^2} = K \tag{4.30}$$

$$\theta = \sqrt{\frac{a}{2R_i K}} \tag{4.31}$$

where K is a constant. The usefulness of Eqs. (4.30) and (4.31) are that only one set of a and θ are needed to find $R_i K$. Once $R_i K$ is known, the effect of any change in a on θ can be predicted.

4.2.1 Saltatory Conduction

A problem with using active propagation to transmit an impulse over long distances is that the signal will take time to traverse the axon. In fact, in some regions of the body an axon can be up to one meter long. To compensate, the nervous system has developed a clever solution to speed up propagation. Schwann cells (a type of glial cell) create a *myelin sheath* around the axon. The presence of this sheath makes it nearly impossible for current to cross the cell membrane. The impact is that membrane directly under the sheath has a much higher R_m and the thicker membrane decreases C_m. Therefore, membrane covered in myelin is effectively *passive*.

As shown in Fig. 4.7, the sheath does leave small regions of the neuron cell membrane exposed, called *Nodes of Ranvier*, which may fire an action potential. If an active patch of membrane fires at a Node of Ranvier, current will flow to the right but will not be able to easily cross the cross the membrane because of the high R_m. Instead, most current will follow the path of least resistance and jump farther down the cable. So, in effect, the high resistance of the myelin creates a short circuit that skips quickly from one Node of Ranvier to the next. Multiple Sclerosis (MS) is a disease that causes inflammation and scaring of the myelin sheath leading to degradation of neural impulse propagation. The symptoms are changes in sensations, muscle spasms, a lack of coordination and balance, pain, and eventually cognitive and emotional impairment.

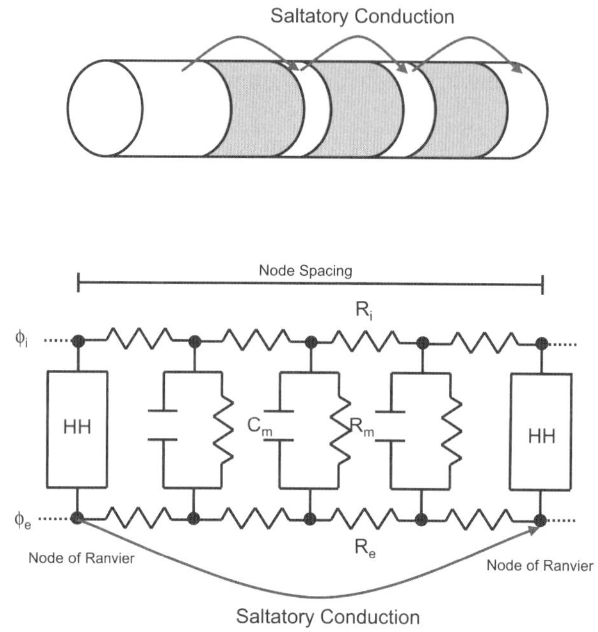

Figure 4.7: Schematic of saltatory conduction.

4.3 PASS THE PAPER PLEASE

Propagation occurs in many other systems, using chemical, mechanical, fluid, or bits of information as the medium of propagation. For an alternative perspective on propagation, consider arranging you and ten of your classmates in a line all shoulder to shoulder. To the right of the line is a large bag filled with shredded paper. The person on the far right end then grabs a large handful of shreds in their right-hand. The rules for each person will be to receive paper in their right-hand, pass it to their left-hand and then pass to the right-hand of the next person in the line. Using these simple rules, the shreds will *propagate* down the line from right to left. If your line tries to propagate shreds quickly, you will notice that most of the shreds are lost by the time propagation ends at the left end of the line. In the activity, the shreds are analogous to charges and the movement of shreds is a current flow. The size of each person's hands arc analogous to a capacitance and how much paper they lose is similar to a leakage resistance. You may also notice that much of the paper has fallen to the ground. In fact, this may the exact situation in the neuron as change that leaves the cell disappears into the extracellular space which acts as a large electrical ground. Next, consider that most of the shreds were lost in the first few transitions with much less being lost as propagation continued. Therefore, the way shreds are lost as propagation continues down the cable may be approximated by a decaying exponential.

As the shreds are lost during propagation, our analogy is to passive propagation. A small change, however, could turn passive propagation to active propagation. If *everyone* in the line has a handful of shreds in their left hand, and the propagation is reinitiated, you will discover that the person at the end of the line will have a considerable amount of shreds. Furthermore, the active propagation activity could be modified for saltatory conduction by simply skipping every other (or every three) people in the line.

4.4 NUMERICAL METHODS: THE FINITE AND DISCRETE CABLE

In all but the most simple situations, the continuous cable model can only be solved numerically using a computer. It is therefore required that the cable be of finite length and divided into small patches of membrane. Beginning with the continuous Eq. (4.11) the *finite difference* approximation may be made for the second derivative in space.

$$\frac{V_m^{k-1} - 2V_m^k + V_m^{k+1}}{(r_i - r_e)\,dx^2} = (r_i - r_e)\left[c_m \frac{dV_m}{dt} + \frac{V_m}{r_m}\right] \tag{4.32}$$

where k is a variable to represent any node in the middle of the cable. Note that the equation able is similar to Eq. (4.10). Rearranging

$$\frac{dV_m}{dt} = \frac{1}{c_m}\left[\frac{V_m^{k-1} - 2V_m^k + V_m^{K+1}}{(r_i - r_e)\,dx^2} - \frac{V_m}{r_m}\right] \tag{4.33}$$

and using the Euler method of Sec. 2.5

$$dV_m = \frac{dt}{c_m}\left[\frac{V_m^{k-1} - 2V_m^k + V_m^{k+1}}{(r_i - r_e)\,dx^2} - \frac{V_m}{r_m}\right] \tag{4.34}$$

$$V_m^{\text{new}} = V_m^{\text{old}} + dV_m \tag{4.35}$$

an every V_m is not discrete in both time and space. Note that in active propagation the gating variable must also be integrated in time, however, they do not require information from their neighbors. Therefore, the simple Euler method may be used.

A problem is encountered, however, with the above formulation. Consider evaluating the left most node on the cable (e.g., $k = 1$). An update of the end node requires information at $k = 0$ which does not exist. The same problem occurs at the right most node of the cable. To overcome the problem at the endpoints, we must enforce a boundary condition. The most common assumption in neural propagation is that no current can leave either end of the cable. This is called a *sealed-end* boundary condition. Mathematically,

$$\frac{d\phi_i}{dx} = -I_i r_i = 0 \tag{4.36}$$

$$\frac{d\phi_e}{dx} = -I_e r_e = 0\,. \tag{4.37}$$

The most straightforward way to enforce this boundary condition is to define a *ghost* node that extends past the end of the cable (see Fig. 4.8). Then, at the two cable ends

$$\phi_{i,e}^{gleft} = \phi_{i,e,}^{2} \tag{4.38}$$

$$\phi_{i,e,}^{gright} = \phi_{i,e}^{n\ \ 1} \tag{4.39}$$

and ensures that at the first and last node, Eqs. (4.36) and (4.37) are satisfied.

Combining Eq. (4.34) and the sealed-end, we can write the equation for node 1 as

$$dV_m = \frac{dt}{c_m}\left[\frac{V_m^{gleft} - 2V_m^1 + V_m^2}{(r_i - r_e)\,dx^2} - \frac{V_m}{r_m}\right] \tag{4.40}$$

$$dV_m = \frac{dt}{c_m}\left[\frac{2V_m^2 - 2V_m^1}{(r_i - r_e)\,dx^2} - \frac{V_m}{r_m}\right]. \tag{4.41}$$

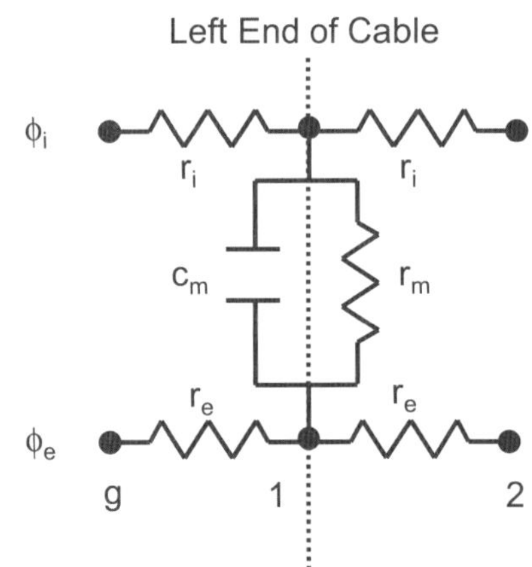

Figure 4.8: Sealed-end boundary condition.

Other possible boundary conditions for the end of the cable are to allow current to leave (*leaky end*) or to clamp the voltage (*clamped end*).

4.5 NUMERICAL METHODS: TEMPLATE FOR CABLE PROPAGATION

In Sec. 2.5, a way of numerically solving a differential equation was outlines. In the active membrane models, we need to keep track of several differential equations as well as compute rate constants, steady-state values and currents. Below is a template for how to write a program to solve the active equations.

```
Define constants (e.g., Ri, Re, a, Cm,
other membrane variables)
Compute initial αs and βs
Compute initial conditions for state variables
(e.g., Vm^rest, m, h, n)

for (time=0 to time=end in increments of dt)
   for (i=1 to i=Last Node in increments of (1)
```

```
        Compute Currents at node i
        Compute αs and βs at node i

        if (i=1)
             Update Differential Equations at left boundary
        else if (i=last node)
             Update Differential Equations at right boundary
        else
             Update Differential Equations at all middle nodes
        end

        Save values of interest into an array (e.g., Vm(i))

    end
end

Store values of interest to a file
```

Homework Problems

(1) Work out the units for Eqs. (4.13)–(4.16).

(2) Show the steps between Eq. (4.22) and Eq. (4.23).

(3) Show how Eq. (4.24) becomes Eq. (4.23) as $t \to \infty$.

(4) Using Eq. (4.24), show that at $x = \lambda$, the time constant, τ_m, can be found as 63% of the steady-state value. Show that at $x = 0$, τ_m can be found as 84% of the steady-state value.

(5) A passive nerve of radius $50\mu m$ is stimulated with a current pulse of $10nA$ in the middle ($x = 0.0cm$) of and cable. Using the plots below to answer the questions:

a) Find τ in $msec$.
b) Find λ in cm. Start with the full solution to the cable equation.
c) What is the R_m in Ωcm^2?

(6) In Fig. 4.6:
a) What is the propagation velocity in cm/s?
b) Explain what would happen to the spatial distribution if the propagation velocity was slower.
c) Explain what would happen to the spatial distribution be is the propagation direction was reversed?
d) Explain what would happen to the spatial distribution if the propagation was saltatory.

(7) In Fig. 4.6, assume $a = 10\mu m$, and predict the propagation velocity if a is changed to $a = 13\mu m$.

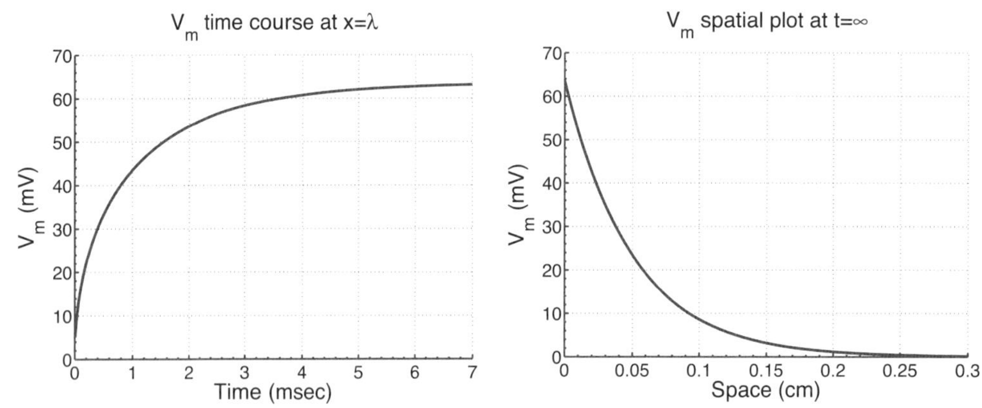

Simulation Problems

(1) Write code to simulation a passive $3cm$ cable with $R_m = 1k\Omega cm^2, C_m = 1\mu F/cm^2, dx = 100\mu m$, $a = 10\mu m$, $R_i = 100\Omega cm$ and assume that the extracellular space is large and acts as a ground and the boundaries are sealed-ends. Demonstrate that your problem is working by reproducing the left and right panels of Fig. 4.5.

(2) Modify the program from 1 above for active propagation of a Hodgkin-Huxely action potential. Stimulate the left end of the cable with a strength and duration sufficient to begin propagation. Demonstrate that your program works by creating a figure similar to 4.6.

(3) Use the code from 2 above to demonstrate the impact of changing the cable radius on propagation velocity.

(4) Use the code from 2 above to demonstrate the impact of changing R_i.

(5) Modify the program from 2 to demonstrate how saltatory conduction speeds propagation velocity. Let $R_m = 10k\Omega$ and $C_m = 0.01\mu F/cm^2$ for the sheath.

CHAPTER 5

Neural Branches

In Ch. 4, we considered propagation in uniform and straight cables. Figure 5.1 shows the complex web of dendrites that is collectively called the *dendritic tree*. It is clear that the tree contains branching cables, changes in diameter and curves. In this chapter we will generalize cable theory to handle changes in membrane properties and branching. We will first examine a method of simplifying the tree by *lumping* the dendrites into one large compartment. We will then turn to the more sophisticated method of *multicompartment* models.

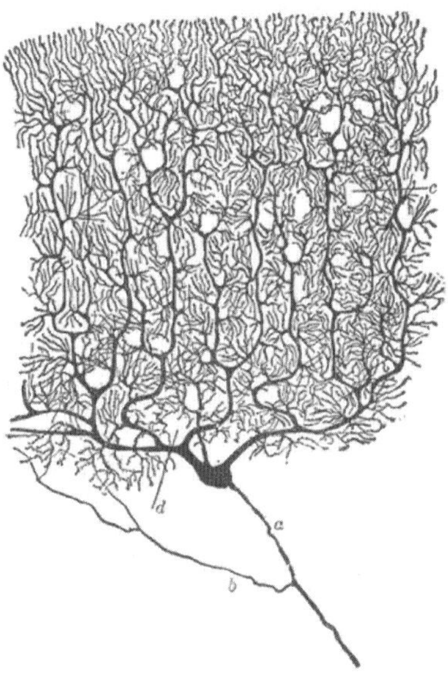

Figure 5.1: Dendritic tree.

5.1 LUMPED MODELS

Lumped models were developed before computers became available as a way to simplify a complex neuron to a single unit. The goal was to make analytical predictions about the behavior of neurons.

5.1.1 The Rall Model

To model the complex dendritic tree, Rall recognized that the branching dendrite was simply acting as a receiver for many inputs and then attenuating the input. He therefore proposed that the dendritic tree could be collapsed down to one signal passive cable of varying diameter. This would mean that the dendritic tree could be represented by an *equivalent* cable as shown in Fig. 5.2. Likewise, the various stimuli to the ends of the dendrites would be approximated by a single stimulus. The advantage of this idea is that computation would become fast and efficient and the actual geometry of the tree would not matter. It also enabled the analytical solution of attenuation down the dendritic tree. The disadvantage was that to achieve this collapse of the tree, Rall had to assume a relationship between the radii of all branches in relation to their main trunk

$$a_{\text{main}}^{3/2} = a_{\text{branch1}}^{3/2} + a_{\text{branch2}}^{3/2} + \dots . \tag{5.1}$$

This relationship was assumed to apply to *every* branch in the tree. Although Eq. (5.1) appears to be very restrictive, Rall showed that for many dendritic trees is was a reasonable approximation.

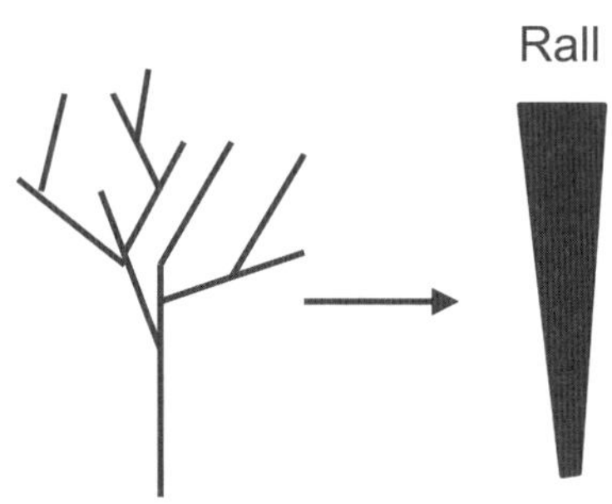

Figure 5.2: Rall's equivalent cable.

5.1.2 The Ball and Stick Model

Another advantage of the collapsed dendritic tree is that the equivalent cable can be attached to a soma and axon to create a *ball and stick* model of a neuron. In Fig. 5.3, the dendrites have been collapsed using the method outlined by Rall and the soma is modeled as single compartment which may be either active or passive. The axon is in fact not present at all and is simulated by a time delay. As we found in Ch. 4, once an action potential has fired in the soma, it will propagate down the axon unattenuated to the axon terminal. So, the delay depends only upon the speed of propagation and length of the axon. All of these simplifications allow for the behavior of a neuron to solved numerically with a minimum of computing power.

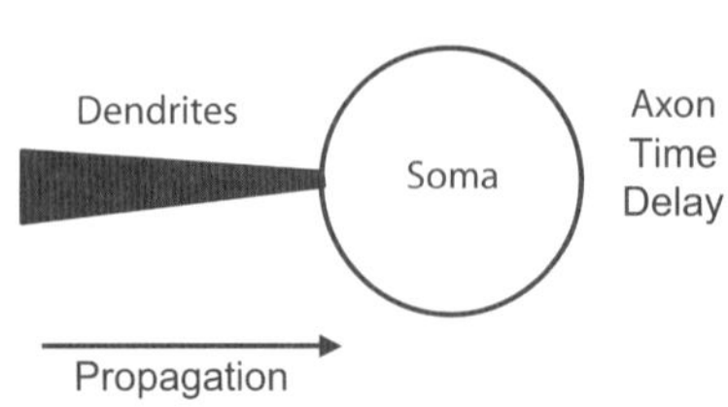

Figure 5.3: Ball and stick model.

5.2 MULTICOMPARTMENT MODELS

It is clear from Fig. 5.1 that not all neurons will follow the assumption of Eq. (5.1). To lift the assumptions of the lumped models, we can consider that the dendrites and axons are made up of many small compartments. The general concept is shown in Fig. 5.4 and is a similar idea to when we derived the original cable equation using many small patches of membrane. Again, current is passed from one compartment to another in one direction only. At a branch, the current is simply split and passed into two (or more) compartments. Below we will derive equations to describe how this current split is achieved. *Multicompartment models* not only remove the Rall restriction but also allow for many different types of post-synapses to be incorporated into the model. The post-synapse will be considered in more detail in Ch. 6. The disadvantage of the multicompartment models is that compared to the Rall model, they must be solved numerically and required considerable greater computing power. Below we will introduce the basic elements of how to construct a dendritic tree of any complexity.

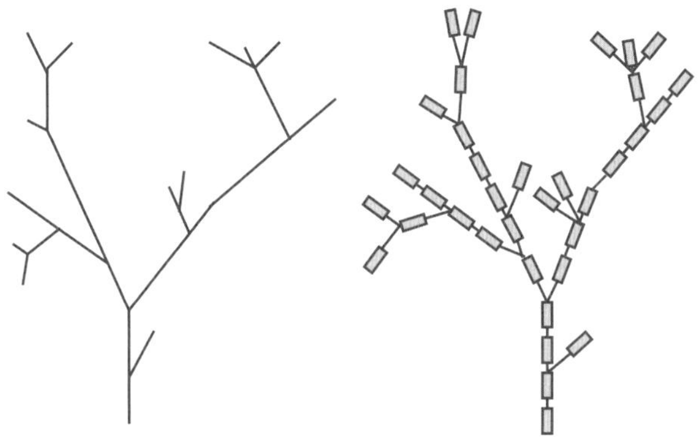

Figure 5.4: Schematic of dendritic compartments.

5.2.1 A Simple Compartment

To begin, we will make a slight change to the definition of a compartment by placing a node at the *center* of each patch as in Fig. 5.5. Each compartment will therefore have a length of $2dx$ and each half resistor in a compartment will be defined as

$$R = \frac{dx \cdot R_i}{\pi a^2} . \tag{5.2}$$

5.2.2 Change in Fiber Radius

Since the composition of the cytoplasm does not change drastically, the intracellular resistivity (R_i, a material property) does not change. Therefore, the determining factor of the half resistance with a radius, a, is

$$R(a) = \frac{dx \cdot R_i}{\pi a^2} . \tag{5.3}$$

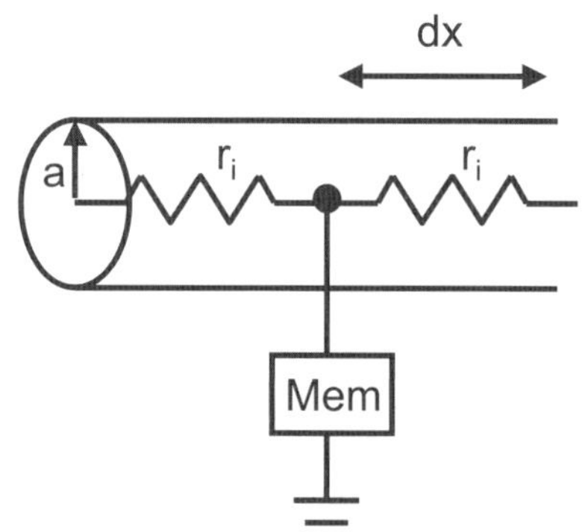

Figure 5.5: Model of a single compartment.

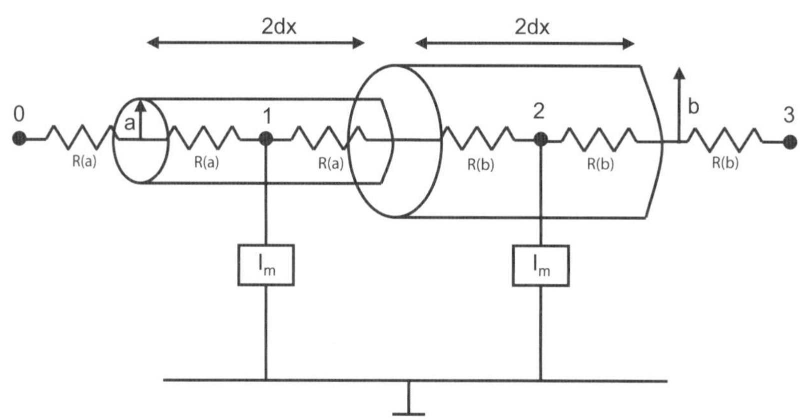

Figure 5.6: Impact of a radius change.

Using the function $R(a)$, we can consider the impact of coupling a patch with a small radius (a) to a patch with a larger radius (b) as in Fig. 5.6. If we assume that all current flows from left to right and that extracellular space is conductive (e.g., $V_m = \phi_i$), then at Node 1

$$\frac{V_m^0 - V_m^1}{(R(a) + R(a))} - \frac{V_m^1 - V_m^2}{(R(a) + R(b))} - I_m = 0 \tag{5.4}$$

and at Node 2

$$\frac{V_m^1 - V_m^2}{(R(a) + R(b))} - \frac{V_m^2 - V_m^3}{(R(b) + R(b))} - I_m = 0\,. \tag{5.5}$$

Although these equations are not compact, they are easy to compute numerically. Because R is also dependent upon dx, it is simple to extend our result to compartments of different lengths.

5.2.3 Branches

It is clear from Fig. 5.1 that we also must describe the points at which the dendrite branches. For simplicity, we will consider a single branch, as shown in Fig. 5.7, and assume that dx and a remain constant. The node labeled b is known as a *branch* node. There is no membrane at the branch node because it is not at the center of a compartment and is only used to make computation easier. At Node b,

$$\frac{V_m^1 - V_m^b}{(R+R)} - \frac{V_m^b - V_m^2}{(R+R)} - \frac{V_m^b - V_m^3}{(R+R)} = 0\,. \tag{5.6}$$

To simplify, we can multiply through by the denominator:

$$V_m^1 - V_m^b - V_m^b + V_m^2 - V_m^b + V_m^3 = 0 \tag{5.7}$$

$$V_m^b = \frac{V_m^1 + V_m^2 + V_m^3}{3}\,. \tag{5.8}$$

Note that if dx or a were not the same everywhere, this equation would become much more complicated but still computable. We can now use V_m^b to derive equations for Nodes 1, 2, and 3:

$$\frac{V_m^0 - V_m^1}{2R} - \frac{V_m^1 - V_m^b}{2R} - I_m = 0 \tag{5.9}$$

$$\frac{V_m^0 - 2V_m^1 + V_m^b}{2R} = I_m\,. \tag{5.10}$$

Similarly:

$$\frac{V_m^b - 2V_m^2 + V_m^4}{2R} = I_m \tag{5.11}$$

$$\frac{V_m^b - 2V_m^3 + V_m^5}{2R} = I_m\,. \tag{5.12}$$

The remainder of the extension of these branches (e.g., Nodes 4, 5, 6, ...) can be formulated as uniform cables. Therefore, the key to writing the equations for a branch is the branch node.

5.2.4 The Soma

Although the Soma is considered to be the center of the neuron, it is often modeled as a single large passive compartment. The special role of the soma, however, is to integrate all impulses entering the dendritic tree. It is also directly connected to the axon hillock which *is* capable of generating an action potential. So, if the dendrites charge up the soma above threshold, the axon hillock will fire and send an action potential propagating down the axon. The soma therefore acts as a relay center between the dendrites and axon. Figure 5.8 shows a simple model of a neuron composed of two active axon compartments of

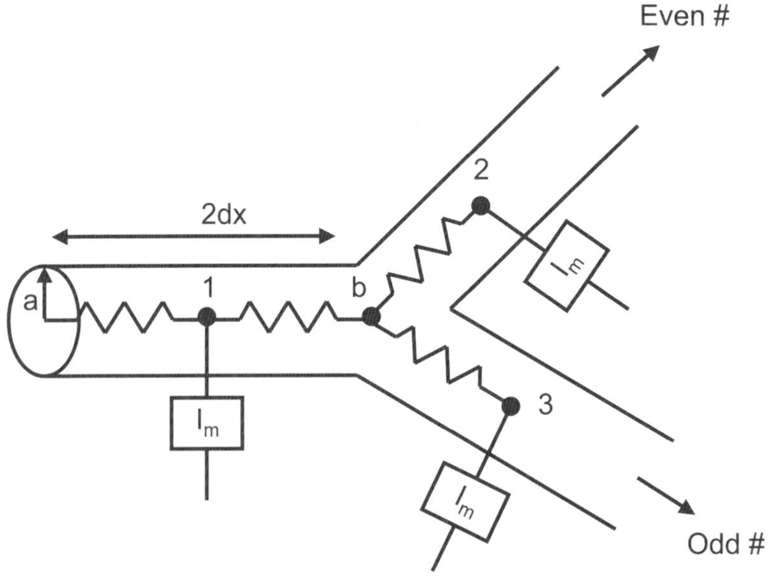

Figure 5.7: Circuit analog of a neural branch.

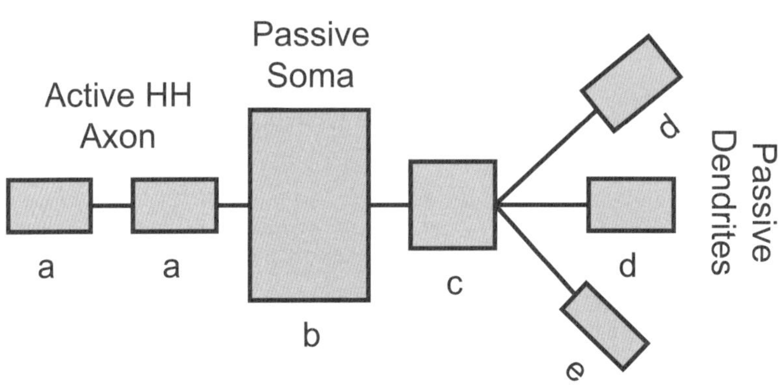

Figure 5.8: A structural model of a simple neuron.

radius a, a passive soma of radius b, a main dendritic trunk of radius c, two branches of radius d, and a third branch of radius e.

By combining the ideas of active and passive cables with the ability to create branches and changes in diameter, we have all the concepts needed to create compartment models of any complexity. In fact, some researchers have created neuronal models, based upon the histology of real neurons that contain hundreds (or even thousands) of compartments. Although most of these models are two-dimensional, the technology exists to create three-dimensional images of the branching structures of real dendrites. You may wish to think about how the analysis above needs to be changed to allow for a 3D structure.

5.2.5 Axon Collaterals and Semi-Active Dendrites

To simplify the discussion above we have made a number of assumptions that may not be true in a real neuron. First, some specialized neurons have either multiple axon projections or a single axon that branches many times. The geometry of a branching axon can be modeled in the same way as a branching dendrite with the exception that the ionic current, I_{ion} will be more complex.

Recent data has also shown that dendrites are not simply passive cables. In other words, there are some ion channels embedded in the membrane that have nonlinear behavior so $I_m \neq \frac{V_m}{R_m}$. To model a *semi-active* dendrite we can simply replace the passive I_{ion} term with the appropriate nonlinear currents. The presence of these channels allows currents and potentials to *back propagate* in the orthodromic direction. Back propagation may impact future pulses in the antidromic direction by changing the concentrations of ions (in particular $[Ca^{2+}]_i$) and transiently change R_m of the dendrite.

Lastly, anatomical studies have shown that small protrusions, called *spines*, appear on some dendrites. Although it is not clear what role spines play, there have been a number of suggestions. One possibility is that spines increase the amount of membrane area for synaptic inputs from other neurons. Another view is that spines store $[Ca^{2+}]_i$ and other ions which modulate the strength of synaptic inputs. Lastly, some consider the spines to serve an electrical role by modulating the attenuation of a voltage headed toward the soma.

5.3 NUMERICAL METHODS: MATRIX FORMULATION

A typical node, k, in an unbranched cable of length n with uniform membrane properties is connected to two neighbors. The governing equation for this node is therefore of the form:

$$\frac{V(k-1) - 2V(k) + V(k)}{R} = I_m(k) \,. \tag{5.13}$$

At the ends of the cable we can assume seals ends as in Sec. 4.3

$$\frac{-2V(1) + 2V(2)}{R} = I_m(1) \tag{5.14}$$

$$\frac{2V(n-1) - 2V(n)}{R} = I_m(n) \,. \tag{5.15}$$

Using matrices we can write this system of equations as

$$\begin{bmatrix} \frac{-2}{R} & \frac{2}{R} & \cdots & & & & \\ \frac{1}{R} & \frac{-2}{R} & \frac{1}{R} & \cdots & & & \\ 0 & \frac{1}{R} & \frac{-2}{R} & \frac{1}{R} & \cdots & & \\ \vdots & \vdots & \vdots & & \ddots & & \\ & & & \cdots & \frac{1}{R} & \frac{-2}{R} & \frac{1}{R} & 0 \\ & & & & \cdots & \frac{1}{R} & \frac{-2}{R} & \frac{1}{R} \\ & & & & & \cdots & \frac{2}{R} & \frac{-2}{R} \end{bmatrix} \begin{bmatrix} V(1) \\ V(2) \\ V(3) \\ \vdots \\ V(n-2) \\ V(n-1) \\ V(n) \end{bmatrix} = \begin{bmatrix} I_m(1) \\ I_m(2) \\ I_m(3) \\ \vdots \\ I_m(n-2) \\ I_m(n-1) \\ I_m(n) \end{bmatrix}$$

or including a more detailed representation for I_m

$$
\begin{bmatrix}
\frac{-2}{R} & \frac{2}{R} & \cdots & & & & & \\
\frac{1}{R} & \frac{-2}{R} & \frac{1}{R} & \cdots & & & & \\
0 & \frac{1}{R} & \frac{-2}{R} & \frac{1}{R} & \cdots & & & \\
\vdots & \vdots & \vdots & & \ddots & & & \\
 & & & \cdots & \frac{1}{R} & \frac{-2}{R} & \frac{1}{R} & 0 \\
 & & & & \cdots & \frac{1}{R} & \frac{-2}{R} & \frac{1}{R} \\
 & & & & & \cdots & \frac{2}{R} & \frac{-2}{R}
\end{bmatrix}
\begin{bmatrix}
V(1) \\ V(2) \\ V(3) \\ \vdots \\ V(n-2) \\ V(n-1) \\ V(n)
\end{bmatrix}
$$

$$
= C_m \frac{d}{dt}
\begin{bmatrix}
V(1) \\ V(2) \\ V(3) \\ \vdots \\ V(n-2) \\ V(n-1) \\ V(n)
\end{bmatrix}
+
\begin{bmatrix}
I_{\text{ion}}(1) \\ I_{\text{ion}}(2) \\ I_{\text{ion}}(3) \\ \vdots \\ I_{\text{ion}}(n-2) \\ I_{\text{ion}}(n-1) \\ I_{\text{ion}}(n)
\end{bmatrix}
$$

or more compactly,

$$
\mathbf{AV} = C_m \frac{d\mathbf{V}}{dt} + \mathbf{I_{ion}} \tag{5.16}
$$

where the bold text indicates a vector or matrix. **A** is called a *coupling* matrix because the location of the entries show exactly how nodes are connected. Each row of **A** corresponds to one node. Although we have assumed R is constant, in reality we know that the resistance will depend upon a and dx. In principle, R could vary in the matrix **A** and will indicate the strength of the connections.

To solve Eq. (5.16) we can perform the same rearrangement as in Sec. 2.5

$$
d\mathbf{V} = \frac{dt}{C_m}\left[\mathbf{AV} - \mathbf{I_{ion}}\right] \tag{5.17}
$$

$$
\mathbf{V_{new}} = \mathbf{V_{old}} + d\mathbf{V} \, . \tag{5.18}
$$

Besides being a compact way of writing the simultaneous equations of a cable, there is a practical reason for the *vectorized* form of Eq. (5.17). There exist many sophisticated methods of multiplying, adding, and factoring vectors and matrices that can drastically speedup the simulations.

Homework Problems

(1) Create the **A** coupling matrix for a cable where $R_i = 100\Omega cm$.

(2) How does the form of the matrix derived in problem 1 change if the cable branches?

(3) Create the coupling matrix for the simple neuron in Fig. 5.8. Assume sealed end boundaries.

(4) Explain how to create the matrix for the right side of Fig. 5.4 assuming all of the radii are equal.

(5) The Traub model of a Pyramidal neuron is shown in Fig. 7.1. Create the coupling matrix for this neuron assuming the diameter of all compartments is the same.

Simulation Problems

(1) Write a computer program to simulate the simple neuron in Fig. 5.8 where the passive properties are as given in Ch. 4, simulation problem 1, and the active properties are determined by HH parameters. Demonstrate that by stimulating one branch with a large enough current that an action potential is fired in the soma and propagates down the axon.

(2) In the program create above, does the stimulus threshold change if you stimulate the branch with radius e versus stimulating one of the branches with radius d.

(3) There are two primary ways that the threshold at a soma could reach threshold. First, a single large current may depolarize one branch, which although attenuated, may still be above the threshold of the soma. Second, many smaller currents may enter many branches simultaneously and be *integrated* (added up) to bring the soma to threshold. Use the program in 1 to demonstrate both ways of firing the soma.

CHAPTER 6

Synapses

Previous chapters have described how electrical activity propagates down the dendrites, integrates in the soma, and propagates down the dendrites. Most of the action, however, occurs at the connections between neurons. There are two primary ways that neurons connect. First are *gap junctions* which are proteins that directly connect the intracellular space of two neurons. Typically, gap junctions are modeled as linear resistor. In this chapter, we will focus primarily on the second type of connection, the synapse. Using engineering terms, the electrical signal that reaches the axon terminal is *transduced* to a chemical signal by the pre-synapse. This chemical signal then diffuses across the synaptic cleft and is transduced back to an electrical signal by the post-synapse on the dendrite of another neuron. A molecule that can transduce an electric signal across a synapse is called a *neurotransmitter*. In reality, these three steps are much more complex and can be broken down in the list below and shown graphically in Fig. 6.1.

1. Neurotransmitters are packaged into vesicles in the soma.
2. Depolarizations in the dendrites charge the soma and cause the axon hillock to fire an action potential.
3. The action potential propagates unattenuated down the axon.
4. The action potential reaches the pre-synaptic terminal.
5. Ca^{2+} channels in the pre-synapse open and Ca^{2+} enters the pre-synapse.
6. Ca^{2+} in the pre-synapse triggers fusion of vesicles (with neurotransmitters inside) with the membrane.
7. The neurotransmitter is released into the synaptic cleft by exocytosis.
8. The neurotransmitter diffuses across the synaptic cleft.
9. The neurotransmitter binds to receptors on the post synaptic membrane.
10. Post-synaptic ion channels open by either a direct or indirect (second messenger) mechanism.
11. The post-synaptic membrane potential changes (either depolarizes or hyperpolarizes).
12. The neurotransmitter is released by the post-synapse, inactivated and finally returned to the pre-synapse.

We have already quantified steps 2–4 and we will not quantify 1 or 7. In this chapter, we will first consider action at the pre-synapse (steps 5 and 6) and then turn to action at the post-synapse (steps 8–12). Although it is typical to examine synapses that connect an axon of one neuron to the dendrite of another, synapses also may form directly on the soma.

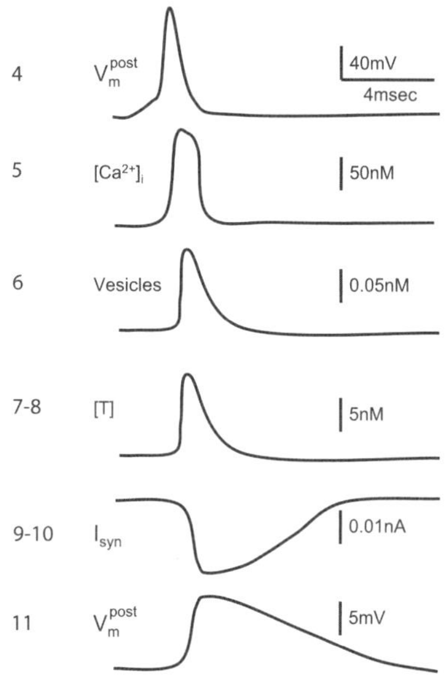

Figure 6.1: Voltages and concentrations of the synapse.

6.1 NEUROTRANSMITTERS

Neurotransmitters are the chemicals that relay electrical signals from the pre-synapse of one neuron to the post-synapse of another neuron. There are approximately ten known neurotransmitter that are classified in two different ways. The first classification is by molecular makeup where amino acids (e.g., glutamate, GABA), amines (e.g., epinephrine, dopamine, serotonin) and peptides (e.g., acetylcholine) are the dominate types. However, there are many other molecules (e.g., vasopressin, neurotensin, insulin) and ions (e.g., zinc) that can impact synapses indirectly and are sometimes considered neurotransmitters. The second classification is by function, where some neurotransmitters induce depolarization (called *excitatory transmitters*) and some induce hyperpolarizations (called *inhibitory transmitters*). The action of one type of neurotransmitter may, however, be excitatory in one region of the nervous system while being inhibitory in other regions. Furthermore, some neurotransmitters have other functional roles. For example, in a developing brain, neurotransmitters can act to guide and strengthen synaptic connections.

6.2 THE PRE-SYNAPSE

The release of neurotransmitter at the pre-synapse is commonly modeled in one of three ways that vary in complexity. The most complex is to model the binding of Ca^{2+}, activation of the vesicle, and release as separate steps. For example,

$$4Ca_i + X \leftrightharpoons k_b k_u X^*$$

$$X^* + V_e \leftrightharpoons k_1 k_2 V_e^* \rightarrow k_3 n T$$

X and X^* represent inactive and active Ca^{2+} binding proteins, V_e and V_e^* represent inactive and active vesicles, T is a molecule of neurotransmitter, and n is the number of neurotransmitters in a vesicle. You will note that it takes four Ca^{2+} ions to activate a single binding protein, a number that has been observed experimentally. There are much more complex models of the Ca^{2+} modulated release of neurotransmitter that include ion channels for Calcium influx as well as pumps to return Ca^{2+} to the synaptic cleft. These models are often of a similar form to I_K and I_{Na} discussed in Ch. 3.

A simpler model can be created by defining an equation for neurotransmitter molecules released as a function of the pre-synaptic voltage. A typical function is the Bolzmann distribution:

$$T(V_m^{\text{pre}}) = \frac{T_{\max}}{1 + e^{-(V_m^{\text{pre}} - V_p)/k_p}} \tag{6.1}$$

where $T_{\max}$ is the maximum concentration that can be released, V_m^{pre} is the pre synaptic voltage, V_p is called the half activation potential, and k_p is a slope factor.

While kinematic and function models attempt to capture some aspects of the physiology, two related and more simple models are sometimes used in computer simulations. First, if an action potential is detected at the pre-synapse, a short pulse of T will be released into the synaptic cleft. Second, the steps 5–10 of synaptic transmission may be replaced by a simple time delay. In other words, if an action potential is present at the pre-synapse it will directly cause a depolarization or hyperpolarization, after some delay, in the post-synapse.

6.3 NEUROTRANSMITTER DIFFUSION AND CLEARANCE

The diffusion of neurotransmitter across the synaptic cleft is typically not modeled. Instead, it is assumed that the released T molecules raise the concentration in the tiny cleft faster than the reaction time of the pre or post synapse. Neurotransmitter *clearance*, on the other hand, may be relative slow. Clearance is accomplished either by reuptake of the molecule by the pre-synapse or by an enzyme that reverts the neurotransmitter into an inactive state. Both of these mechanisms can be modeled using differential equations, the most simple of which is a first order decaying exponential.

$$\frac{dT}{dt} = -k_t \ . \tag{6.2}$$

Typically, the neurotransmitter (either in an active or inactive form) is taken back into the pre-synaptic axon and transported back to the soma. Here the molecules are reassembled, repackaged into vesicles, and sent back to the pre-synapse for future release. It is worth noting that if a neuron is undergoing bursting it is possible for release and uptake of a neurotransmitter to be occurring at the same time.

6.4 THE POST-SYNAPSE

6.4.1 The Post Synaptic Current

The post-synapse translates a concentration of neurotransmitter into a change in the post-synaptic potential (V_m^{post}). The neurotransmitter binds to a post-synaptic docking site and either directly or indirectly causing ions to rush either into or out of the post-synaptic membrane. The current generated by this ion flow across the post-synaptic membrane is defined the same way as other membrane currents.

$$I_{\text{syn}}(t) = g_{\text{syn}} O(t) \left[V_m^{\text{post}} - E_{\text{syn}} \right] \tag{6.3}$$

where V_m^{post} is the post-synaptic membrane potential, g_{syn} is the maximum conductance, O is the probability of an open channel, and E_{syn} is the reversal potential.

6.4.2 Excitatory, Inhibitory and Silent Synapses

Given the formulation of Eq. (6.3), E_{syn} will determine whether the synapse will be *excitatory*, *inhibitory*, or *silent*. To classify E_{syn}, we will assume that the post synaptic potential is at rest ($V_m^{\text{post}} = V_m^{\text{rest}}$). If $E_{\text{syn}} > V_m^{\text{post}}$, I_{syn} will be negative. Remember that a negative membrane current will depolarize the membrane so I_{syn} will *excite* the cell membrane. The resulting depolarization is called a often called an *Excitatory Post Synaptic Potential* (EPSP). Likewise, if $E_{\text{syn}} < V_m^{\text{post}}$, I_{syn} will be positive, hyperpolarize V_m^{post} and *inhibit* the cell membrane. The hyperpolarization of the post-synapse is called an *Inhibitory Post Synaptic Potential* (IPSP). If $E_{\text{syn}} = V_m^{\text{rest}}$ the synapse is said to be *silent*. The role of the silent synapse will be explored further in Sec. 6.4.5.

6.4.3 Neurotransmitter Gating

The gating mechanism for a synaptic current, O in Eq. (6.3), may be described in the same way as the gating of any other ionic channel (see Ch. 3). The exception is that instead of being dependent upon the membrane voltage, synaptic gating is typically dependent upon the concentration of neurotransmitter in the synaptic cleft. There are many possible formulations for O so we will cover only the most significant. A generic neurotransmitter gated reaction is

$$C + nT \rightarrow \alpha\beta O$$

governed by a differential equation

$$\frac{dO}{dt} = \alpha[T](1 - O) - \beta O \tag{6.4}$$

where $[T]$ is the concentration of neurotransmitter present. Similar to other gating mechanisms we could rewrite Eqs. (6.4)

$$\frac{dO}{dt} = \frac{O_\infty - O}{\tau_O} \tag{6.5}$$

$$O_\infty = \frac{\alpha T_{\max}}{\alpha T_{\max} + \beta} \tag{6.6}$$

$$\tau_O = \frac{1}{\alpha T_{\max} + \beta} \, . \tag{6.7}$$

More simply, $O(t)$ could be described by an analytic function, for example a bi-exponential

$$O(t - \tau_s) = K \left[e^{-(t-t_s)/\tau_2} - e^{-(t-t_s)/\tau_1} \right] \tag{6.8}$$

where K is the maximum amplitude of O (usually 1), τ_s is the time of the presynaptic spike, τ_1 is the rising time constant, and τ_2 is the falling time constant.

6.4.4 Multiple Gating Mechanism

It is possible for a synaptic current to be gated by a more complex combination of neurotransmitters, V_m^{post}, ionic concentrations and secondary chemicals. As an example, the model below is for the NMDA synaptic current that is dependent upon $[T]$, the extracellular concentration of magnesium ($[Mg^{2+}]_0$), and V_m^{post}:

$$I_{NMDA} = g_{NMDA} \cdot B\left[V_m^{post}, [Mg^{2+}]_o\right] \cdot O(t)[V_m^{post} - E_{NMDA}] \tag{6.9}$$

where $O(t)$ has a similar formulation, and dependence on $[T]$, as Eq. (6.4) and

$$B\left[V_m^{post}, [Mg^{2+}]_o\right] = \frac{1}{1 + exp\left[\frac{-(0.062 V_m^{post})[Mg^{2+}]_o}{3.57}\right]} . \tag{6.10}$$

6.4.5 Second Messenger Gating

Although we will not show the differential equations, the follow reaction illustrates how a G-protein second messenger is used in the gating of the $GABA_b$ receptor

$$R_o + T \leftrightharpoons \quad R \leftrightharpoons \quad D$$

$$R + G_0 \leftrightharpoons \quad RG \leftrightharpoons \quad R + G$$

$$G \leftrightharpoons \quad G_o$$

$$R \leftrightharpoons \quad R_o$$

$$K^+ + nG \leftrightharpoons \quad O .$$

First, $[T]$ binds to a free receptor R_o, which becomes active (R). At some later time, R can become deactivated (D). An inactive g-protein (G_o) can bind with R to become active (G) through a two step process. n activated g-proteins can then bind with a single K^+ ion to open the channel. In the background, the active R and G are inactivated (reset) and ready for another reaction.

6.5 SYNAPTIC SUMMATION

Above, we modeled I_{syn} as a single generic synaptic current. In reality, a single post synapse may have more than one type of receptor, even mixing excitatory and inhibitory receptors. Likewise, the pre-synapse may be capable of generating different types of neurotransmitters. To model a synapse with multiple types of receptors we can sum up the effects as in the parallel conductance model.

$$I_{\text{syn}}(t) = I_{\text{syn}}^{GABA} + I_{\text{syn}}^{\text{Dop}} + I_{\text{syn}}^{\text{Ser}} + I_{\text{syn}}^{\text{silent}} \tag{6.11}$$

where I_{syn}^{GABA}, $I_{\text{syn}}^{\text{Dop}}$, $I_{\text{syn}}^{\text{Ser}}$, and $I_{\text{syn}}^{\text{silent}}$ are formulated using some form of Eq. (6.3). In this way, one synapse can respond to many different molecules.

Given this summation we can now understand the role of the silent synaptic current. Since a silent synapse has a reversal potential at the resting potential, it will try to depolarize the membrane if $V_m^{\text{post}} < V_m$ and hyperpolarize the membrane if $V_m^{\text{post}} < V_m$. In other words, the direction of $I_{\text{syn}}^{\text{silent}}$ will be such as to bring the post synaptic potential back to rest. This is a very important feedback mechanism since it prevents the potential from deviating far from rest. You can also think of the silent synapse as a type of inertia that the excitatory and inhibitory receptors must overcome to be effective.

Three additional properties of synapses are important. First, it is also possible for a post-synapse to form directly on the soma in which case there will be no attenuation due to the dendrites. Second, the left-hand side of the reactions in Sec. 6.4.4 reveal how dependent an open channel is on all of the reactions. For example, if there are not enough inactive g-proteins (G_o), then no more channels can open. This is a problem called *availability* and is a rate-limiting step in any step-wise opening of the post-synaptic channels. Third, the opposite problem may occur if $O \to 1$, meaning that all the post-synaptic channels are open. When this is the case the post-synapse has become *saturated* and additional neurotransmitter will not have any effect.

6.6 SIMPLIFIED MODELS OF THE SYNAPSE

There are many ways to simplify the action of the synapse. Typically, the simplification is to remove any dynamics of the neurotransmitter and create a direct electrical connection between the pre and post synapse. One simplification is to model the conductance in Eq. (6.3) as an *alpha function* (top panel of Fig. 6.2).

$$G(t) = g_{\max}\frac{t}{\tau}e^{1-t/\tau} \tag{6.12}$$

$$I_{\text{syn}} = G(t)\left[V_m^{\text{post}} - E_{\text{syn}}\right] \tag{6.13}$$

where t is the time after the action potential reaches the axon terminal, τ is a time constant, and $g_{\max}$ is the maximum possible conductance. A slightly more complex model is a bi-exponential (bottom panel of Fig. 6.2)

$$G(t) = \frac{A g_{\max}}{\tau_1 - \tau_2}\left(e^{-t/\tau_1} - e^{-t/\tau_2}\right). \tag{6.14}$$

An even more simple model would be to skip the conductance term all together and simply model the synaptic current, I_{syn}, as a pulse that follows the arrival of a pre-synaptic action potential. In any of these simple synapse models it is possible to control the onset time and therefore easy to introduce any delays that may occur.

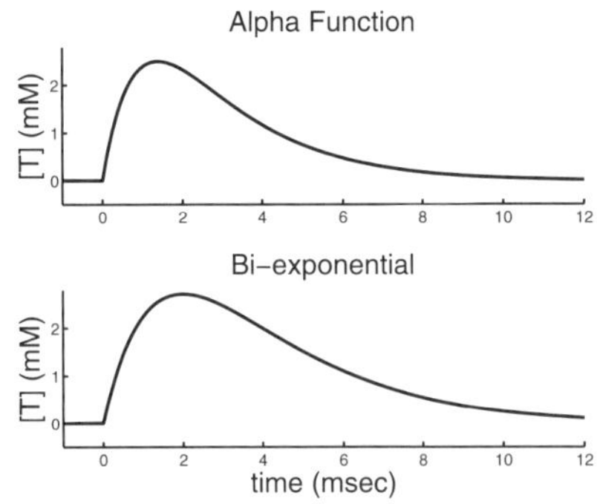

Figure 6.2: Alpha function and bi-exponential model for synaptic conductance.

6.7 THE MANY MECHANISMS OF DISEASES AND DRUGS

The effect of many drugs and diseases in the brain can be generalized as either decreasing or increasing the effect that the pre-synaptic potential has on the post-synaptic potential. There are, however, many methods by which these effects may be achieved. Below are five different mechanisms.

1. If the neurotransmitter uptake mechanism is not present due to a mutation or is blocked by a drug, the neurotransmitter will not be cleared and continue to have an impact. Alternatively, the uptake could be too effective and the neurotransmitter will be cleared before it can have its full effect.

2. If drugs or mutations either block Ca^{2+} channels or bind to Ca^{2+} ions (a buffer) in the pre-synapse, the mechanism for vesicle release will be compromised. The impact is that no neurotransmitter will be released into the cleft.

3. The structure of the neurotransmitter itself could be compromised either by a mutation or by a drug that binds to the neurotransmitter before it can reach the post-synapse.

4. The post-synaptic channel could be compromised either by disallowing neurotransmitter binding (e.g., docking proteins on the I_{syn} channel are mutated) or by a change in the channel itself that limits the passage of ions. A more complicated target for drugs or diseases is to alter a second messenger which relays the message of neurotransmitter binding.

5. Diseases and drugs can induce long-term changes on the number of vesicles available, number of neurotransmitters per vesicle, or number of post-synaptic binding sites. These changes may impact availability and saturation.

Most drugs have more than one effect because the same drug can impact different regions of the brain in different ways. Furthermore, the most powerful drugs often act through more than one of the mechanisms on the list above.

6.8 SYNAPTIC PLASTICITY AND MEMORY

In 1949, Donald Hebb published a book, *The Organization of Behavior*, which introduced the idea that the "strength" of a synapse could change over time in response to external factors. His idea was that *local memory* could be encoded in the strength of the connections between neurons. By strength, Hebb meant that the pre-synapse could have a greater impact (either excitatory or inhibitory) on the post-synaptic membrane. Given all of the ways that drugs and diseases can act on the synapse, it is possible for natural changes in synaptic strength to be accomplished by any or all of these mechanisms.

The ability of a synaptic strength to change is called synaptic *plasticity*. The mechanism of plasticity is one of the most elusive questions in neurobiology because of the many factors that could play a role. What is known is that some molecular level change occurs at either the pre-synapse, post-synapse or neurotransmitter clearance. For example, if a receptor channel on the post-synapse is over expressed, i.e., more channels waiting to bind with a neurotransmitter) the post-synapse will become more sensitive to the concentration of neurotransmitter and be less susceptible to saturation. The most simple way to model synaptic plasticity is to change g_{syn} in Eq. (6.3). The interpretation is that the value of g_{syn} is related to the number of synaptic ion channels. The opposite is true if the receptor channel is under expressed.

Hebb also proposed a mechanism by which synapses would be triggered to change, elegantly expressed in the phrase *neurons that fire together are wired together*. In other words, the more two neurons fire at the same time, the more tightly they will be connected. These types of synapses were in fact found to exist and were given the name *Hebbian synapses*. Recent studies, however, have shown that there are other reasons that neurons may either connect or disconnect, including pH, ion concentration, or drug use.

One of the most interesting ways for synapses to change is the phenomenon of *Long Term Potentiation* (LTP) discovered by Jerje Lomo in 1966. A typical experiment has three phases that are shown in Fig. 6.3. First, an external stimulus is applied to one neuron, which fires an action potential. This action potential releases neurotransmitter which crosses the synaptic cleft and induces a depolarization in the post-synapse of a second neuron. The recorded EPSP, V_1^{post} serves as a baseline. Second, a rapid train of stimuli are applied to the first neuron. The number of stimuli and rate is variable but is typically performed over a period of minutes. Third, at some later time a single stimulus is applied to neuron one and V_2^{post} is recorded. Surprisingly, the response has been *potentiated*, or $V_2^{post} > V_1^{post}$. The interpretation is that the post-synapse has become more sensitive to activity at the pre-synapse, or in other words, the connection has become strengthened. Even more surprising is that this effect may last many weeks or even months. Understanding the mechanism of LTP is a very active area of research since is a controlled way to study how synaptic strength changes.

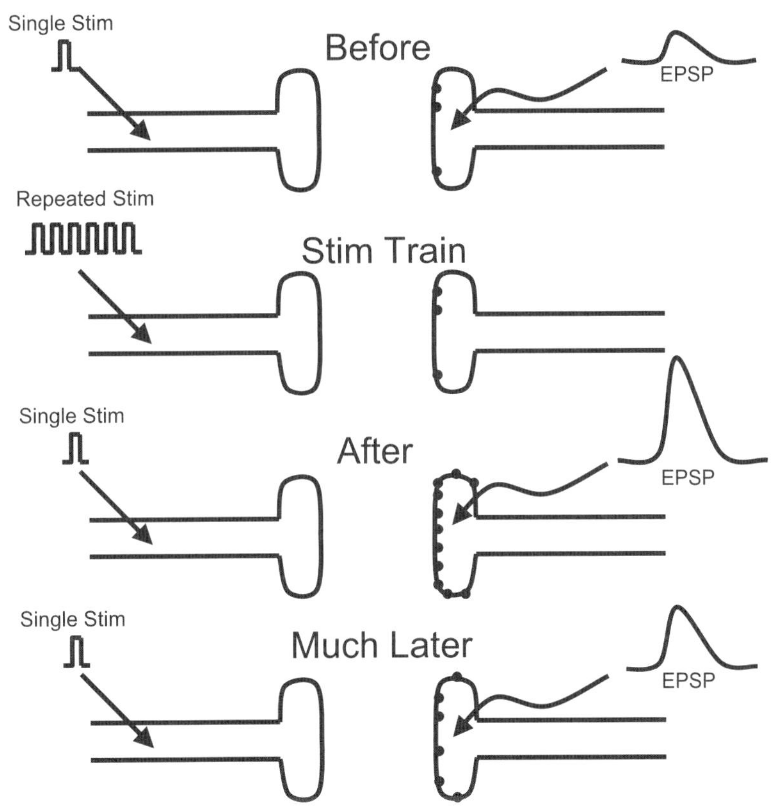

Figure 6.3: Long-term potentiation of EPSP.

CHAPTER 7

Networks of Neurons

It has been well established that an individual neuron does very little in the way of information processing or learning. The current hypothesis is therefore that it is the complex connections and interactions between neurons that gives rise to the enormous range of functions and behavior of the brain. The first half of this chapter will focus largely on the work of Roger Traub's research group on connecting thousands of multicompartment neurons. The second half of this chapter will introduce some basic concepts of artificial neural networks.

7.1 NETWORKS OF NEURONS

Previous chapters built up the theory necessary to create a multicompartment neuron with any complex geometry as well as the connections between neurons at the synapse. This section will demonstrate how many neurons may be connected together to perform some basic functions of networks of neurons.

7.1.1 The Traub Pyramidal Neuron

Based upon experimental studies in slices of the rat hippocampus, Traub and Miles created a generic Pyramidal neuron. A schematic of the neuron is shown in Fig. 7.1 and is composed of 28 compartments,

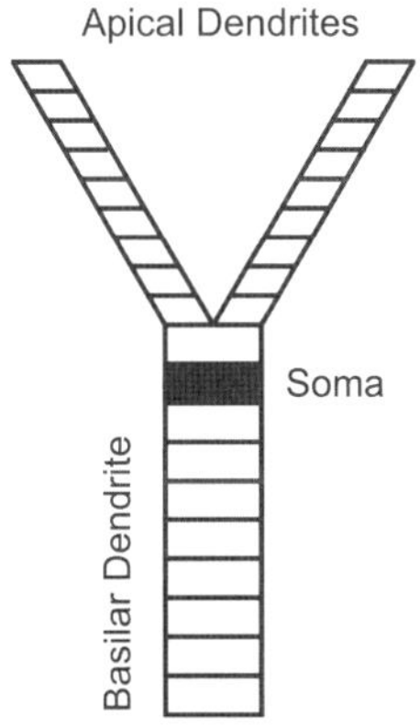

Figure 7.1: Traub pyramidal cell geometry.

divided into apical and basilar dendrites separated by the soma. The dendrites are modeled as passive cables and the soma is modeled using the bursting membrane model described in Sec. 3.4.2. For a complete description of the model, see *Neuronal Networks of the Hippocampus* by Traub and Miles (1991).

7.1.2 Conduction down the Axon and Synaptic Connections

The axon of the Pyramidal cell is not modeled directly because it is assumed that any action potential at the soma will reach the axon terminal after some delay. The axon is therefore modeled simply as a delay that depends upon the conduction velocity and length of the axon.

Once an action potential is delayed and reaches the pre-synapse, the impact on the post-synapse is modeled using the generic synaptic current equation first introduced in Eq. (6.3).

$$I_{\text{syn}}(t) = c_e O(t) \left[V_m^{\text{post}} - E_e \right] \tag{7.1}$$

where the maximum conductance term is c_e and the gating variable, O, takes the form

$$\frac{dO}{dt} = \text{rect}(t_{\text{on}}, t_{\text{off}}) - O\ . \tag{7.2}$$

The rect function is a pulse that is equal to 1 between t_{on} and t_{off} and 0 everywhere else. In Traub's model, if the soma voltage is greater than $20mV$ and hasn't fired in $3msec$ (i.e., refractory period), then t_{on} is set to the delay for the axon. t_{off} determines how long the pulse will last and is governed by the type of synapse.

7.1.3 Interneurons

Real networks of neurons are rarely composed of a single cell type. Traub's group therefore incorporated *interneurons*, also known as inhibitory neurons, into their model by making three changes to the Pyramidal cell model. First, the multicompartment geometry was reduced. Second, the active membrane at the soma was changed to eliminate the Calcium current. Third, synaptic connection from interneurons were made to be inhibitory by changing the Nernst potential and t_{off} in Eqs. (7.1) and (7.2) at the post-synapse. The strength of the inhibitory connection was called c_i.

7.2 MODEL BEHAVIOR

Traub's research lab was able to simulate nearly 10,000 individual cells and made some ground-breaking observations that agreed well with experimental recordings. Before explaining the large-scale trends, they demonstrated three interesting phenomenon, in networks of two or three cells, that formed the basis for the more complex behavior. We repeat these simulated experiments below.

7.2.1 All or None Synaptic Induced Firing

The strength of an excitatory synapse may be modulated through c_e and was found to initiate an action potential burst in an all-or-none fashion. Figure 7.2 shows Pyramidal cell 1 connected to Pyramidal cell 2 through an excitatory synapse at the soma of cell 2. A burst was initiated at the soma of cell 1 and the strength of the synapse connection (c_e) was varied. It is clear from the figure that bursting in cell 2 is a governed by the strength of the connection.

In large scale models, it is difficult to always draw the shape of every neuron, so many presentations of *neural circuits* use some combination of circles to represent cells and triangles to represent synapses. In Fig. 7.2, the excitatory connection between two Pyramidal cells is represented by two open circles (e.g., excitatory cells) connected together by a line terminating in an open triangle (excitatory synapse). The direction of the triangle also indicates that cell 1 is driving cell 2.

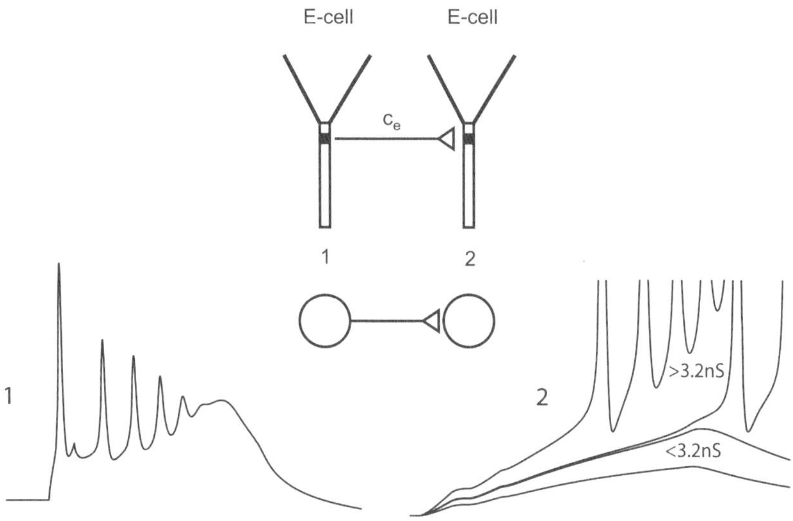

Figure 7.2: All-or-none synaptic transmission of bursting.

7.2.2 Suppression of Neuronal Firing

The bursting in the second neuron may be inhibited by an interneuron. Figure 7.3 shows the case where Pyramidal cell 1 will tend to cause Pyramidal cell 2 to fire (e.g., $c_e > 3.2nS$) in the absence of inhibition. Interneuron cell 3 (closed circle), on the other hand, will tend to prevent Pyramidal cell 2 from firing. Here it is the strength of the inhibitory connection (c_i, closed triangle) that determines whether cell 2 will fire or not. For example, if $c_i > 5nS$ the strength of the inhibitory connection will prevent an action potential from firing in cell 2. If $c_i < 5nS$, the inhibitory connection is not strong enough to prevent an action potential in cell 2, but it may delay when cell 2 fires.

7.2.3 The Importance of Delays

The impact of an inhibitory cell is governed not only by the strength of the connection, but also the timing. Figure 7.4 show a similar three cell network as Fig. 7.3, but c_e and c_i have been fixed so that cell 2 does not fire. The timing of the interneuron action potential, however, was delayed relative to the firing of Pyramidal cell 1. As this delay becomes large, Pyramidal cell 1 has already committed to bursting and the late inhibition has little effect.

7.2.4 Cyclic Activity in the Brain

The brain is known to generate a number of rhythms that vary depending upon the waking state and mental tasks being performed. Although the function of these rhythms is not known, computational and theoretical work have demonstrated how the brain may use the same neural circuits to generate different rhythms. Typically, these circuits consist of tens to thousands, or even millions of cells. As a simple demonstration, consider the left-hand panel of Fig. 7.5. The single pulse may activate Cell 1, which will activate Cell 2 and on to Cell 3. If the delay for passing the electrical impulse through the network is long

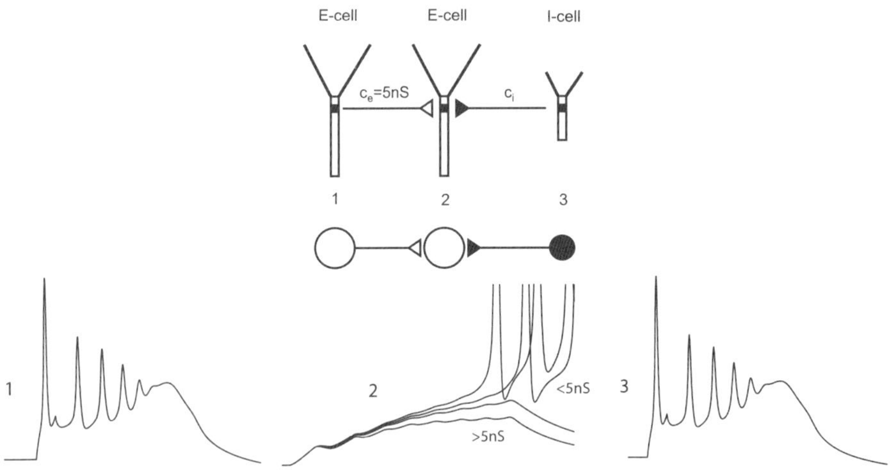

Figure 7.3: All-or-none inhibition of bursting by strength of inhibition.

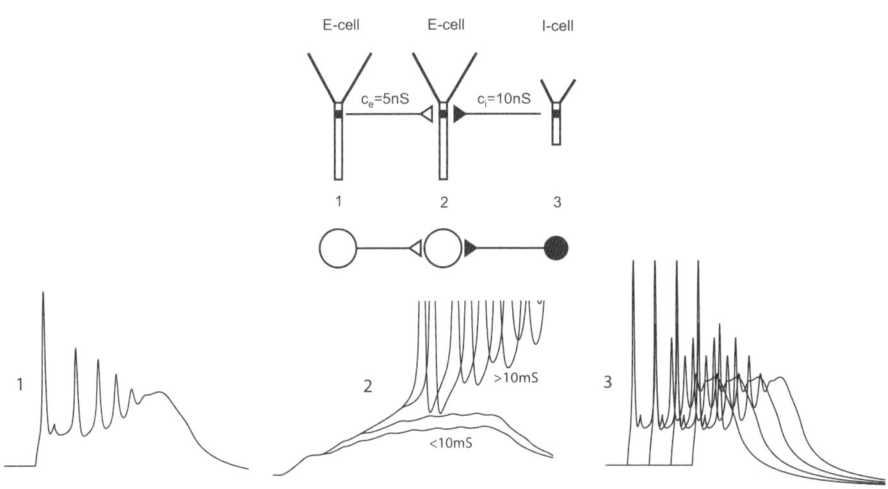

Figure 7.4: All-or-none inhibition of bursting by delays.

enough, Cell 3 could then reinitiate firing in Cell 1 after it has repolarized. In this way, a simple neural circuit, possibility composed of many cells, could have some natural cycle of firing.

To modulate the rate of the cycle, consider adding an interneuron, as in the right panel of Fig. 7.5. If c_i is high, it may end the cycle all together. However, c_i could also be a way to delay how quickly Cell 3 reinitiates firing in Cell 1, thus slowing down the rate of the cyclical firing. This situation could be even more interesting if c_i itself was modulated by other synapses, hormones or ionic concentrations.

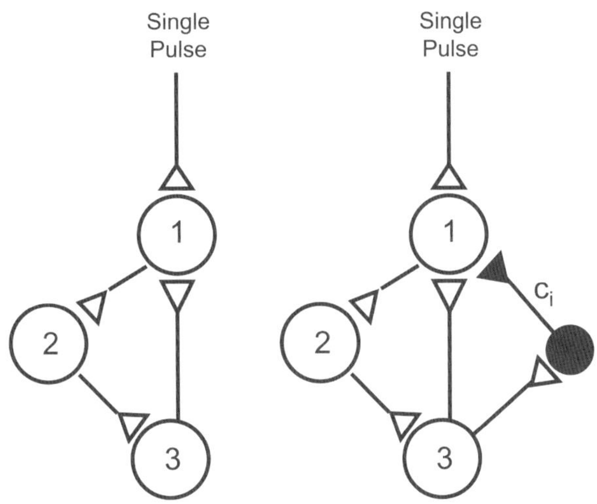

Figure 7.5: Cyclic firing in a network of neurons.

7.3 ARTIFICIAL NEURAL NETWORKS

In 1943, McCulloch and Pitts published an article that showed how the basic actions of a neuron could be generalized. Furthermore, they showed that these general units, called *perceptrons*, could be used to perform logical functions such as *and* and *or*. Each perception is a system with a simple input/output transfer function that roughly models the behavior of a neuron. Because of the simple transfer function, many neurons (thousands or even millions) may simulated quickly. The general philosophy, unlike much of this text, is that the higher-level functions of the brain are governed by the *connections* between neurons and not the neurons themselves. So, to model the higher-level functions of the brain (learning, distributed memory, planning, pattern recognition), it is not necessary to model the details of the neuron. Rather, it is more computationally efficient to model neurons as simple units that are connected together in complex ways as in Fig. 7.6. In the following section we will give a brief overview of neural networks composed of perceptrons.

7.3.1 The Perception and Single Layer Networks

In general, a perception, i, may only be in one of two *states*, firing (1) or not firing (0). This state is the output of the perception. But the perception may also take as inputs, the output of other perceptions. Computing the output of a perception given the inputs is a two step process. First a weighted sum of the inputs is computed as

$$\sum = \sum_{i=1}^{N} w(i, j) a_i \tag{7.3}$$

where a_i are the N inputs and $w(i, j)$ are weights between perceptions i and j. The second step is to apply an *activation function*

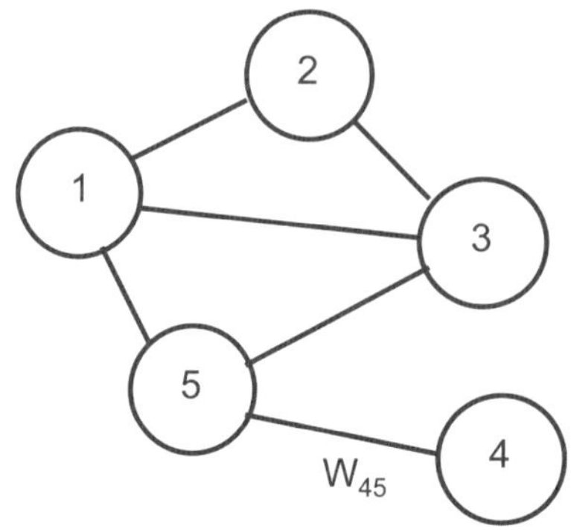

Figure 7.6: Weighted connections between perceptrons.

$$a_j = g\left(\sum\right) = g\left(\sum_{i=1}^{N} w(i, j)a_i\right). \tag{7.4}$$

Often, the function $g()$ is a threshold function as in the left panel of Fig. 7.7 and defined by

$$a_j = g(\sum) = \begin{cases} 1 & \text{if} \sum \geq \theta \\ 0 & \sum \leq \theta \end{cases}$$

but $g()$ may also take on a more smooth function such as a sigmoid

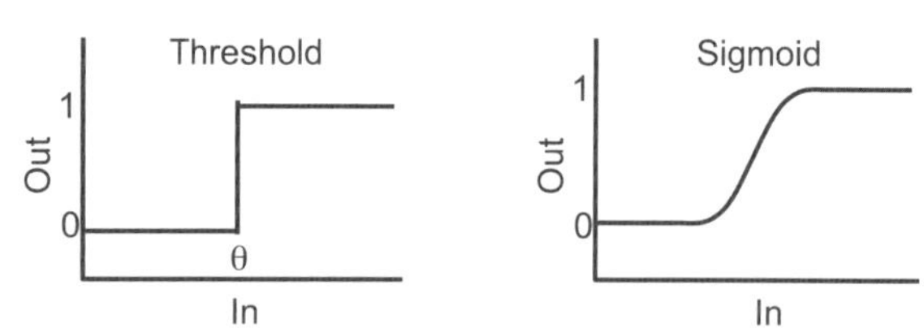

Figure 7.7: Threshold and sigmoid activation functions.

Using the threshold $g()$ function and appropriate choice of weights, McCulloch and Pitts were able to show how to create **NOT**, **AND**, and **OR** logical functions as shown in the top panel of Fig. 7.8. The input to the gates (e.g., x_1 and x_2) are always either 0 or 1.

As shown in the bottom panel of Fig. 7.8, the logic function can also be considered graphically. Inputs are graphed on the axes and the squares represent the desired outputs, filled for a logical 1 and open for a logical 0. The dotted line represents a *decision line* that is determined by the weights and threshold. For a two-input perceptron, consider a group of inputs and weights that when summed are exactly at the threshold

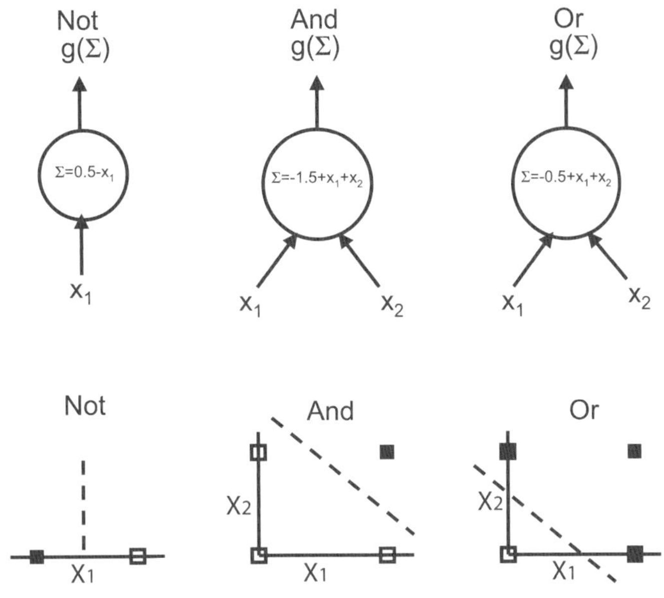

Figure 7.8: Perceptrons computing logical NOT, AND, and OR.

$$\sum_{i=1}^{2} w_i a_i = \theta = w_1 a_1 + w_2 a_2 \,. \tag{7.5}$$

If we solve for a_2

$$a_2 = -\frac{w_1}{w_2} a_1 + \frac{\theta}{w_2} \,. \tag{7.6}$$

Equation (7.6) is in the form of a decision line, with a slope of $-w_1/w_2$ and intercept of θ/w_2, that separates the inputs into two categories. These two categories are assigned the output of 1 or 0 depending upon whether they are above or below the decision line. A decision line is therefore the most simple form of pattern recognition. Figure 7.8 shows the decision lines for the AND, OR, and NOT gates where the thresholds have been set to 1.

Although the math will not be presented here, given some set of inputs that need to be classified, the appropriate decision line may be created as in the left panel of Fig. 7.9. Finding the decision line is a very simple *learning rule* by which the weights and thresholds are adjusted. Graphically, this is equivalent to changing the slope and intercept of the line in an iterative process until the line effectively separates to two inputs.

The right panel of Fig. 7.9b demonstrates a case where a simple line is not capable of separating two sets of points. Here, open squares represent inputs which should be classified together and closed squared are in another category. The situation becomes even more complex if there are more than two classification categories. The problem is that the points in the right panel of Fig. 7.9 are not *linearly separable*, meaning they can not be separated by a line. The most simple example of a logical function that is not linearly separable is the **XOR** function shown graphically in Fig. 7.10. The problem of building and trained neural networks capable of classifying nonseparable points nearly ended active research in neural networks in the 1970s.

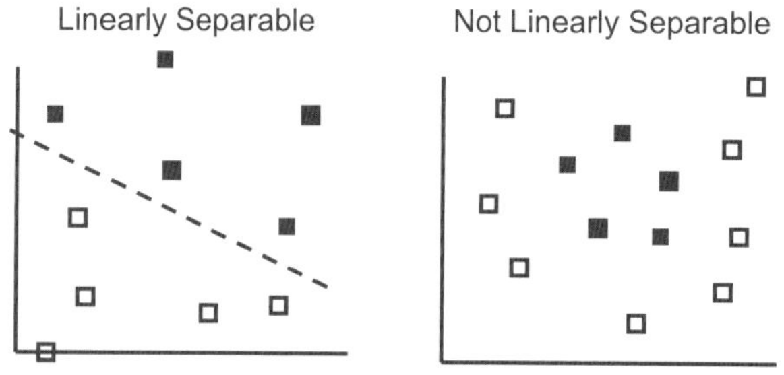

Figure 7.9: Example of linearly separable and not linearly separable data.

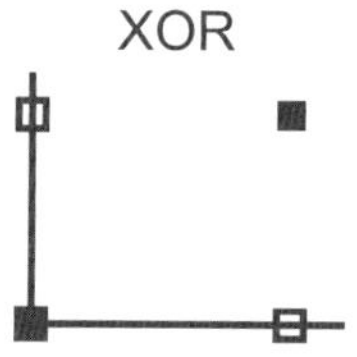

Figure 7.10: XOR logical function is no linearly separable.

7.4 MULTI-LAYERED NETWORKS

The points in Figs. 7.9 and 7.10 points can be classified if many lines are used. For example, to classify the XOR points, only two lines are needed. The practical meaning of adding more lines is the addition of *hidden layers* of perceptions in the network as shown in Fig. 7.11. Each unit receives inputs only from the preceding layer and can only send outputs to units in the layer ahead. Hidden layers allow for multiple types of points, separated by nonlinear regions, to be grouped together. Furthermore, adding more hidden layers, in principle allows for any complex set of regions to be defined.

7.4.1 Backpropagation

Initially there was no way to systematically *tune* the weights to separate any generic set of groupings. Graphically, this is equivalent to changing the slope and intercept of many decision lines in a coordinated way. It was not until the early 1980's that a method was found. First, an ideal target output, T (a vector

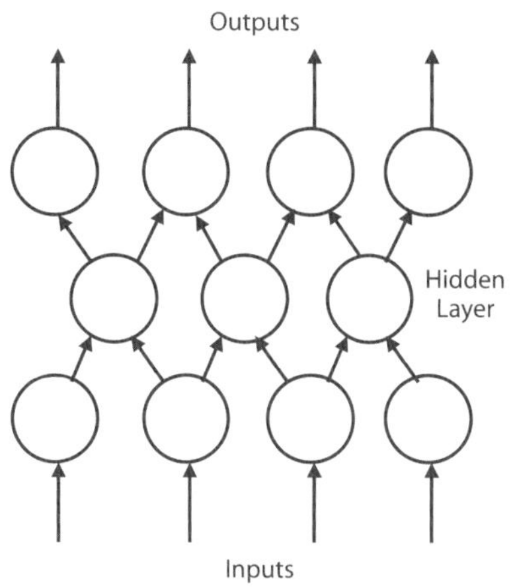

Figure 7.11: Generic feedforward multilayer network.

of outputs), is defined that corresponds to known inputs. Given an actual output, a_j, an error can be computed between the actual and target output. The outputs can then be adjusted by tuning the weights in a direction such that the sum of squared output errors, E_p, will be minimized.

$$E_p = \frac{1}{2}\left(T - a_j\right)^2 . \tag{7.7}$$

Substitution of Eq. (7.4) into Eq. (7.7) yields

$$E_p = \frac{1}{2}\left(T - \sum_{i=1}^{N} w(i,j)a_i\right)^2 . \tag{7.8}$$

Another interpretation of E_p is that it defines an *energy function* or energy surface. The surface is at a minimum when the error between a_i and T is smallest. In theory we might imagine the error being zero, however, this generally not possible. Finding a practical solution is therefore is equivalent to finding the set of weights that minimizes Eq. (7.8). For an arbitrary case there is no analytical way to find the minimum. Instead the most common approach is the iterative method of *gradient descent*. First a function is defined that characterizes how the energy surface (E_p) will change if the weights (w) change.

$$\frac{dE_p}{dw} = \left[T - g\left(\sum w(i,j)a_i\right)\right] g'\left(\sum w(i,j)a_i\right) a_i \tag{7.9}$$

which is simply the derivative of Eq. (7.8) with respect to w. Evaluating Eq. (7.9) at a point gives the slope of the energy surface with respect to w. The negative of this multidimensional slope is the direction w should be changed to cause the most rapid drop in E_p. This process is repeated in an iterative fashion to arrive at a minimum. It should be noted that for the gradient decent method to work, g must be a differentiable (e.g., a sigmoid). The new weights can then be computed as:

$$w^{\text{new}}(i, j) = w^{\text{old}}(i, j) + \alpha \frac{dE_p}{dw} \tag{7.10}$$

where α is called the *learning rate*. The updating of weights in this way is called the *delta rule* and is specific case of a more general method called *backpropagation*. Backpropagation is a family of learning rules where blame is assigned to individual connection weights for errors. Once the blame is targeted, the weight may be moved in a direction that will decrease the error.

7.4.2 Momentum and Thermal Noise

The delta rule in Eq. (7.10) can be slow if α is small. On the other hand, if α is too large, the numerical solution may oscillate or even diverge from the minimum. A second *momentum* term may be added to Eq. (7.10):

$$w^{\text{new}}(i, j) = w^{\text{old}}(i, j) + \alpha \frac{dE_p}{dw} + \nu dw^{\text{old}}(ij) \tag{7.11}$$

where ν is a momentum constant and dw^{old} is the previous change in w. Here, α may remain small, but the total change to w may be adjusted each iteration.

It is also possible for this iterative learning rule to converge on a local minimum and become stuck. To help reach a global minimum some neural networks use *simulated annealing*. Noise is added to Eq. (7.10), large at first but then slowly decreasing toward zero. The effect is that at some intermediate noise level, the current solution will not settle into any of the local minimums but only the global minimum. A further decrease in the noise level, will not be enough to move the solution outside of the basin that forms the global minimum.

7.4.3 Training and Use

Most neural networks undergo two phases of use. The first we have already discussed and is called the *training* phase. It is during training that the weights of the network are adjusted by applying known inputs and specifying target outputs. After the network is performing well on the training set, the *usage* phase may begin. In other words, the network is ready for real and unknown inputs.

7.4.4 Energy Surfaces and Sheets

A multidimensional energy surface is not easy to visualize, but a three-dimensional analogy may make some concepts more clear. Imagine a sheet that is stretched across a number of chairs and tables. The sheet will naturally have some peaks that represent unstable points and valleys that represent local minima. If a ball is then dropped on the sheet, it will roll toward the nearest local minima and come to rest at the bottom. Here the topography of the sheet acts as an energy surface and the ball rolling toward the local minimum is similar to the method of gradient decent.

For the analogy to be useful, however, the minima should correspond to some desired output and the location where the ball dropped must correspond to an input. Consider placing into each minima a perfectly scripted capital letter (e.g., 'E' in one minima, an 'F' in another minima). The initial location onto which the ball drops would then represent an imperfect letter that may be close to an 'E' but does not exactly match the perfectly scripted 'E'. In this case, the gradient decent method would classify the input as an 'E' by sending the ball into the local 'E' minima. Likewise, if the imperfect letter was closer to the ideal 'F', the ball would fall into the 'F' minima.

The analogy we have built up so far is for a network that has already been trained. To train our sheet-ball system, we would begin with a perfectly straight sheet and then present a perfect 'E' as an input. To ensure that the sheet would remember the 'E' in the future we would change the contour to make a local minima at exactly this point. The change in sheet topology would is analogous to changing the weights in a neural network. The effect would be to warp the sheet around the local minimum such that even a distorted 'E' input would roll to the lowest point. The same type of process could be used to generate other local minima to correspond to 'F', 'G', and 'H'. In the end, the sheet would be trained to recognize distorted letters.

7.4.5 Network Structure and Connectivity

A consideration when building a neural network is how many perceptrons are needed to make the desired classification. If the network is too small, the error even at the global minimum may be unacceptable, and the network is an *underfit*. On the other hand, if the network is very large (an *overfit*), the increased computation may only result in a small gain in accuracy. A more recent solution to this problem is to use *tiling*. A tiling algorithm begins with a small network which is trained to classify inputs using the methods discussed above. When the global minimum is reached, the error is assessed. If the error is not acceptable, more perceptrons are added and new connections are made. The training then resumes on the larger network. This iterative process can be continued until the network produces results that are within a desired error tolerance.

7.4.6 Recurrent and Unsupervised Networks

Thus far, we have only considered *supervised* training of a neural network. Although supervision may have different interpretations, generally it means that the user specifies some known inputs paired with desired T outputs. A learning algorithm, such as Eq. (7.10), can then be used to adjust the weights. In an *unsupervised* network, T may not be given and it is up to the network to develop classification patterns that pair similar inputs. The neural networks discussed thus far have been of a class called *feedforward* because information flows from one layer to the next in one direction. For a network to develop classification patterns, it must be able to adjust weights through feedback connections. Networks with built in feedback are often called *recurrent* networks and can be used to not only classify patterns, but also store distributed and adaptable memories of information and processes. The two most mentioned types of recurrent networks are shown in Fig. 7.12 but the number of network variations has grown rapidly and is still growing. Although we will not discuss these types of networks, they have been shown to perform some incredible feats, such as unaided pattern recognition, storage of short-term memory and planning.

7.5 NUMERICAL METHODS

Although it is possible to generate the computer code needed to simulate networks of neurons and artificial neural networks, a number of software packages exist which drastically reduce the start up time. To simulate networks of neurons, two programs have established themselves as the gold standards of neural compartment simulations: GENESIS and NEURON. GENESIS (`http://www.genesis-sim.org`) and NEURON (`http://neuron.duke.edu`) were born out of the need for a common platform for

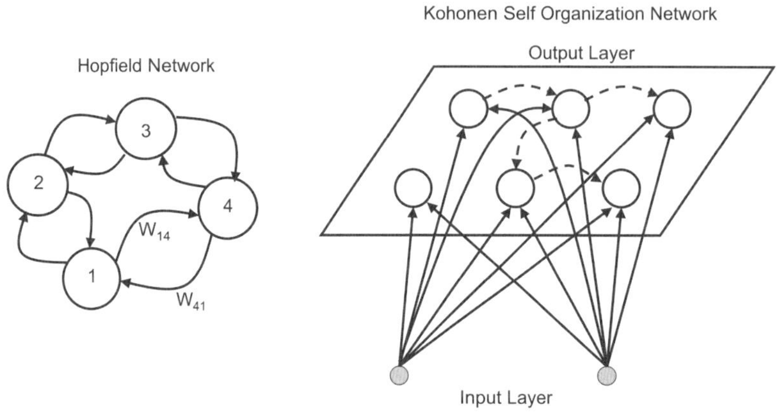

Figure 7.12: Hopfield and Kohonen recurrent networks.

neural simulations. Both allow for stand-alone simulations on a single CPU or massively parallel simulations on supercomputers. Both have graphical user interfaces that show neural geometry, recording, and stimulating electrodes and built in functions for viewing the time courses of voltages, currents, and concentrations. To simulate artificial neural networks, MATLAB has a Neural Network tool box that contains many optimized functions for building, simulating visualizing, and analyzing neural networks.

CHAPTER 8

Extracellular Recording and Stimulation

The theory we have derived so far is largely based upon transmembrane potentials and currents. Recording V_m or I_m in a real preparation, however, is nontrivial and nearly impossible in the clinic. In most situations it is, therefore an extracellular potential that is recorded. In this chapter we will use Maxwell's equations to derive a relationship between extracellular potentials and membrane currents.

8.1 MAXWELL'S EQUATIONS APPLIED TO NEURONS

Maxwell's equations describe how an electric field creates a distribution of electric potentials that can be recorded everywhere in space. One mechanism by which an electric field may arise is if a current source is present in a large conductive bath of fluid. In the physics and electrophysiology literature, a large bath is called a *volume conductor*. If we assume a small current, I_o, emanates from a point in a uniform conducting medium of infinite extent, the current will flow radially in all directions. By placing a small sphere of radius, r, around the current source, we can compute the flux, J, through the sphere as

$$\mathbf{J} = \frac{I_o}{4\pi r^2}\mathbf{a_r} \tag{8.1}$$

where $4\pi r^2$ is the surface area of the sphere and $\mathbf{a_r}$ is a unit vector in the radial direction.

Although an electric field can theoretically change at the speed of light, biological systems are much slower. Therefore, all analysis of biological system can assume changes in the electric field to be *quasi-static*. The practical meaning is that if we record a potential, even if it is far from the source, it will reflect the current at exactly that time. In reality, this is not true but the field changes much faster than the biological current can change. Given this assumption

$$\mathbf{E} = -\nabla\phi_e \tag{8.2}$$

where $\mathbf{E}$ is the electric field vector and $\nabla\phi_e$ is the gradient of extracellular potentials in the bath. To relate the gradient of ϕ_e to flux, we can use the general form of Ohm's Law:

$$\mathbf{J} = -\sigma_e\nabla\phi_e \tag{8.3}$$

where σ_e is the conductivity of the bath. Far from any current sources, Eq. (8.3) reduces to Laplace's Equation.

$$\nabla^2\phi_e = 0\,. \tag{8.4}$$

Combining Eqs. (8.1), (8.2), and (8.3),

$$\mathbf{J} = \frac{I_o}{4\pi r^2}\mathbf{a_r} = -\sigma_e \nabla \phi_e \ . \tag{8.5}$$

We know, however, that components of $\nabla\phi_e$ can only be in the radial direction, so

$$-\sigma_e \frac{d\phi_e}{dr} = \frac{I_o}{4\pi r^2}\mathbf{a_r} \tag{8.6}$$

integration with respect to r yields

$$\phi_e = \frac{I_o}{4\pi\sigma_e r} \ . \tag{8.7}$$

Note that in Eq. (8.7), ϕ_e is the same for any surface where r is constant (concentric spheres).

We have mentioned potentials before but it is worth giving a formal definition here. The electric potential at a point is simply the work required to move a single charge (e.g., e^-, K^+) from infinity to that point against the electric field. Therefore, as $r \to \infty$, ϕ_e must equal zero.

Equation (8.7) is useful when there is one current source. When more than one current source is present, we can sum up the effects of all the currents. First, we can assume each current is generated within some small volume, dV, and has some incremental effect on the overall ϕ_e. Differentiating Eq. (8.7)

$$d\phi_e = \frac{I_o dV}{4\pi\sigma_e r} \tag{8.8}$$

and to sum up the effects

$$\phi_e(x', y', z') = \frac{1}{4\pi\sigma_e}\int \frac{I_m(x, y, z)}{r} dV \tag{8.9}$$

$$r = \sqrt{(x - x')^2 + (y - y')^2 + (z - z')^2} \tag{8.10}$$

where r is the distance from each current source, (x, y, z), to the recording point, (x', y', z'). Note that we have assumed that the I_o of Eq. (8.8) is the membrane current, I_m, emanating from a small patch of membrane. In this way, we can theoretically calculate potentials in a bath surrounding an active neuron, given all of the I_m sources (see Fig. 8.1).

8.1.1 Forward and Inverse Problems

If all of the I_m and r terms are known, ϕ_e may be calculated using Eq. (8.9). This calculation is known as the *forward problem* and has a unique solution. Equation (8.9) also demonstrates a fundamental problem encountered in all extracellular recordings. ϕ_e is an *average* of the effect of many current sources, weighted by the radii. As in any average, information is lost. Therefore, if you record ϕ_e there is no way to back calculate all of the I_m sources. In fact, you may not even know how many current sources compose ϕ_e. Computing I_m sources from ϕ_e is called the *inverse problem*. Despite the limitations of the inverse problem, ϕ_e is much easier to record and is the typical measurement made for both research and clinical purposes. Although we will focus on the forward problem here, there exists a large body of research on gaining the maximum amount of information from ϕ_e. We will explore some of these methods in Ch. 9.

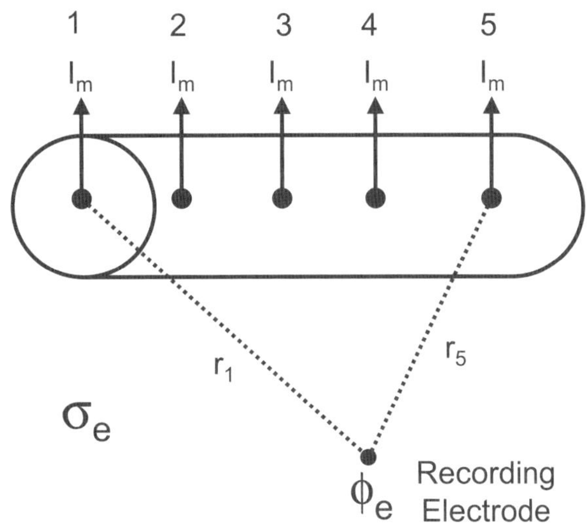

Figure 8.1: Summation of dipoles in calculating an extracellular potential.

8.2 EXTRACELLULAR POTENTIAL RECORDINGS

8.2.1 Extracellular Potentials Generated by an Active Cable

In our derivation of propagation in a cable, we assumed that the cable was very thin. If we also assume that the cable is surrounded by a large bath and an electrode far from the cable, the current sources, I_m, will appear to be point sources. Furthermore, if we align the cable with the x-axis, then the infinitesimal volume in Eq. (8.1) because an infinitesimal length, dx:

$$\phi_e(x', y', z') = \frac{1}{4\pi\sigma_e}\int \frac{I_m(x, y, z)}{r}\,dx \tag{8.11}$$

where I_m at each point in the cable may be computed in one of two ways.

$$I_m = C_m\frac{dV_m}{dt} - I_{\text{ion}} \tag{8.12}$$

or

$$I_m = \frac{\pi r^2}{R_i}\frac{d^2V_m}{dx^2}\,. \tag{8.13}$$

Note that in the second formulation ϕ_e can be computed at a particular moment in time by knowing only the spatial variation in V_m along the cable.

8.2.2 Current Sources in a Complex Domain

Although a section of an axon or dendrite may be approximated by a straight cable, in reality axons and dendrites are not straight. Therefore, the Eq. (8.11) is not valid. Furthermore, given the complex tangle of neurons, the assumption of infinite bath of conductive fluid is not valid either. It is possible to numerically solve for ϕ_e in a complex domain by building a circuit analog as in Fig. 8.2. Although the circuit is only a very small portion of a much larger circuit network, it should be clear that we can represent intracellular and extracellular space by passive resistors and membranes as nonlinear current sources. The advantage of this arrangement, besides allowing for complex geometries, is that extracellular and intracellular potentials can be recorded directly at any point (e.g., Node 2). It also naturally allows for the membrane of the cell to influence currents and potentials in the extracellular space. At the boundaries we can make the simple assumption that no current may leave, leading to sealed ends as described in Sec. 4.3. To model a 3D section of neuronal tissue, it is conceptually simple to extend our resistor network into a third dimension.

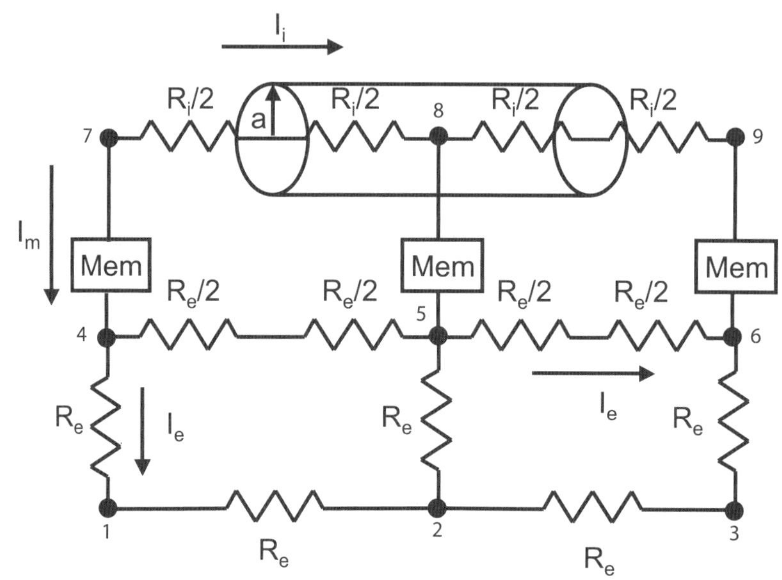

Figure 8.2: Extracellular potentials in a complex network of neurons.

The obvious disadvantage of a circuit network is that it is computationally very slow because potentials and currents must be computed at *every* point. The more subtle disadvantage is that the technology does not currently exist to efficiently image the complex web of neurons and directly translate the image to a circuit model.

8.2.3 The Electroenchephogram (EEG)

From the previous section and Ch. 7 is should be clear that large populations of neurons can generate a significant extracellular potential that may be detected at some distance. The more active and synchronized the neurons, the larger the signal. Of course, any behavior will be greatly averaged but a general idea of the degree of activity can be determined. At least in principle, we have described precisely the situation in recording the *electroenchephalogram* or EEG. The EEG is the potential recorded from the surface of the scalp. The deflections, although in the μV range, correlate to the activity of the brain in regions under the electrode. Therefore, the EEG is typically recorded using many electrodes positioned directly above known brain structures.

8.3 PRACTICAL ASPECTS OF RECORDING NEURAL POTENTIALS

The typical process of recording and storing electric potentials is summarized in Fig. 8.3. In the section below, we review the key components of a recording system. Processing will be the subject of Ch. 9.

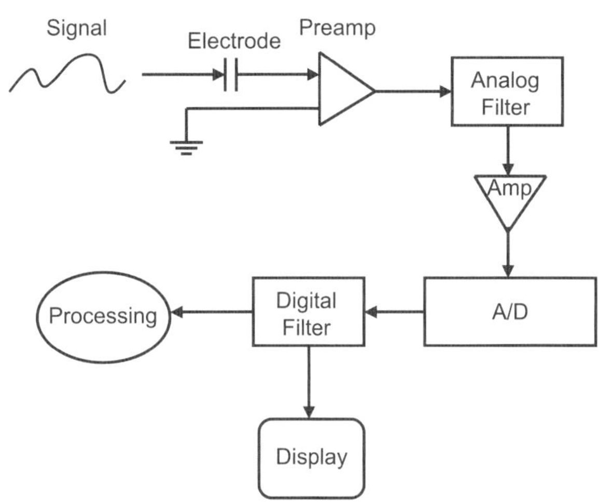

Figure 8.3: Recording, filtering, amplification, and digitization of an extracellular potential.

8.3.1 Electrodes

Currents in the brain are carried by ions. Currents in computers and wires, however, are carried by electrons. The role of an electrode is therefore to transduce ionic current to electric current. The chemical reaction most typically used is:

$$Ag \rightleftharpoons Ag^+ + e^-$$
$$Ag^+ + Cl^- \rightleftharpoons AgCl\ .$$

Although many metals could be used to create an electrode, the *Silver-Silver Chloride* electrode is easy to fabricate, has a fast electrical response, and is compatible with biological tissues.

Electrodes for recording potentials from tissue are typically small ($< 1mm$). They can be placed directly on the surface of neural tissue or, since neural tissue is somewhat spongy, can easily be pushed deep inside of a three-dimensional section of tissue. Recordings made in the brain are typically called *cortical recordings* if the electrodes are one the surface of the brain, or *sub-cortical* if they are pushed into the depth of the tissue. Sub-cortical recordings can in fact be made from awake patients because the brain has no sensory neurons.

In EEG recordings we are not concerned with uncovering individual neuronal firings but rather the general activity below the electrode. Therefore, many electrodes are placed on the scalp so comparisons can be made between the relative activity of different brain regions. Since an average is desired, the electrodes often cover a large area (e.g., $1cm$ in diameter). The amplitude of the EEG is small because the skull and scalp have a low conductivity (σ_e). To decrease the impedance of ion flow to the electrode, the scalp may be *abraded* and a gel with free Cl^- ions is often applied.

8.3.2 Recording Preparations

In most tissue recordings there are at least three electrodes present. The *recording electrode* will detect the potentials resulting from the electric field generated by membrane currents. In the true definition of a potential, however, another electrode at infinity would be necessary. As this is impossible in practice, most recordings have a second electrode, called the *ground*, that is far from any electrical activity. Common locations for the ground would be a large slab of metal in the room or some location in the body that is very poorly electrically connected to the recording electrode. The third electrode is called the *reference* and is typically located in an electrically connected region of tissue but not close to the recording electrode. To understand the role of the reference, consider that the recording electrode will detect *all* current sources scaled by the radii. So, current sources close by will have a large impact but there will be few of them. On the other hand, the distant current sources will individually have a small contribution but there could be many of them. The role of the *reference* electrode is to record these distant sources so they may be subtracted from the recording signal.

8.3.3 Filtering, Amplification and Digitization

The amplitude from a neural recording (EEG or tissue recording) may be as small as $1\mu V$ and contain noise from the recording device, room, and motion artifacts. Therefore, neural signals are generally filtered and amplified. In high-end recording systems, amplification and filtering are performed in a series of steps. Even after filtering and amplification, the signal is still continuous. To be stored in a computer, the signal must be *digitized.* Digitizing will transform a continuous signal (with infinitely many values) to a discrete signal (with some finite number of values). The specifics of how to amplify, filter, and digitize a neural signal will not be covered here but can be found in texts on biomedical instrumentation.

8.4 EXTRACELLULAR STIMULATION

If a stimulus is applied to one of the extracellular nodes in Fig. 8.2, the potential at that point (ϕ_e) will become more negative. According to Laplace's Equation [Eq. (8.4)], this negative potential will fall off with distance from the stimulus point. However, it is possible that even after the attenuation, the extracellular potential outside of an active section of membrane could be driven more negative. Recall that $V_m = \phi_i - \phi_e$ and that Sodium channels will open if $V_m > V_m^{th}$. So, if ϕ_e outside the membrane decreases enough, the membrane potential may reach threshold. In this way, an extracellular stimulus can be used to activate a patch of membrane.

In a real preparation, a stimulating electrode will be surrounded by a complex tangle of axons, dendrites, somas, and glial cells, so the exact amplitude and duration to induce an action potential will not be known. Furthermore, a single action potential propagating down a signal axon may not be enough to have any global impact, as it will be lost in the constant *chatter* of the neurons. Therefore, the goal of most neural stimulation is to cause all or most of the neurons around the stimulus electrode to fire and then recruit even more neurons. This idea of recruitment is called *capture*. To ensure that a given stimulus is achieving capture, it is common to record a potential at some distance from the stimulus, and then turn up the stimulus amplitude until the distant neurons are firing in response to the stimulus.

In addition to a stimulating electrode, a second electrode, call the *return*, is often used as well. The purpose of the return electrode is to provide a sink (or return) for the injected current. The advantage of using a return electrode is that current will find the shortest pathway(s) from the stimulus to the return. By placing the stimulus and return relatively close together it is possible to focus current into a particular location.

8.5 NUMERICAL METHODS: COMPUTING ϕ_E IN A CABLE

To compute ϕ_e(t) numerically, at each time step

$$\phi_e(t) = \frac{1}{4\pi\sigma_e} \sum \frac{I_m(t)}{r} dx \tag{8.14}$$

must be computed.

```
Define Propagation and Membrane Constants constants
Define σe
Define location of recording electrode
Compute initial conditions
Compute radii (r(i)) for each patch to recording electrode.

for (time=0 to time=end in increments of dt)

   Imsum=0.0
   for (i=1 to i=Last Node in increments of (1)

       Update Currents
       Update Differential Equations
       Imsum = Imsum + (Im(i)/r(i))*dx

   end

   φe(time) =  1/(4π σe)*Imsum
   Save φe(time)

end

Store values of interest to a file
```

Homework Problems

(1) Estimate $\phi_e(t)$ at the electrode in Fig. 8.5, where three current sources are propagating down a cable at $300mm/s$. Assume $dV = 2mm^3$, all distances are in mm, and all currents are in $\mu A/mm^2$.

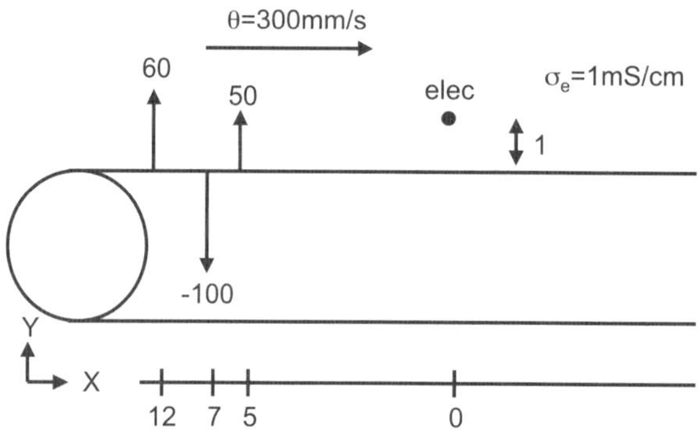

(2) Derive the equations for Fig. 8.2.

Simulation Problems

(1) Modify the program in Ch. 4 simulation problem 2 to compute $\phi_e(t)$ at $x = 1.5cm$ and $0.2cm$ above the cable. Assume that $\sigma_e = 1mS/cm$.

(2) Show how the shape of ϕ_e changes as the location of the recording electrode is change. Explain the changes you see.

(3) Show how the shape of ϕ_e changes as R_i is changed. Explain the changes you see.

(4) Program Fig. 8.2 assuming $R_e = 10\Omega cm$, $R_i = 100\Omega cm$, and I_{ion} is described by the Hodgkin-Huxely model.

(5) Use the program in problem 2 above to create a strength-duration curve for a stimulus at node 2 in Fig. 8.2.

CHAPTER 9
The Neural Code

Over the past several decades, two primary thrusts have driven neuroscience research. The first thrust has been to understand how neurons *encode* information and communicate with other neurons. The second thrust has been to *decode* neural messages by listening in on the chatter of neurons with extracellular electrodes. These two thrusts are by no means separate, and many researchers work in both areas. As the encoding and decoding of information in the brain is a rich field of study, we will only cover the most widely used techniques here. In this chapter, we will first examine how neurons use firing rates to encode information and then turn to the basics of how to decoded the conversation.

9.1 NEURAL ENCODING

Experiments have shown that the shapes of action potentials do not vary significantly from neuron to neuron. It is assumed, therefore, that information is not transmitted from neuron to neuron encoded in the shape of the action potential. What does vary significantly from neuron to neuron is the rate at which they may fire. Furthermore, the firing rate, or frequency, of one neuron can be modulated by the synaptic inputs to that neuron. The firing of a neuron can also affect the firing rate of the neurons to which it is connected. A central principle of neuroscience is therefore that neural information is encoded in *firing frequency*.

9.1.1 The Membrane as a Frequency Detector

To understand how it is possible for neurons to encode information in the frequency of firings, consider a single current pulse applied to a membrane. In Sec. 2.3 we found that if a large stimulus was applied for a short time, the membrane may not fully charge to threshold. When the current was turned off, it will took take time for the membrane to discharge back to rest. Now, consider that a train of short-duration stimuli are applied to the membrane. If the duration between stimuli is long, the membrane will have time to fully discharge after each stimulus is turned off. If the stimulus interval time is decreased, the membrane will not fully discharge before the next stimulus is applied. The result is that the membrane potential will *drift* toward more depolarized potentials. This phenomenon is called *temporal integration* or *temporal summation*. In Fig. 9.1, the stimulus duration and amplitude remained constant and only the frequency was changed. The red plots show a low-frequency stimulus where the drift does not reach threshold. The black plots are for a higher frequency where the drift does reach threshold. These two different behaviors demonstrate how the frequency of stimulation is capable of modulating neural behavior at the cellular level. A secondary way that frequency can alter neural behavior is in the timing. Even when the rate is fast enough to induce an action potential to fire, the frequency can alter *when* the action potential fires. For example, in Fig. 9.1 a higher stimulus frequency caused an action potential to fire earlier.

9.1.2 The Synapse as a Frequency Detector

A similar phenomenon of temporal integration can occur at the synapse but at a slower rate. For example, if a number of action potentials reach the end of the axon terminal in rapid succession, short bursts of neurotransmitter will be released into the cleft faster than it can be cleared. Therefore, the concentration

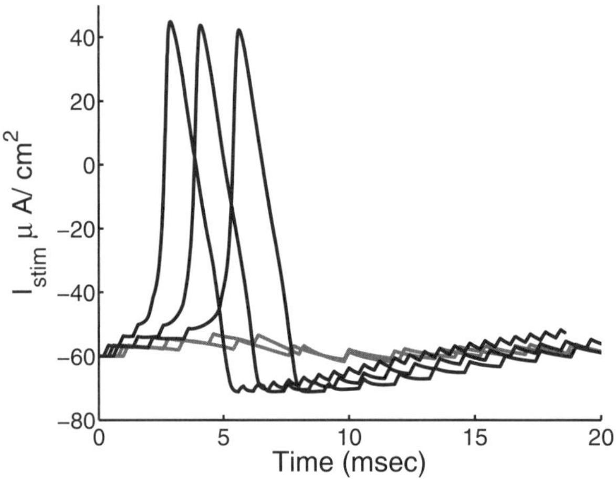

Figure 9.1: Effect of stimulus frequency on charging and action potential firing.

of neurotransmitter will slowly rise, leading to a sustained post-synaptic current. Recall that the post-synaptic current may either depolarize (excite) or hyperpolarize (inactive) the post synaptic membrane.

9.2 NEURAL DECODING

The purpose of neural decoding is to listen in on many neurons to understand how they are passing information to one another. As it is the timing of action potentials that is important, the extracellular potentials, ϕ_e, contain all the information needed to decode neural messages. From Ch. 8, we know that as an action potential approaches an extracellular electrode, ϕ_e becomes positive. As the action potential is moving away from the electrode, ϕ_e will become negative. Therefore, if an electrode is placed very close to an axon, the *biphasic* deflections in ϕ_e (as in Fig. 9.2) will reflect a single-action potential.

9.2.1 One Electrode, Many Recordings

When an extracellular electrode is very close to a single axon, other surrounding neurons contribute very little to ϕ_e and the recording will reflect only one neuron. In reality, neurons form a complex web and we would not know the exact location of our recording electrode in the web. Therefore, whenever one action potential fires and propagates down an axon, a corresponding deflection will be observed in the recording. Further complicating the extracellular potential time course is the fact that axons are not straight cables and may be at any distance or angle relative to the electrode. The result is that ϕ_e is no longer biphasic.

A somewhat more challenging problem is to distinguish between the many neurons that may contribute to a single ϕ_e recording. Figure 9.3 shows the relatively simple case of two neurons being recorded by two electrodes. Notice that the deflections are not biphasic because of the orientation of the electrode relative to the axon. Furthermore, because of the geometry, the firing of cell 1 will appear as a different deflection at electrode 1 and electrode 2. This situation becomes even more complicated if many neurons are firing. Lastly, consider how difficult it would be to separate the firing of cell 1 and 2 if they fired at nearly the same time.

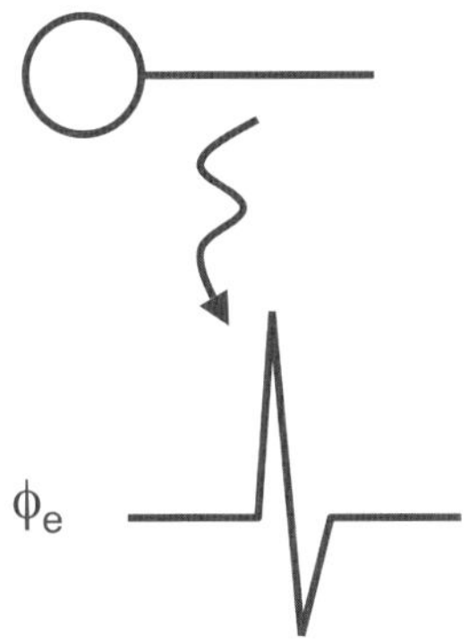

Figure 9.2: Example of a biphasic electrogram.

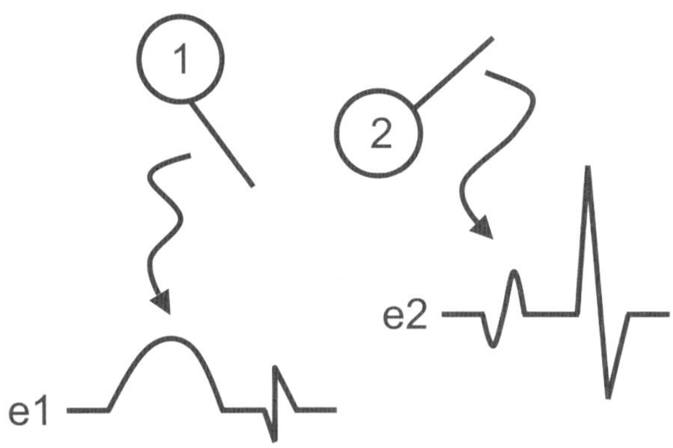

Figure 9.3: Electrograms from multineuron preparations.

The key to neural decoding is to assume that the electrodes do not move and that the action potentials propagate at the same speeds down the axons. With these assumptions the firing of cell 1 will always give the same pattern of deflections at electrodes 1 and 2. Likewise, the firing of cell 2 will always generate the deflections in ϕ_e. The deflection in ϕ_e corresponding to a particular neuron will therefore always have the same shape. In principle it is possible to use the extracellular shapes to sort or *bin* the firings of individual neurons and therefore record the neural code from many neurons simultaneously.

9.2.2 Eavesdropping

An analogy to the recording and interpreting neural information is to imagine entering a room filled with people speaking a foreign language that you do not understand. Upon entering the room you are far from every individual conversation and so you hear mostly noise. In fact, all you may be able to infer is how passionate the conversations are on the whole. If you move closer to any particular conversation you may be able to gain more information on that particular group. Again, you do not speak their language, but you may be able to gain some information about who is a leader, who is a follower, when arguments occur and when they are resolved. All the while, the noise is still in the background. Now, consider that you perform this task, day after day, week after week. Eventually, you would begin to recognize patterns in the language and pick up on subtleties of the conversations.

Eavesdropping on a foreign language class is similar to our current status in decoding neural patterns. We understand parts of the neural language, and can gain some idea of when a population of neurons is working on a problem, but on the whole they speak a language that is very foreign to us.

9.2.3 Spike Sorting

The prospect that a recording from one electrode contains information from many neurons is another step toward decoding neural information. The next step is to separate out the firing of individual neurons. Typically this separation is called *spike sorting*. The idea is to create a bin for each neuron and then sort the timing of firings into that bin. While there are many methods for accomplishing binning the most straightforward method is to define a number of *template deflections* that correspond to specific neurons firing (Fig. 9.4). This is an important step for two reasons. First, the number of template deflections will be the same as the number of bins and therefore makes an assumption about how many neurons are contributing to ϕ_e. If there are two few or two many bins, some spikes will be misclassified. Second, the templates will serve as the ideal to which other deflections will be compared. Although the templates may be determined by eye, there do exist some methods of automating the selection of templates using statistical differences in the signals.

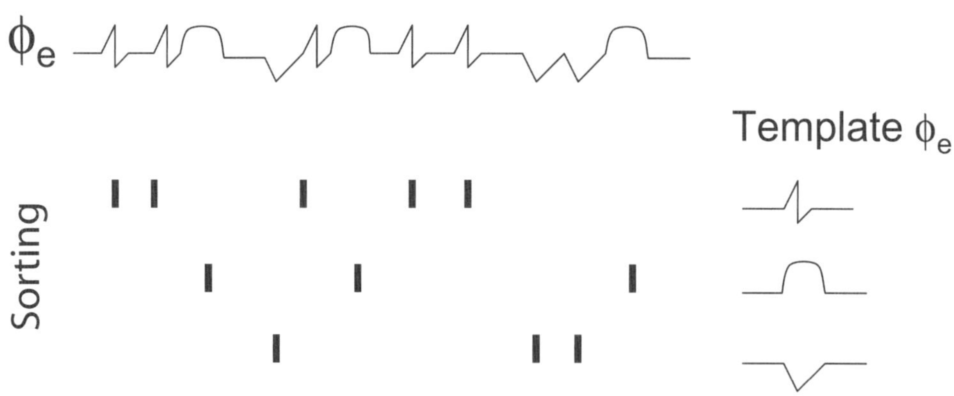

Figure 9.4: Spike sorting of extracellular deflections.

9.2.4 Time Binning and Firing Frequency

Once the timing of each neural firing has been extracted, the firing rate (frequency) can be found. It is often the case, however, that a neuron fires in a burst and then remains quiet, followed by another burst. To detect the change in frequency, the spikes may be binned in time. In *time binning*, all spikes are counted within a bin that spans a given length of time. By counting the number of spikes that occur within the time bin, an average firing rate may be computed. Changes in firing frequency can therefore be tracked over time. For example, the sequence of firings in the right side of Fig. 9.5 have been separated into four bins.

9.2.5 A Graphical Representation

As extracellular potentials are relatively simple to record, it is often the case that many hundreds of electrodes are used. The result is hundreds of ϕ_e recordings from potentially thousands of neurons. After sorting and binning, information is often represented graphically in one of two ways. First, the individual spikes may be shown in a matrix, or *raster*, where each row represents a single neuron (Fig. 9.5, right panel). A raster plot shows all of the raw data but can be difficult to interpret. As information is encoded as a firing frequency, the timing of an individual spike is less important. The second graphical representation, therefore displays firing rate and as a color (Fig. 9.5, right panel). Again, each row corresponds to a single neuron. The advantage of the color coded map is that the firing frequency of different neurons may be correlated more easily. It is often the case that many trials may be recorded of the same phenomenon and the frequency results averaged before displaying the data graphically.

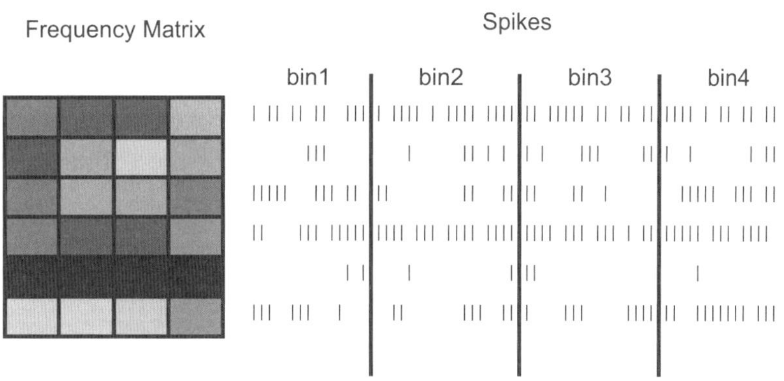

Figure 9.5: Graphical binning and frequency matrix.

9.2.6 Windowing

There are a number of potential problems with time binning. For example, consider the impact of slightly offsetting the four bins in Fig. 9.5. If the firing frequency changed suddenly, the change may be averaged out over two neighboring bins. A simple solution would be to add more time bins, each with a shorter duration. As the bin size is made smaller, however, there will be less possible spikes within each bin to count and the range of possible frequencies will be limited. On the other hand, if the bin size is large, the frequency curve will be much more discrete. A more clever method of binning is to wait for some number of spikes (e.g., 10 or 20) and count the time that elapses. In this way, the bin size adapts to the rate of firing. For fast firing the bin is small and for slow firing the bin expands. The problem with adjustable bin sizes is that it becomes difficult to correlate the firing patterns of different neurons.

A more dynamic method of computing firing frequency is to define a single bin of some duration (Δt) and slide it along the spike train. This technique is known as *windowing*. For each slide, the number of spikes can be counted and used to compute a frequency for that particular slide time. The simplest window may be defined as

$$w(t) = \begin{cases} 1/\Delta t & \text{if } -\Delta t/2 \leq t \leq \Delta t/2 \\ 0 & \text{Otherwise} \end{cases} .$$

If the slide is continuous rather than discrete then we can rewrite the frequency estimate as

$$f(t) = \int w(\tau)s(t-\tau)\,d\tau \tag{9.1}$$

where the window, w remains stationary and the spike train is moved. When presented in this way, the window is sometimes called a *kernel.*

It is also possible to use windows with different shapes that will weight some spikes more than others. For example, the Gaussian kernel is defined as

$$w(t) = \frac{1}{\sqrt{2\pi}\sigma_w} \exp\left(-\frac{\tau^2}{2\sigma_w^2}\right) \tag{9.2}$$

and so spikes closer to the evaluation time are more heavily weighted in the frequency estimate.

Thus far we have assumed that the entire spike train is known, and so times both before and after the evaluation time may be used. These kernels are called *noncausal* because they involve spikes that will occur after the evaluation time. In any real-time application, only spikes that occurred before the evaluation time are known and therefore a *causal* window must be used. In principle, the same window shapes can be used, but instead of evaluating at the center of the kernel the evaluation is performed at the right edge.

9.3 TUNING CURVES

While Sec. 8.4 described how an extracellular stimulus may induce neural firing, the previous sections of this chapter described how to characterize a neural response. The combination of an controlled input (stimulation) and recorded output (neural firing rate) enables a transfer function to be created that characterizes a network of neurons. This type of transfer function is typically called a *tuning curve* and is represented graphically as a plot of a stimulus parameter versus firing rate. The stimulus parameter can be any number of possible input parameters. As there are too many possible types of stimuli we will broadly classify them as either direct electrical stimulation or indirect stimulation. Direct electrical stimulation can take the form of cortical or sub-cortical stimulation through electrodes or even magnetic stimulation (see Sec. 10.3.2). The stimulus parameters that may be varied are the stimulus strength, duration, or pacing rate. Indirect stimulation may take the form of light, smell, taste, touch, sound, hormones, or drugs injected into the blood stream.

Once the empirical data are collected, it is often the case that an analytic function, called a *tuning function*, is fit to the data. There are many mathematical forms for tuning functions but two deserve special mention here. First are a group of Gaussian-like curve that have a maximum firing rate at some particular value of the stimulus parameter with a lesser response for deviations from that value. Networks of neurons that have a Gaussian tuning curve have a sensitivity to a particular range of the stimulus. Second are a group of sigmoid-like functions that do not respond for some large range of stimuli, then rapidly transition to a strong response as the stimulus is changed. In other words, networks of neurons that have a sigmoidal tuning curve exhibits threshold behavior.

Figure 9.6 shows how cone cells, eye cells that detect color, respond to different wavelengths of light. The solid line shows the firing frequency of cones that detect red light. The dashed line shows cone cells that detect green light. The dotted line shows cone cells that detect blue light. Note that each type of cell has a background firing frequency but they will respond most strongly, i.e., increased firing rate) in response to a small range of wavelengths.

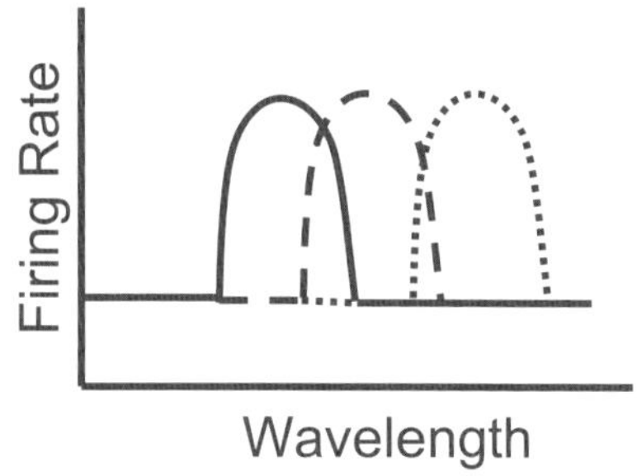

Figure 9.6: Example tuning curve from a cone cells in the eye.

CHAPTER 10

Applications

10.1 SCIENCE AND SCIENCE FICTION

A compact definition of science fiction is *science that hasn't happened yet*. Using this definition we can ask when some science fiction of the brain will become science. A quick overview of some of the most popular science fiction books and movies will show that there are (and have been) some incredible leaps of imagination in what people and machines may be able to do in the future. Although at the time these ideas were far fetched, the pace of breakthroughs is such that some of these wild ideas may not be far from realization.

The purpose of this text is to provide a basic quantitative background from which to jump off to more advance areas of study. Although we cannot go into the details of each of these very interesting research areas, this chapter is an overview of some of the most promising new directions in neuroscience and neuroengineering. As you read each section, try to consider the technology from two perspectives. First, how will the quantitative background of the previous nine chapters inform and supplement these new lines of research? Second, given these new technologies, what ideas do *you* have for future experiments or devices?

10.2 NEURAL IMAGING

The electrical function was uncovered using primarily patch clamp electrodes and extracellular electrodes. These two techniques are very efficient and can detect small changes over a short period of time. Over the past four decades, however, three new imaging techniques have enabled a fresh and complimentary view of the function of the brain.

10.2.1 Optical Recordings

In the late 1960s and early 1970s, the work of the three research groups of Tasaki et al. found that small molecules could be embedded in the cell membrane (Fig. 10.1) that would absorb light at one wavelength (color) and emit light at a different wavelength (left panel of Fig. 10.2). Furthermore, for some molecules, called *fluorescent dyes*, the intensity of the emitted light was proportional to V_m within a range that spanned the normal action potential (right panel of Fig. 10.2). Some subset of these molecules could also respond to changes less than $1msec$, short enough to resolve the action potential upstroke.

A typical imaging system is shown in Fig. 10.3 and consists of two light pathways. In the first pathway (solid line), a laser is used to excite the tissue at a high frequency through a dichromatic mirror. The function of the dichromatic mirror is that it will pass light of a high frequency but reflect light of lower frequencies. The cell then absorbs high frequency photons and emits low-frequency photons. In the second pathway (dotted line), the low-frequency light emitted from the cell is reflected off of the mirror, through a microscope and then to a collection device (usually a CCD camera or photodiode array).

The optical imaging method provides at least three advantages over the use of extracellular electrodes. First, as demonstrated in Ch. 8, a single ϕ_e recording is a weighted average of all surrounding electrical activity. On the other hand, optical mapping enables V_m to be recorded from many locations simultaneously. Second, when coupled with a high power microscope, the field of view can be adjusted

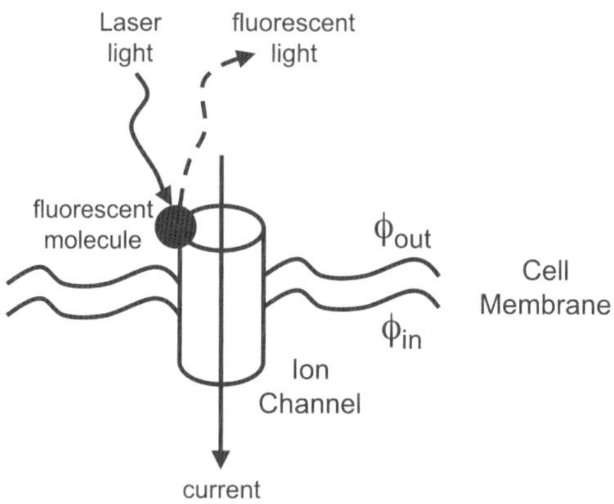

Figure 10.1: Molecular binding of a fluorescent molecule to the cell membrane.

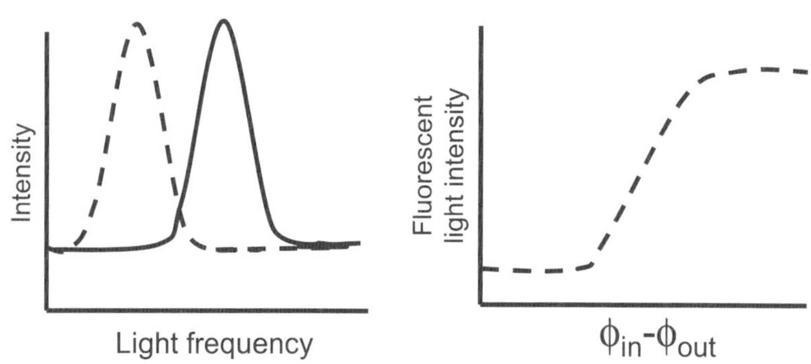

Figure 10.2: Left: Example absorption (dotted) and emitted spectra for a fluorescent molecule. Right: Example of proportionality between V_m and emitted light intensity.

to examine populations of neurons, an individual neurons or even regions of a single neuron. Lastly, new dyes have been discovered that are dependent on $[Ca^{2+}]$ instead of V_m, opening up the potential to better understand bursting and synaptic transmission.

The are, however, a number of disadvantage of optical mapping. First, the equipment is more expensive than using Ag-AgCl electrodes. Second, the dye itself is somewhat toxic and therefore can not be used for long periods of time or in humans. Third, as light can only penetrate a short distance into tissue, optical imaging is generally limited to thin slices of tissue. Lastly, the response times of fluorescent dyes are orders of magnitude slower than electrodes. Therefore, resolving fast propagation can be difficult.

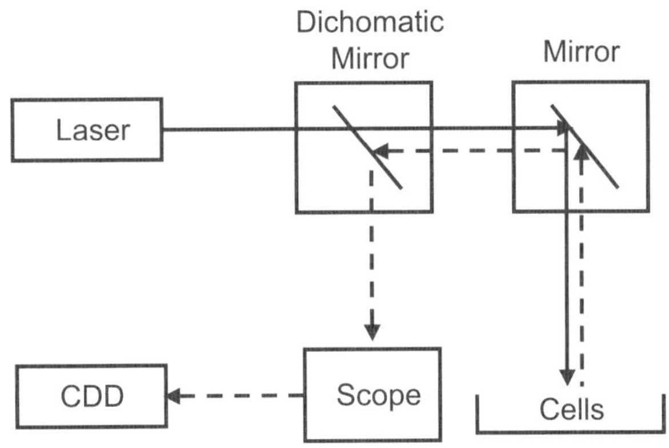

Figure 10.3: Optical mapping hardware.

10.2.2 Functional Magnetic Resonance Imaging

Magnetic Resonance Imaging (MRI) is a technique widely used in neurology clinics for its ability (unlike X-rays) to image soft tissue and (unlike ultrasound) penetrate the bone of the scull. At its most fundamental level, MRI is a three-step process as shown in Fig. 10.4 . First a very large (1-3 Tesla) magnetic field, B_0, is applied. The effect is to align all protons (H^+) to the B_O field. Since H^+ ions have a natural spin, they continue to spin like a top even under the influence of the B_0 field. Second, a fast radio frequency pulse, B_1, is introduced perpendicular to the B_0 field. The effect is to knock the aligned spinning molecules into a wobbly orbit. Lastly, the image is created by detecting how these wobbly orbits decay back to being aligned with the B_0 field. The decay rate is called the T_1 time and can be detected by the flow of current induced in a coil of wire. Superimposed upon the T_1 decay is a second decay called T_2. This second decay is a result of interactions between the spin of neighboring molecules. Most importantly in creating medical images, the T_1 and T_2 times are dependent upon the type of material.

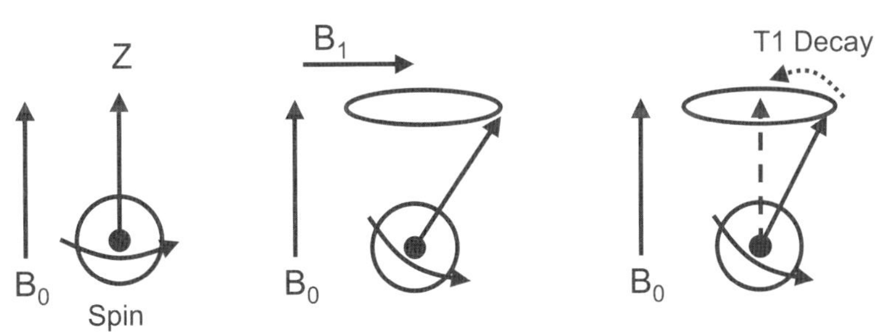

Figure 10.4: Overview of the biophysics of MRI.

There are many subtleties to the MRI technique that have been skipped over, but impressive images of soft tissue can be made quickly in 3D at resolutions approaching $2mm^3$, allowing for fast diagnosis of structural problems. The two most significant disadvantage of MRI are that the technique is expensive and requires very specialized equipment.

Functional MRI (fMRI) is a reinterpretation of traditional MRI that allows for regions with active neurons to be highlighted. Active regions of the brain will demand more O_2, causing oxyhemoglobin (HbO_2) to become deoxyhemoglobin (Hb). Deoxyhemoglobin is paramagnetic and so has a significant impact on spin interactions and T_2 times. Regional changes in Hb concentration, as a result of neuronal activity, can therefore be detected as a change in the T_2 times. Again, real-time 3D maps can be created. The disadvantage of fMRI is that neurons must be firing for many seconds before the drop in deoxyhemoglobin can be detect.

10.2.3 Diffusion Tensor Imaging

A technique that also uses MRI technology is Diffusion Tensor Imaging (DTI). The same physics and equipment are used and the images are similar. But, more information can be added by taking several closely timed images. When the images are averaged together, regions that are static will be reinforced while regions that are in motion will be blurred. Since the brain is not moving, the blurring is caused by the diffusion of water molecules. It has been found that water diffuses much more easily inside the cytoplasm of the axons and less easily across the membrane. In this way, the pathways of axon bundles (tracts) can be identified using DTI as the directions along which the most blurring occurs.

10.3 NEURAL STIMULATION

10.3.1 Cortical Stimuli during Surgery

During neurosurgery, a surgeon will do their best to avoid removing or damaging regions that are vital to the patient's normal cognitive functions. Unfortunately, these areas are not in the same location from one person to the next so the surgeon will often create a *functional map* before beginning the surgery. Creating a function map is often performed by electrically stimulating a region of the cortex and then asking the patient what they experienced. Remember that neurosurgery is often performed on conscious patients who feel nothing because the brain has no sensory neurons.

10.3.2 Deep Brain Pacing

Recently, the symptoms of Parkinson's Disease and some forms of chronic pain have been reduced by injecting current into specific sites deep in the brain. It has also been suggested that repeated pacing may help depression and prevent epileptic seizures. Furthermore, patient controlled stimulation of select locations in the brain stem or spinal cord could lead to better control of basic bodily functions. For example, stimulation of the sacral nerve is an effective means of bladder control.

Building upon the technology of cardiac pacemakers and defibrillators, some companies have already begun to offer deep brain stimulating devices. There are two potential problems with deep brain pacing. First, the blood brain barrier must be compromised to insert the electrodes, potentially opening the way for dangerous bacterial infections and viruses. Second, in some patients an external stimulus can trigger an epileptic seizure.

10.3.3 Magnetic Recordings and Simulation

We have so far only considered *electric* fields. Currents in the brain, however, generate extremely small magnetic fields ($10^{-9} - 10^{-6}$T) that can be recorded on the scalp by a Superconducting Quantum Interference Device (SQUID). An individual SQUID element uses a phenomenon known as the Josephson effect which only occurs when two supercooled superconductors are separated by a thin isolating barrier. The recording of magnetic potentials from the brain is called *magnetoencephalography* (MEG) and is typically performed with an array of SQUIDs.

Just as currents generate a magnetic field, Faraday's law of induction allows a time varying magnetic field to induce current to flow. Therefore, applying a strong time-varying magnetic field to the brain can force current to flow in neurons. Unlike electrical fields which do not easily penetrate the insulating scalp and skull, the magnetic field can extend deep into the brain, attenuated only by distance. The stimulation of the brain through a magnetic field is called *Transcranial Magnetic Stimulation* (TMS). A pulse or pair of pulses can be delivered that will cause a depolarization and firing of neurons. For example, if the depolarization is in the occipital cortex, a patient will sense flashes of light. Repetitive TMS (rTMS) has been shown to induce longer lasting changes and provides some interesting treatment and research options. Clinically, rTMS may help alleviate some of the symptoms of migraines, stroke, Parkinson's disease, and depression but the mechanisms are unknown. In the research laboratory, rTMS is an effective and noninvasive way of "knocking out" a very specific region of the brain. By focusing a magnetic field of a particular orientation and polarity, it is possible to hyperpolarize a small region of the brain, rendering it unexcitable. If the subject is asked to perform a task that requires that region, they will be incapable of completing the task. Unlike the images created by fMRI, rTMS provides much stronger evidence that a particular region of the brain is actually *used* to perform a specific task.

10.4 DRUG AND GENETIC THERAPIES

Neurons make very heavy use of proteins and amino acids that function as neurotransmitters, enzymes and ion channels. It is therefore not surprising that genetic mutations and changes in expression can have an enormous impact on neural conduction. Any breakthroughs in genetic therapies or drug design/delivery will therefore quickly find applications in neuroscience. Two mechanisms will most likely provide the greatest benefit. First, when a mutation has caused a deformation of a protein, an engineered DNA strand can be introduced to the body (typically through a viral vector) with a correction. The virus then spreads the corrected DNA to neurons which begin producing copies of the corrected protein. Alternatively, a synthetic drug can be designed to perform a desired function. In fact, the action of many current drugs, both legal and illegal, function in a very similar manner. Second, the mechanisms that control protein expression can be altered indirectly through the availability and effectiveness of second messengers. Again, correction of defective second messengers may be accomplished through genetic or pharmaceutical means. In the future, drug and genetic therapies may even be used in combination to correct a defective protein and ensure proper expression.

One very serious issue with delivering drugs and genes to the central nervous system is the *blood brain barrier* (BBB), which severely limits the transport of molecules to the brain. Two technologies may lift this barrier. First, a better understanding of transport across the BBB may allow drug designers to attach tags to the drug that fool transport mechanisms into carrying the drug across the membrane. Alternatively, there may be a way to find small molecules that can diffuse through the BBB that have the desired effect indirectly. Second, it should be possible to build implanted devices (similar to an insulin pump) that can control the direct deliver drugs directly to the brain, therefore bypassing the blood brain barrier.

10.5 BRAIN MACHINE INTERFACE

The concept of a Brain Machine Interface (BMI) is an extension of the old idea that a human and a machine can work together to perform a task. When neuroscientists and neuroengineers refer to BMI, however, they mean the direct connection of a human brain to an external machine in a feedback loop. Signals from the brain are recorded either from extracellular electrodes implanted directly into the cortex or from the surface EEG. Although implanting electrodes in the cortex is invasive, it is permanent and allows for many hundreds of simultaneous recordings instead of the 30–50 recordings of the EEG. The signals must then be decoded to parse out relevant information. Many of the techniques of decoding in Ch. 9 are used to sort spikes and generate frequency maps. These data must then be translated into commands to be sent to the external machine. Perhaps the most difficult aspect of BMI is to design a system that performs the translation of the neural code to machine commands. These machine commands, may be the opening of a valve, rotation of a motor or opening of an electrical switch.

The hope of a BMI is that it will be able to adapt to new situations. Similar to neural networks, the BMI must first be trained by sending in inputs to which the desired outputs are known. Using these known input-output pairs, the parameters of the neural-machine translation are adjusted. Next, the model, now tuned, can be sent an arbitrary input and generate an output that is close to desired. In the ideal situation, the person can use their own senses to monitor the progress of the external device. Corrections to the device can then be made by simply thinking (changing neural firing patterns). In a somewhat less ideal situation, the effect may need to be transduced into a signal that the person can evaluate.

BMI was originally envisioned as a therapy to restore motor control to disabled patients, particularly those suffering from ALS, spinal cord injuries, stroke, cerebral palsy, or amputees. The sick patient would be capable of controlling a device using neural firings that correlate to some thought. It is also possible that the "external" device would in fact be contained within the body. For example, patients with incontinence could have a small valve installed on their urethra that would be directly controlled by thoughts from the brain. The concept of thought controlled devices has many other applications. For example, dangerous jobs (fighter pilot, underwater bridge building), exploration (space missions) or rescue missions could be performed by a machine under direct control from a human brain.

There are a number of hurdles to overcome for BMI to become widely applicable. For complex tasks, such as the delicate finger movements needed to tie a shoe, the amount of information processing and coordination in time is well beyond the present capabilities of a wearable device. Information can only be sent at the speed of light, so exploration of space using BMI will always contain fundamental delays. These delays may also cause some series problems in performing tasks that require fast action such as flying a fighter jet. Despite these obstacles, there are some encouraging finds. In an amazing display of the brain's plasticity, early studies using BMI in animals have shown that after prolonged use, the cortex dedicates some area to the external device. In other words, from the perspective of the brain, the external device has become an additional part of the body.

10.6 GROWING NEURAL TISSUE

The goals of tissue engineering are to: (1) stimulate the natural growth of new tissue inside the body; and (2) artificially grow tissue outside the body. Below we highlight two examples of the application of tissue engineering in neuroscience.

10.6.1 Nerve Guides

Some of the most debilitating injuries effect only a small population of neurons. For example, quadriplegics and paraplegics often have a gap in their spinal cord that prevents signals from reaching portions of the body below the gap. If this gap could be spanned, all (or most) function would be returned. In the repair of neural gaps, creating *nerve guides* or nerve channels have shown some promise. The idea is to provide a channel that limits the direction that a nerve can grow. In this way two neurons (or nerve bundles) can be forced to connect. Typically, this method will only work to span gaps less than $1cm$. To span larger gaps a nerves can be grafted from some other part of the body. More recently, the material of the nerve guide has been the focus of much research. In particular, the nerve guide can be made of a bio-degradable substance which serves its purpose and then is naturally broken down by the body. Some new materials can be created with embedded growth factors that will release slowly and stimulate both neuron and capillary growth. Lastly, there is evidence to suggest that Hebb's "fire together wire together" idea applies in the peripheral nervous system. Therefore, a continual stimulus down the nerve gap may promote a better electrical connection.

10.6.2 A Biological Flight Simulator

A second application for tissue engineering is to create biological neural networks that will function similarly to artificial neural networks. These networks of real neurons can take in inputs from some external source, process the inputs and send out a meaningful output. Similar to an artificial neural network, the connections between real neurons are plastic, meaning that they can adapt to identify patterns in inputs.

Perhaps one of the most striking and creative examples of the capabilities of a biological network of neurons is the work of Thomas DeMarse. Using circuit boards created from special materials, neurons were grown and interfaced with recording and stimulating electrodes. The neurons, however, were randomly spread on the circuit and allowed to self-assemble. During this self-assembly, they are being trained to fly a flight simulation program in much the same way a neural network would be trained. After the neurons connected, the network could successfully fly the flight simulator without external input. Even more impressive was that given an unexpected input (e.g., a cross-wind, engine failure), the network adapted the outputs to compensate. Growing neurons on a circuit board may have some interesting applications in creating biologically based computers.

Growing an entire brain is far outside the realm of our current technology. The most significant hurdle is that tissue engineering has not found an effective way to grow three-dimensional tissues. The fundamental limitation is the transport of O_2 and glucose and removal of waste products. In the body, these functions are performed by capillaries which are no more than $100\mu m$ from each neuron. A first step toward overcome this limitation is again in new materials (e.g., gels) which allow for vasculature to develop along side the neural tissue. To complicate matters, neurons in a 3D structure will most likely require glial cells to perform maintenance functions and possible participate in electrical impulse communication.

10.7 ARTIFICIAL INTELLIGENCE

Artificial Intelligence (AI) is an interdisciplinary field composed of computer scientists, mathematicians, psychologists, philosophers and engineers. There are two primary goals of AI. First is to mimic the function of the brain to do something useful (e.g., drive your car for you). Second is to build an artificial intelligence so we can learn more about the way our own brain works. One fundamental stumbling block has been to define what is meant by intelligence. To most, intelligence requires memory, pattern recognition, the ability to generate an orderly sequence of reasoning (logic), the ability to plan ahead and some feedback between these elements. But some have argued that intelligence should include additional elements such as emotions. Still others have argued that intelligence can only arise in the context of a body that can interact with (sense) the world. And perhaps the most perplexing philosophical question of all is does intelligence give rise to consciousness or the other way around?

As the answers to the questions above have been debated for millennia with little progress made, many engineers and computer scientists have approach artificial intelligence from the perspective of utility. Early artificial intelligence research attempted to reduce thinking to the manipulation of symbols by rules. Although this *symbolic* approach produced some interesting applications (e.g., expert systems), it quickly became clear that the number of rules needed to produce anything resembling intelligence was enormous. The second approach focuses more on the ideas of connected neurons to build something that resembles the brain in both structure and function. This *connectionist* approach includes neural networks and has made some steps toward matching patterns, storing memories in many distributed locations and the ability to plan ahead. All indications are that as better mathematical models of neuron and brain function are developed, they will find their way into the field of artificial intelligence.

10.8 CONCLUSION

Although it may not be possible to predict which of the above technologies will change the way we understand and use our brain in the future, it is nearly certain that some will influence our lives in profound ways. The road forward will not be easy and as is typical in science, not all of the obstacles to progress are known to us at this time. The other certainty is that the brain still has many more secrets waiting to be discovered. With continued study and hard work, you may have a date with scientific history and even add your name to the list of prize winners in the preface.

Biography

Joseph Tranquillo is an assistant professor of biomedical engineering at Bucknell University where he has been a faculty member since 2005. He received his Doctor of Philosophy degree in biomedical engineering from Duke University (Durham, NC) and Bachelor of Science degree in engineering from Trinity College (Hartford, CT). His teaching interests are in biomedical signals and systems, neural and cardiac electrophysiology, and medical device design. His research interests are in electrophysiology of the heart and brain, high performance biocomputing, and the non-linear dynamics of coupled oscillators. He is an elected member of Sigma Xi and Heart Rhythm, a member of the IEEE Engineering in Medicine and Biology and Biomedical Engineering Society, and serves as a member-at-large on the Biomedical Engineering Division of the American Society of Engineering Education. When not teaching or doing research, he enjoys improvisational dance and music, running marathons, backpacking, brewing Belgian beers and raising his two children Laura and Paul.

Index

Tremor

From Pathogenesis to Treatment

Giuliana Grimaldi
University of Palermo

Mario Manto
Université Libre de Bruxelles

MORGAN & CLAYPOOL PUBLISHERS

ABSTRACT

Tremor represents one of the most common movement disorders worldwide. It affects both sexes and may occur at any age. In most cases, tremor is disabling and causes social difficulties, resulting in poorer quality of life. Tremor is now recognized as a public health issue given the ageing of the population.

Tremor is a complex phenomenon that has attracted the attention of scientists from various disciplines. Tremor results from dynamic interactions between multiple synaptically coupled neuronal systems and the biomechanical, physical, and electrical properties of the external effectors.

There have been major advances in our understanding of tremor pathogenesis these last three decades, thanks to new imaging techniques and genetic discoveries. Moreover, significant progress in computer technologies, developments of reliable and unobtrusive wearable sensors, improvements in miniaturization, and advances in signal processing have opened new perspectives for the accurate characterization and daily monitoring of tremor. New therapies are emerging.

In this book, we provide an overview of tremor from pathogenesis to therapeutic aspects. We review the definitions, the classification of the varieties of tremor, and the contribution of central versus peripheral mechanisms. Neuroanatomical, neurophysiological, neurochemical, and pharmacological topics related to tremor are pointed out. Our goals are to explain the fundamental basis of tremor generation, to show the recent technological developments, especially in instrumentation, which are reshaping research and clinical practice, and to provide up-to-date information related to emerging therapies. The integrative transdisciplinary approach has been used, combining engineering and physiological principles to diagnose, monitor, and treat tremor. Guidelines for evaluation of tremor are explained.

This book has been written for biomedical engineering students, engineers, researchers, medical students, biologists, neurologists, and biomedical professionals of any discipline looking for an updated and multidisciplinary overview of tremor. It can be used for biomedical courses.

KEYWORDS

tremor, oscillations, generator, signal processing, Fourier analysis, nonlinear, inertia, sensors, accelerometer, electromyography, haptics, fuzzy logic, brain computer interface, movement disorders, essential tremor, Parkinson's disease, cerebellum, muscle spindles, synapses, neurotransmitters, clinical scales, neuroimaging, blood studies, drugs, animal models.

Contents

CHAPTER 1

Introduction

1.1 DEFINITION OF TREMOR

Tremor is defined as a rapid back-and-forth movement of a body part (McAuley and Marsden, 2000). Tremor is one of the most common movement disorders encountered in clinical practice and is a readily apparent motor phenomenon in most instances. It occurs both in normal individuals (the so-called physiological tremor) and as a symptom of a disorder, most often of neurological origin.

Pathological tremor is usually a rhythmic and roughly sinusoidal oscillatory movement. However, tremor is a nonlinear and nonstationary phenomenon. It is distinct from other involuntary movement disorders such as chorea, athetosis, ballism, tics, and myoclonus (see Table 1.1) by its repetitive and stereotyped feature (Bhidayasiri, 2005). The different tremors are grouped according to their frequency, amplitude, topographical distribution, and task or position-dependence. The most commonly used clinical classification is to distinguish tremor into *rest tremor*, *postural tremor*, and *kinetic tremor* (see also Chapter 4 for the clinical characterization).

1.2 PHYSIOLOGICAL TREMOR

In physiological tremor, two distinct oscillations (mechanical reflex and central neurogenic) are superimposed upon a background of irregular fluctuations in muscle force and limb displacements (Elble, 1996, 2003). Frequency studies show, in the majority of cases, similar frequencies on both sides.

The *mechanical reflex component* is the largest of the two oscillations. Its frequency is governed by the inertial and elastic properties of the body (Elble, 1996). Damped oscillations are generated in response to pulsatile perturbations, such as those produced by irregularities in motor unit firings (a motor unit includes a motoneuron and the depending muscle fibers, see Chapter 2) and by blood ejection during cardiac systole.

The frequency (ω) of these passive mechanical oscillations depends directly upon the stiffness (K) of the joint and inversely upon the inertia (I) according to the equation:

$$\omega = \sqrt{(K/I)}.$$

TABLE 1.1: Differential diagnosis of involuntary movements

INVOLUNTARY MOVEMENTS	DEFINITION/FEATURES	DISEASES COMMONLY ASSOCIATED WITH THE MOVEMENT DISORDER
Tremor		
Rest	See text	Parkinson's disease
Postural		Essential tremor
Kinetic		Cerebellar tremor
Dystonia	Prolonged muscle contractions leading to abnormal postures; may be repetitive; twisting movements	Drug-induced
		Genetic
		Idiopathic
Chorea	Irregular; often hidden in voluntary movement; generates a dance-like movement	Huntington's disease
Athetosis	Continuous slow hyperkinesia of distal segments of limbs; causes an octopus-like movement	Stroke
Ballism	Fast and ample movement of proximal segments of limbs; gives a "throw away"-like movement; more severe in upper limbs	Stroke
		Inflammatory diseases
Tics	Fast and short hyperkinetic movements usually with a facial or head topography	Gilles de la Tourette syndrome
Myoclonus	Sudden, short (20–150 ms) movement; may cause a pseudorepetitive muscular contraction	Essential myoclonus
		Myoclonic epilepsy
		Symptomatic myoclonus

Consequently, tremor frequency will increase from proximal to distal segments. Physiological tremor of the elbow has a frequency of 3–5 Hz, wrist tremor 7–10 Hz, and metacarpophalangeal joint tremor 12–30 Hz (Elble, 2003).

The *central neurogenic component* of physiological tremor is invariably associated with the modulation of motor unit activity. Rhythmic motor unit activity is not just a simple passive response to sensory feedback, but is driving the limb oscillation. Moreover, regardless of their mean frequency of discharges, participating motor units are entrained at about 8–12 Hz. The frequency of the central neurogenic tremor shows no response to modification of inertia or stiffness and is independent of the length of the stretch reflex. For these reasons, the central neurogenic tremor is believed to originate from an oscillating neuronal network within the central nervous system (Elble, 1996).

The normal tremor behavior when an inertial load is added to the limb is a decrease of frequency, according to the equation $\omega = \sqrt{(K/I)}$, when there is no significant contribution from the stretch reflex or central oscillations. This is confirmed by surface or needle electromyographic (EMG) studies showing no rhythmic motor unit entrainment despite the rhythmic oscillations of the limb (Figure 1.1a). In some subjects, a prominent motor unit entrainment with and without mass loading is identified (Figure 1.1b). In pathological cases (for instance, essential tremor, see Chapter 5), lower- and higher-frequency oscillations can be identified. These last ones are associated with motor unit entrainment and do not decrease with inertial loading. For these reasons, high-frequency oscillations are interpreted as central neurogenic oscillations, whereas low-frequency oscillations correspond to the mechanical reflex resonance frequency (Elble, 2003).

In addition to the 8- to 12-Hz frequency band, frequency oscillations in the 15- to 30-Hz range and around 40 Hz are commonly found in the kinematic and EMG recordings in the upper limbs. Based on the fact that changes in the amplitude of specific frequency bands of tremor as a function of added mechanical load are informative of the central or peripheral modulation of tremor, some authors have studied the effects of increments of load on the intensities of the 8- to 12-, 20- to 25-, and 40-Hz neural rhythms (Vaillancourt and Newell, 2000). Their findings support the view that the 8- to 12-Hz oscillation resides within the central nervous system with its amplitude remaining independent of mechanical resonance oscillations and that the 20- to 25-Hz tremor oscillations are related to the mechanical properties of the finger. This means that the 20- to 25-Hz band is related to cortical activity, but its amplitude is modulated by mechanical reflex oscillations. Concerning the 40-Hz peak found in EMG recordings, it is likely originating in the central nervous system before being low-pass filtered by the muscle–tendon complex. The tendency of human limb segments to exhibit rhythmic oscillations around 40 Hz is observed in EMG recordings during strong voluntary contractions ("Piper rhythm"; Piper, 1907). Data of surface EMG traces simultaneously recorded with magnetoencephalographic signals (see Chapter 6) suggest that Piper rhythm of human muscles is linearly correlated with focal activity in the controlateral motor cortex both in

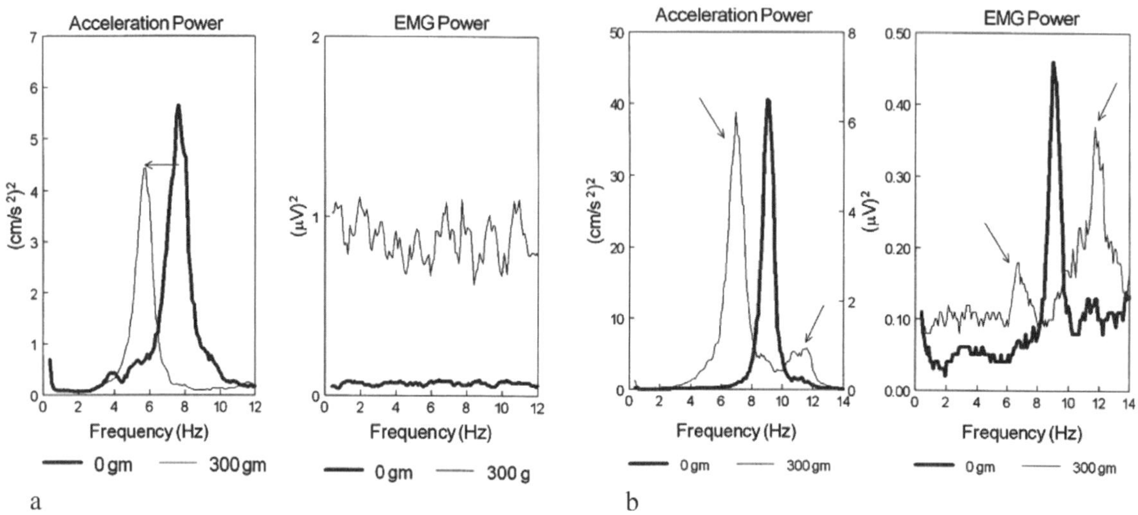

FIGURE 1.1: (a) Wrist tremor and rectified–filtered electromyography (EMG; Extensor carpi radialis bilateral: ECRb) from a 27-year-old woman with no evidence of motor unit entrainment without (thick trace) and with 300-g loading (thin trace). Note the 2-Hz reduction in the acceleration spectral peak with 300-g loading (arrow). (b) Wrist tremor and rectified–filtered EMG (ECRb) from a 21-year-old man with prominent motor unit entrainment with and without mass loading. With no mass loading, the acceleration and EMG spectra contained a single coherent peak at 9.4 Hz. Mass loading produced coherent peaks in the EMG and acceleration spectra at 6–7 and 11–12 Hz (arrows). Left vertical axis of the acceleration spectrum: 0-g load; right vertical axis: 300-g load. From Elble (2003), with permission from Elsevier.

isometric contractions and during phasic movements. A longer lag to tibialis anterior than that to forearm extensor muscle may be detected. This interval is linked to the conduction in fast pyramidal pathways (see Chapter 2). Because Piper rhythm can be picked up from most muscles, including those with few or no muscle spindles, it is unlikely that the coherence between cortical activity and the Piper rhythm is due to a simple reafference mechanism. Piper Rhythm is related to the degree of force exerted in tonic and phasic contractions probably because of a stronger excitation of the motor cortex during forceful contractions. Alternatively, it may be related to a greater attention demand in tasks requiring forceful contraction (Brown et al., 1998).

1.3 SOURCES OF TREMOR: NEURONAL NETWORKS

We can thus summarize the sources of tremor into three groups (Figure 1.2; Hallet, 1998):

I—Mechanical oscillations: Motion of joints and muscles obey the laws of physics, and tendon–muscle–joint complexes can be compared with masses and springs. Therefore, oscillations can be interpreted as related to movements of these masses and springs (Figure 1.3).

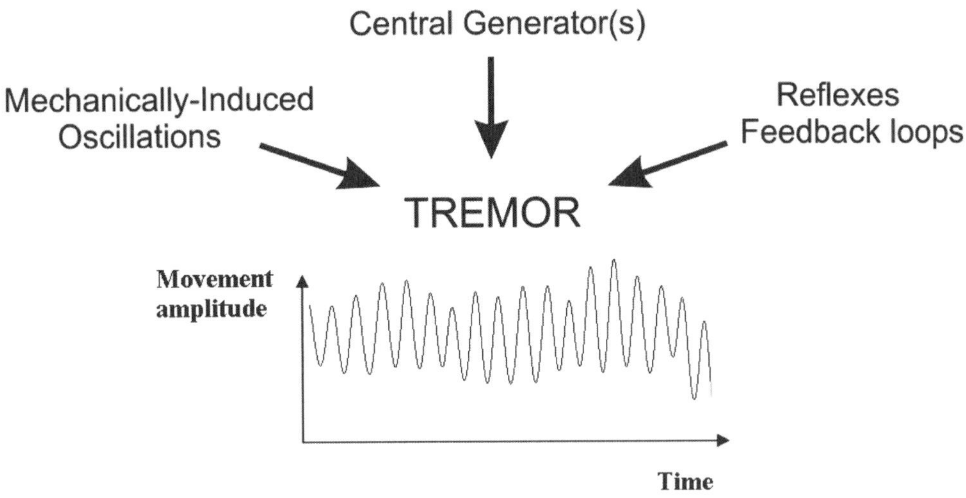

FIGURE 1.2: The three main sources of tremor.

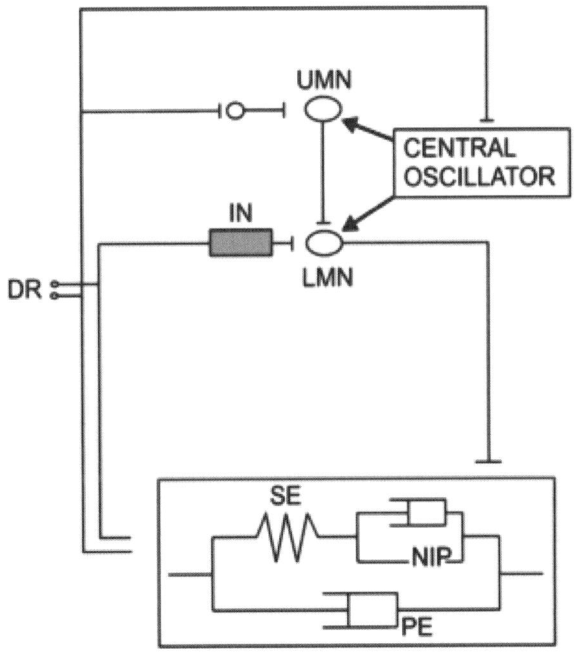

FIGURE 1.3: Central and peripheral loops in the nervous system. The figure illustrates the interaction between the central oscillator and the upper motoneuron (UMN)/lower motoneuron (LMN). IN indicates the pool of interneurons in the spinal cord. DR corresponds to the dorsal root ganglia. The rectangle in the bottom represents Hill's muscle model (SE: series elastic component; NIP: neural input processor in parallel with a viscous component PE).

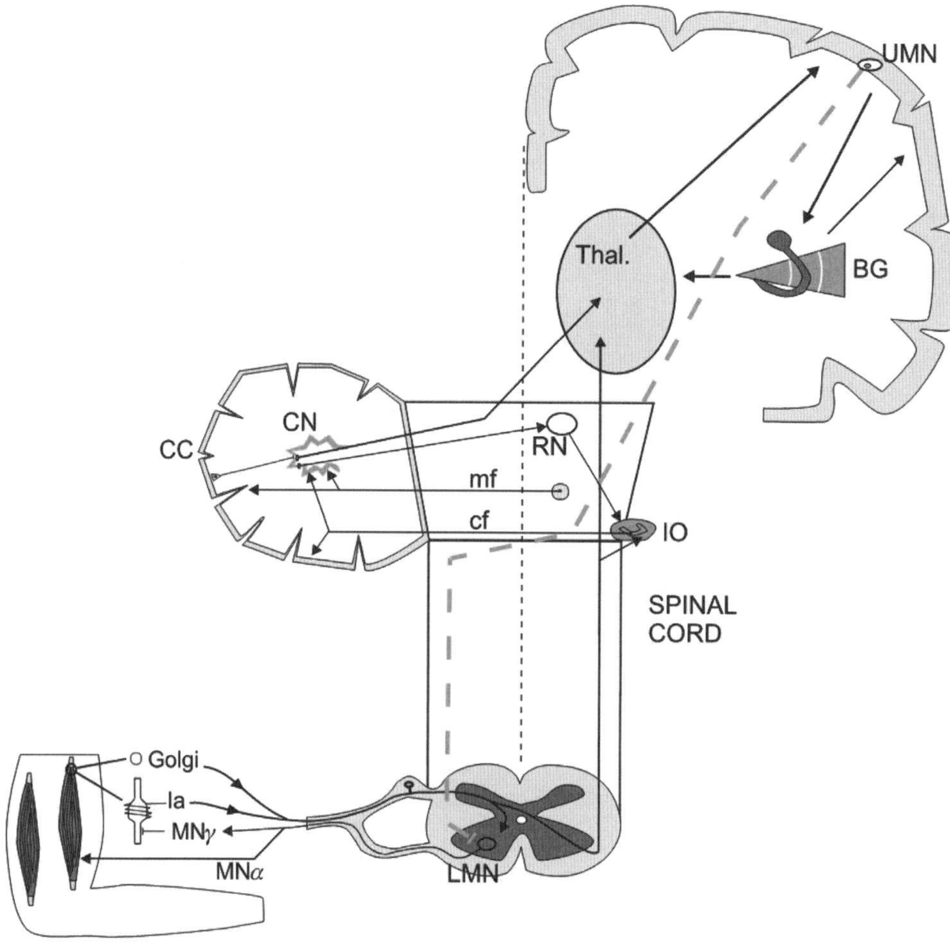

FIGURE 1.4: Pathways involved in tremorgenesis. Muscle spindles are receptors located inside muscles, made up by fibers sensitive to changes in length. In response to lengthening, the signals reach spinal and supraspinal centers where motor commands are generated and sent back to the extrafusal muscle fibers. UMN: upper motoneuron; Thal: thalamus; BG: basal ganglia; CC: cerebellar cortex; cf: climbing fibers; mf: mossy fibers; CN: cerebellar nuclei; IO: inferior olive; LMN: lower motor neuron; MNα: alpha motoneuron; MNγ: gamma motoneuron; RN: red nucleus; Ia: Ia sensory afferent fibers. Note that afferents project to the cerebellum ispilaterally.

II—Reflex oscillations: They are related to loops: *peripheral loops* from muscles to the spinal cord and back again (spinal level) and *central loops* from the periphery to the spinal cord and segments at the supraspinal level including the brainstem, cerebellum, basal ganglia, and cerebral cortex (Figures 1.3 and 1.4).

- The simplest loop is from the muscle spindle. Ia afferents are connected monosynaptically to the motoneuron, and the motor axon projects to the extrafusal muscle fibers.
- A classical example of a central loop is given by the role of comparator and movement controller of the cerebellum.

III–Central oscillations: Since the first electroencephalographic (EEG) recordings, it is obvious that the neural activity follows rhythmic behaviors. Cerebral cortex, basal ganglia, cerebellum, and brainstem nuclei are all involved in tremogenesis (see Chapter 3).

REFERENCES

Bhidayasiri R. Differential diagnosis of common tremor syndromes. *Postgrad Med J.* 2005;81: 756–762.

Brown P, Salenius S, Rothwell JC, Hari R. Cortical correlate of the Piper rhythm in humans. *J Neurophysiol.* 1998;80:2911–2917.

Elble RJ. Central mechanism of tremor. *J Clin Neurophysiol.* 1996;13:133–144.

Elble RJ. Characteristics of physiologic tremor in young and elderly adults. *Clin Neurophysiol.* 2003;114:624–635.

Hallet M. Overview of human tremor physiology. *Mov Disord.* 1998; 3(3):43–48.

McAuley JH, Marsden CD. Physiological and pathological tremors and rhythmic central motor control. *Brain.* 2000;123(Pt 8):1545–1567.

Piper H. Über den willkürlichen muskeltetanus. *Pflügers Gesampe Physiol Menschen Tiere.* 1907;119: 3001–3338.

Vaillancourt DE, Newell KM. Amplitude changes in the 8–12, 20–25 and 40 Hz oscillations in finger tremor. *Clin Neurophysiol.* 2000;111:1792–1801.

CHAPTER 2

Anatomical Overview of the Central and Peripheral Nervous System

This chapter provides an overview of the relevant anatomy for the understanding of the organization of the central and peripheral nervous system. This anatomical framework is the substratum of tremor in human and must be kept in mind when assessing tremor, as several key features of tremor are directly explained by the organization of these anatomical structures. Moreover, several therapies [such as deep brain stimulation (DBS), see Chapter 8] are based on this organization.

2.1 CEREBRAL CORTEX

The cerebral cortex is the layer of gray matter in the cerebral hemispheres' external coat. The surface of the cerebral cortex is folded in sulci. The cerebral cortex is divided into four lobes: frontal, parietal, temporal, and occipital lobes. Two additional structures are included in the cerebral cortex: the insula and the limbic lobe.

Cortical areas are linked to subcortical structures through afferent and efferent fibers. Moreover, commissural tracts connect both cerebral hemispheres. From a functional point of view, cortical areas are divided into motor, sensory, and associative areas. According to Brodmann classification, each area is identified by a number (Figure 2.1).

1. Frontal lobe
    - Area 4 corresponds to the frontal ascendant circonvolution. This primary somatomotor area issues motor commands along the pyramidal tract. Commands reach the brainstem nuclei and spinal cord. As a consequence of the somatotopical distribution of the neurons controlling the limbs, face, and trunk movements, a representation of the contralateral hemibody is discernible (the so-called motor homunculus). Motor areas receive main projections from the thalamus.
    - Area 6, or premotor area, sends information mainly to Area 4, pontine nuclei, red nucleus, striatum, and mesencephalic substantia nigra.
    - Area 8 allows conjugated eye movements in response to head rotation.

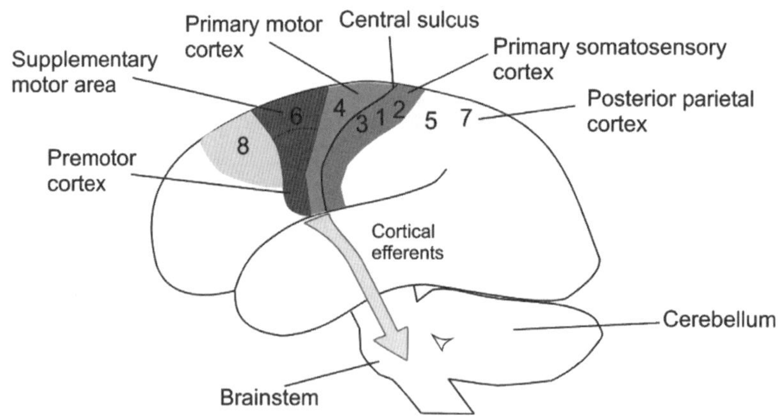

FIGURE 2.1: Representation of the brain. Main Brodmann's areas involved in tremor are illustrated.

2. Parietal lobe
   - Areas 3, 1, 2 are the somatosensitive areas. They are located in parietal ascending circonvolution, behind Rolando's scissura. A sensory homunculus is represented. These areas are highly connected with the motor areas, so that motor activity is always performed in a context of a sensory afferent according to the cortical reflex arch, except for deafferentation states.
3. Occipital lobe
   - Area 17 is the visual primary area. Vision plays an important role in the modulation of tremor intensity.

Associative areas integrate information from different systems. The temporo-parieto-occipital cortex is crucial for a complex sensorial elaboration that integrates somatic, acoustic, and visual inputs. The *prefrontal area* is involved in the planning of voluntary movements. The *posterior parietal area* coordinates somatic and visual inputs and integrates them into movement engrams.

2.2 BASAL GANGLIA AND THALAMUS

The basal ganglia are made up of five deep subcortical nuclei: caudate and putamen (forming the striatum), globus pallidus, subthalamic nucleus, substantia nigra, which is divided into pars reticulata and pars compacta (Figure 2.2). In each cerebral hemisphere, the basal ganglia receive information from the cerebral cortex and project it to the thalamus, prefrontal, premotor, and motor cortex. Therefore, motor functions of the basal ganglia are mediated partly by the frontal cortex. In addition to the motor functions, the basal ganglia also play key roles in cognitive operations, emotions, and learning.

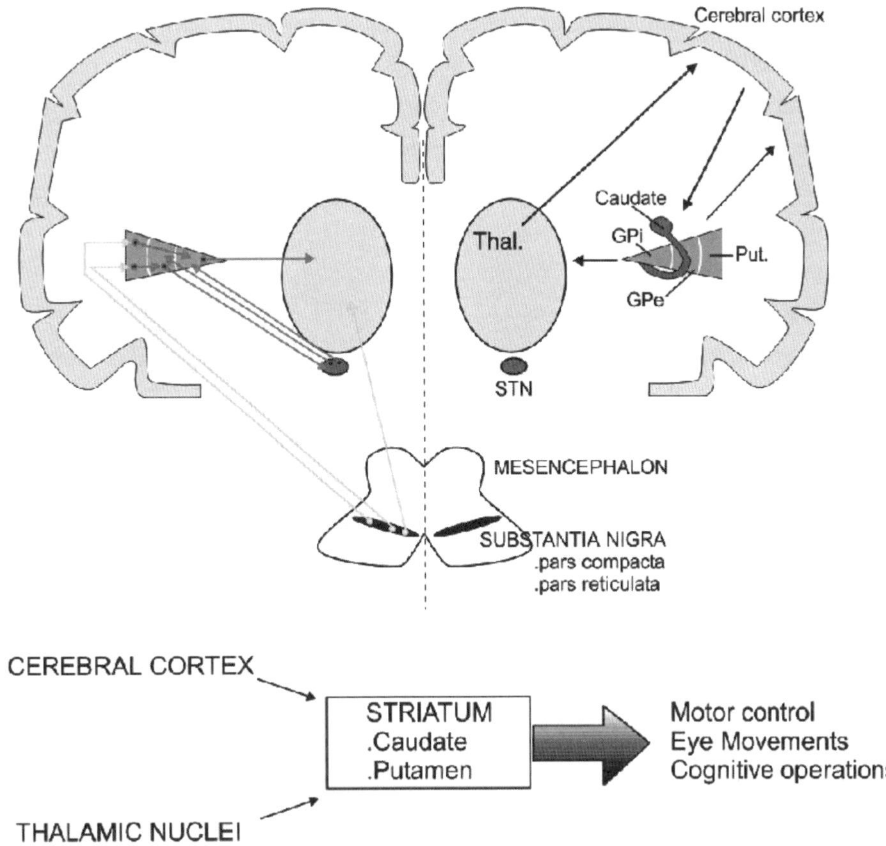

FIGURE 2.2: Coronal representation of the motor circuits of the basal ganglia. Main anatomical connections are illustrated. Nearly every input to the basal ganglia reaches the caudate nucleus and the putamen (Put.). These two nuclei form the striatum. The subthalamic nucleus (STN) receives projections from the external portion of the pallidum (GPe) and projects back to the globus pallidus and to the pars reticulata (substantia nigra). In addition, the STN receives projections directly from the primary motor cortex/premotor cortex, allowing a direct control to the output of basal ganglia. The striatum receives an important dopaminergic projection from the pars compacta. Thalamic nuclei are the main targets of the basal ganglia.

The striatum receives the majority of the afferents from the cerebral cortex (motor, sensitive, associative, and limbic areas) and the thalamus. Inputs from the cortex to the striatum are glutamatergic and therefore excitatory (see also Chapter 3). Projections maintain a topographical organization, so that putamen is principally involved in motor control, caudate in the ocular movement control and in some cognitive operations, and their ventral part in limbic functions. From the globus pallidus and substantia nigra pars reticulata, the basal ganglia fibers reach the thalamic nuclei [nuclei ventral lateral (VL), ventral anterior (VA), mediodorsal (MD), and centromedian (CM)], which project back to the prefrontal, premotor, supplementary motor area, and motor cortex. The globus pallidus and substantia nigra pars reticulata are the two main outputs of the basal ganglia. The basal ganglia influence motor descending systems (corticospinal and corticobulbar). The connection with the superior colliculus is involved in the control on ocular movements. All these interconnections give the basal ganglia an important role in the planning and performance of complex motor strategies.

2.2.1 Direct and Indirect Pathways

Projections from the striatum reach the internal segment of the globus pallidus and the substantia nigra pars reticulata and, finally, the thalamus. This is the *direct pathway.* The information leaving the striatum to the external segment of the globus pallidus and the subthalamic nucleus proceeds via the *indirect pathway.* Projections get back from the subthalamic nucleus to the internal segment of the globus pallidus and the substantia nigra pars reticulata. The neurotransmitters of the direct and indirect pathways are distributed as follows:

1. Direct pathway: neurotransmitters are GABA and substance P, which are inhibitory. Therefore, when the striatum is stimulated by cortical projections, it inhibits the globus pallidus and the substantia nigra pars reticulata. Given that these two nuclei have inhibitory effects on the thalamus, the thalamus is released from tonic inhibition, and movements are allowed. Thalamus–cortical activation induces, in turn, activation of premotor and supplementary motor areas projecting upon the motor cortex, brainstem, and spinal cord.
2. Indirect pathway induces a decrease of activation of the neurons of motor areas. Indeed, the striatum inhibits (via GABA and enkephaline) the external segment of the globus pallidus, which, in turn, inhibits the subthalamic nucleus. The subthalamic nucleus' excitatory influence on the internal segment of the globus pallidus and the substantia nigra pars reticulata allows the inhibitory action of these latter nuclei upon the thalamus.

Additional effects are provided from dopaminergic projections of the substantia nigra pars compacta, which has a facilitatory action on movement by exciting the direct pathway and inhibiting the indirect pathway.

The thalamus is thus a crucial synaptic relay both for sensory inputs and motor output. The thalamic nuclei retransmit motor codes—sent by cerebellar circuitry and basal ganglia—toward motor areas of the frontal lobe. Moreover, the thalamus is involved in the maintenance of consciousness and contributes to autonomic responses.

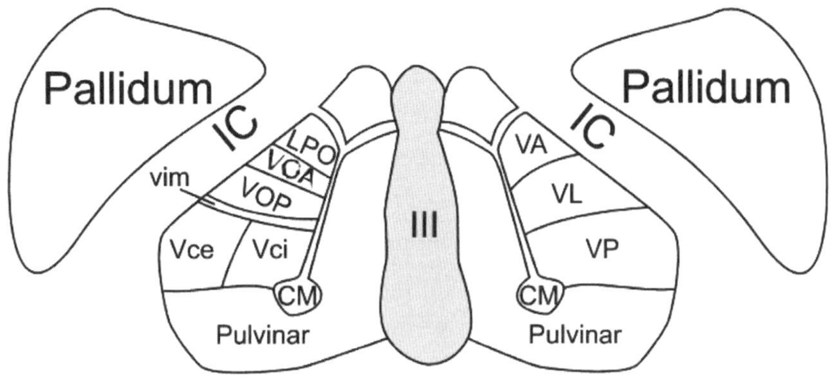

FIGURE 2.3: Illustration of the thalamic nuclei. Most commonly used terminologies are shown. LPO: lateral polaris, VOA: ventralis oralis anterior, VOP: ventralis oralis posterior, Vim: ventralis intermedius, Vci: ventralis caudalis interna, Vce: ventralis caudalis externa, CM: centrum medianum, VA: ventralis anterior, VL: ventralis lateralis, VP: ventralis posterior, IC: internal capsula, III: third ventricle. Adapted from Goldman and Kelly (1995).

2.2.2 Thalamic Nuclei

The thalamic nuclei can be divided into the relay nuclei and the diffuse-projection nuclei. Six groups of thalamic nuclei are described (Figure 2.3):

1. Lateral nuclei (ventral and dorsal tiers): receive restricted sensory and motor afferents and project to specific cortical regions (sensory, motor, or association cortex). They are made up of the following nuclei: *ventral anterior* (VA) and *ventral lateral* (VLN) involved in motor control; *ventral posterior* (VP) critical for somatic sensation; *medial and lateral geniculate* mediating information about hearing and vision, respectively; *lateral dorsal*; *lateral posterior*; *pulvinar* (the biggest thalamic nucleus). The pulvinar is involved in the integration of sensory information.
2. Medial nuclei: most representative is the *medial dorsal.*
3. Anterior nuclei: involved in emotions.
4. Intralaminar nuclei: diffuse-projection nuclei. The largest is the *centromedian nucleus* (CM) that projects to the frontal cortex and the basal ganglia (striatum).
5. Midline nuclei: diffuse-projection nuclei located in the dorsal half of the wall of the third ventricle.
6. Reticular nucleus: located in the lateral side of the thalamus. This nucleus is strictly interconnected with a specific relay nucleus and is the sole nucleus having inhibitory output and no projections towards cerebral cortex.

2.3 CEREBELLUM

The cerebellum overlies the posterior parts of the pons and medulla, occupying a large part of the posterior fossa. Structurally, the cerebellum consists of four pairs of nuclei embedded in white matter and surrounded by a cortical mantle of gray matter (Colin et al., 2002). Although the circuitry is functionally heterogeneous, the cytoarchitecture of the cerebellum is remarkably homogeneous. The cerebellum contains more neurons than any other region of the brain.

The cerebellum is divided into three main parts: the *anterior lobe*, the *posterior lobe*, and the *paraflocculus/flocculus* separated by the fissure prima and the fissure posterolateralis, respectively (Figure 2.4a). Three areas can be considered (Dow, 1942):

1. Flocculonodulus *(vestibulocerebellum)*
2. Vermal anterior and posterior lobes with mainly spinal connections (*paleocerebellum*)
3. Mediolateral part having principally corticopontocerebellar connections (*neocerebellum*)

Afferents enter the cerebellum through three pairs of peduncles: the inferior peduncle (or restiform body), the large middle peduncle (or brachium pontis), and the superior peduncle (or brachium conjunctivum). Efferents from the cerebellar nuclei leave the cerebellum through the superior and inferior peduncles. The number of afferent fibers exceeds the number of efferents by a ratio of about 40:1, hence the enormous computational capabilities of the cerebellar circuitry.

Three types of fibers enter the cerebellum:

1. Climbing fibers: thin, myelinated and slow conductive (20 m/s) fibers firing at low frequency (about 1 Hz). They arise exclusively from the contralateral inferior olive, crossing the midline, and ascending in the restiform body.
2. Mossy fibers: large myelinated fibers. They are fast conducting and fire at high frequency rates. They arise from a large spectrum of ipsilateral and contralateral sources.
3. Cholinergic and monoaminergic afferents: distributed diffusely.

2.3.1 Cerebellar Cortex

The cerebellar cortex (Figure 2.4b) contains *Purkinje cells*, *granule cells*, and *inhibitory interneurons* organized in a trilayer structure with the Purkinje cell layer (ganglionic layer) separating the outer molecular from the inner granular layer.

1. *Purkinje cells* are GABAergic and thus inhibitory (Ito and Yoshida, 1964). Their axons project to the cerebellar nuclei and vestibular nuclei. Purkinje cells receive a glutamatergic

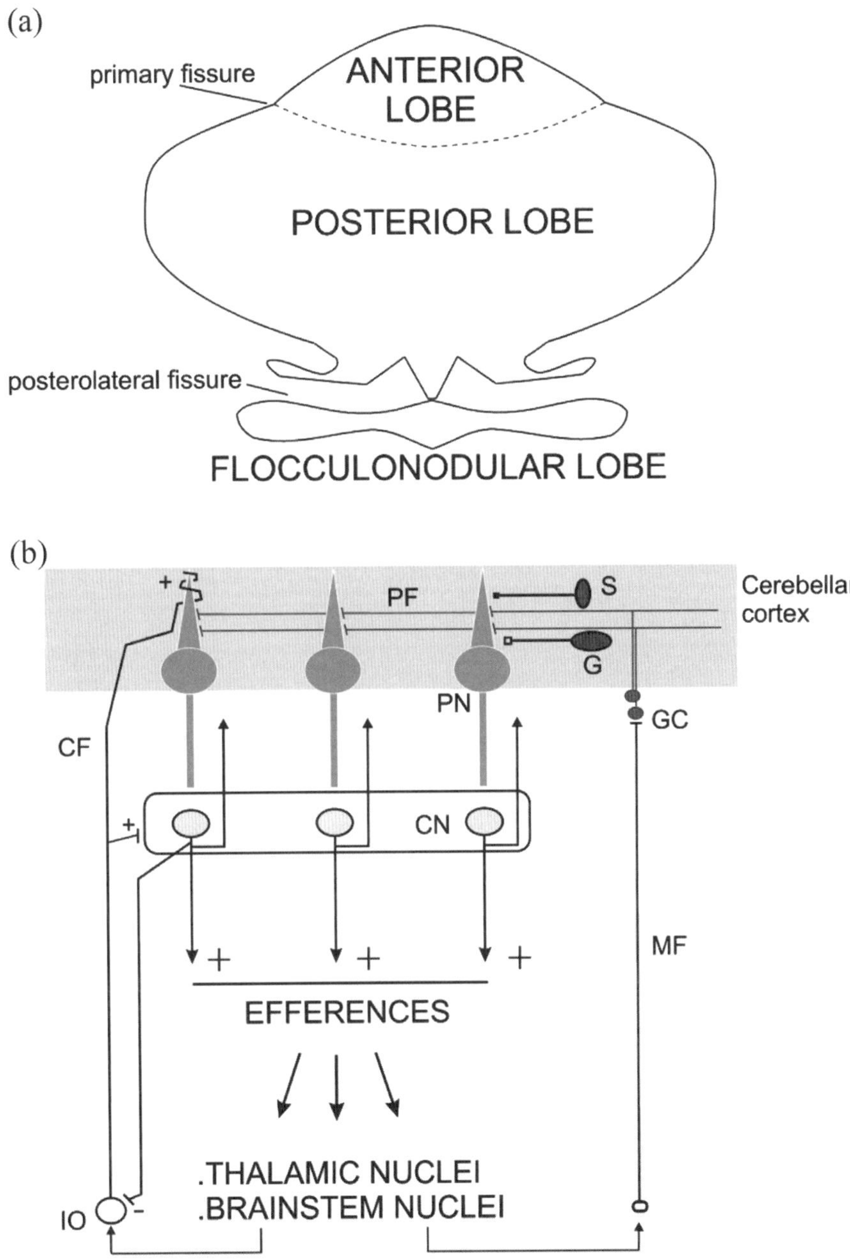

FIGURE 2.4: (a) Illustration of the unfolded cerebellum subdivided in three lobes. (b) Organization of the cerebellar cortex. Purkinje neurons (PN) project upon cerebellar nuclei (CN). Inhibitory interneurons of the cerebellar cortex (S: stellate cells, G: Golgi interneurons) modulate the activity of Purkinje cells. Granule cells (GC) are the source of parallel fibers (PF) branching upon Purkinje cells. Climbing fibers (CF) emerge from the inferior olive (IO). Mossy fibers (MF) emerge from various sources in the brainstem.

projection from the inferior olive. This synapse-climbing fiber/Purkinje neuron is the most powerful excitatory synapse of the brain. The Purkinje cell's dendrites also establish excitatory synapses with the parallel fibers of granule cells (Eccles et al., 1967).

2. *Granule cells* present about four to five dendrites and a thin unmyelinated T-shaped axon whose branches constitute the parallel fibers running between the Purkinje neurons.
3. Inhibitory interneurons are located both in the molecular and in the granular layer. They balance the excitatory activity targeting the Purkinje cells. Inhibitory interneurons in the molecular layer outnumber the Purkinje cells by a ratio of 10:1 (Andersen et al., 1992; Pouzat and Hestrin, 1997). Interneurons residing in this layer are the *basket cells* and *stellate cells*. *Lugaro cells* receive massive innervation from Purkinje cell axon collaterals. In turn, their axon ascends in the molecular layer and makes numerous symmetrical GABA-ergic synapses with the basket and stellate cells (Lainé and Axelrad, 1998). *Golgi cells* receive the main excitatory input from the parallel fibers on the dendritic arbor and are inhibited by Purkinje cell collaterals. Golgi neurons sense and regulate the mossy fibers' input.

2.3.2 Cerebellar Nuclei

Cerebellar nuclei (fastigial, globosus, emboliformis, and dentate nucleus) receive the inhibitory projection from the cerebellar cortex (Ito, 1984). The dentate nucleus is the largest and appears as a convoluted band, having the shape of a folded bag, with the hilus directed medially (Carpenter, 1985). Nucleofugal fibers are excitatory (glutamatergic) anywhere except for the nucleo-olivary projections (Schwarz and Schmitz, 1997). All nucleofugal fibers send collaterals that end in the granular layer. The nucleocortical projections reach mostly the areas from which they receive Purkinje cell axons (Buisseret-Delmas, 1988; Buisseret-Delmas and Angaut, 1988; Trott et al., 1998). This is one example of an internal loop within cerebellar circuitry. Nuclear cells receive a dense Purkinje cell innervation. On overage, the ratio between Purkinje and nuclear cells is 26:1, a given nuclear cell receiving about 14 terminals from the same Purkinje cell. Inhibitory inputs to the nuclear cell represent ~60% of the total synaptic input (Ito, 1984). On the other side, the excitatory input on nuclear cell comes, partly from collaterals of the climbing fibers and, extensively from the mossy fiber afferents.

2.3.3 Climbing Fibers

Arise from the olivary complex. This complex is organized in functional units that tend to fire synchronously. The anatomical and functional unit of the olivocerebellar system consists of a sagittal band of cortex receiving climbing fibers from a small part of the inferior olive and sending its output to a small part of the nuclei. A given area in the inferior olive corresponds to a definite area in the inferior output structure. The inferior olive translates its afferent inputs in to a well-controlled, low-frequency (1 Hz) discharge in time and distributed to a precise cluster of sagittally organized Purkinje cells in space. Inferior olive receives projections from the spinal cord, motor cortex, sensory spinal root of the trigeminal nerve, and red nucleus (Colin et al., 2002).

2.3.4 Mossy Fibers

Mossy fibers have numerous origins, mainly somesthetic, vestibular, acoustic, visual, and cortical. The principal spinocerebellar tracts are the dorsal and the ventral spinocerebellar tracts. The *dorsal spinocerebellar tract* (DSCT) arises from Clarke's column. It conveys information mainly from the hind limb and enters the cerebellum via the inferior peduncle. Its cells are either proprioceptive—stimulated by muscle spindles primary Ia and secondary group endings II, or by tendon Ib endings—or exteroceptive cells—monosynaptically activated by cutaneous low-threshold, fast-adapting hairy, and high-threshold, slowly adapting pressure receptors. The *ventral spinocerebellar tract* (VSCT) arises from the third to the sixth lumbar segments and conveys information from the hind limb. Axons cross the midline, ascend ventrally to the DSCT in spinal cord, cross a second time at the level of the ipsilateral brachium conjunctivum, and end bilaterally in the anterior lobe. All ventral spinocerebellar neurons receive strong polysynaptic input from the ipsilateral flexor reflex afferents (Oscarsson, 1973).

Fibers from vestibular nuclei project diffusely to the vermis, the flocculus, the paraflocculus, the paramedian lobule, the fastigial, and the interpositus nucleus, but not to the lateral cerebellum.

Afferents entering the cerebellum through the brachium conjunctivum emerge also from the reticular precerebellar nucleus. Fibers from lateral reticular precerebellar nuclei (located in the lower medulla) ascend bilaterally through the brachium conjunctivum (Clendenin et al., 1974, 1975; Ekerot, 1990a–c). Other groups of fibers—from the nucleus reticularis tegmenti pontis (NRTP)—relay visual information to the flocculus and are involved in the control of gaze.

Each large Brodmann's area projects to one or two slabs oriented rostrocaudally in the pons (Serapide et al., 1994). Each slab projects mainly contralaterally to the cerebellar hemisphere in a well-defined region.

2.3.5 Mediolateral Subdivision of the Cerebellum

Three zones have been identified: a vermal zone that projects to the fastigial nucleus, an intermediate zone whose Purkinje cells are connected with interposed nuclei, and a lateral zone projecting to the dentate nuclei.

2.3.6 Projections of Cerebellar Nuclei

The cerebellar ouput arises exclusively from the cerebellar nuclei, except for the vermal Purkinje cell axon to lateral vestibular nucleus.

The fastigial nucleus projects bilaterally to the vestibular and reticular nuclei. The fastigial afferents are involved in the control of reticulospinal and vestibulospinal tracts. Their effects on the spinal alpha and gamma motor neuron are complex (Ito, 1984).

The anterior interpositus nucleus projects mainly to the ventral lateral (VLN) and ventral anterior (VA) nuclei of the contralateral thalamus and sends collaterals to the caudal magnocellular part of the red nucleus (mesencephalon), pontine nuclei, superior colliculus. The posterior interpositus nucleus sends projections towards the thalamus (lower density of projections if compared with those from the anterior part) and to the medial part of the red nucleus. The dentate nucleus has similar projections of the interpositus. However, there are two differences: a small number of fibers from the dentate nucleus project to the rostral intralaminar thalamic nuclei, and the collaterals reach the anterior parvocellular part of contralateral red nucleus.

The thalamic nuclei are thus the major sites that receive cerebellar nucleofugal projections. In turn, thalamic neurons project to the cerebral cortex. Through this *thalamocortical relay*, the fastigial nucleus projects bilaterally to the hind limb area of the motor cortex and the parietal cortex. The interpositus nucleus projects contralaterally to the trunk area of the motor cortex and the premotor cortex. The dentate nucleus projects contralaterally to the forelimb area of the motor cortex, premotor cortex, and prefrontal association cortex.

2.3.7 Guillain–Mollaret Triangle (Dentatorubro-Olivary Pathway)

This anatomical notion refers to the projection from the dentate nucleus to the contralateral red nucleus and inferior olive, via the superior cerebellar peduncle, the central tegmental tract, and the inferior cerebellar peduncle. This pathway plays critical roles in tremor genesis.

2.4 SPINAL CORD

The spinal cord is a cylindrical-shaped portion of nervous matter located inside the vertebral bone canal. The rostral extremity is in continuity with the medulla. The caudal extremity forms the conus medullaris ending in the filum terminalis in correspondence to coccygeal bones. The spinal cord is organized into 31 continuous spinal segments divided into four regions: cervical, thoracic, lumbar, and sacral. Spinal nerves enter and leave the lateral medullary zones through the dorsal (sensory) and the ventral (motor) roots. Once compacted, the spinal nervous trunk gets off the vertebral canal through the intervertebral foramina. The area of skin supplied by axons from a single dorsal-root ganglion is defined as a *dermatome*.

In a cross-section, the spinal cord shows the butterfly-like shaped gray matter surrounded by the white matter.

1. The medullar gray consists of cell bodies and dentrites of neurons and glial cells. The gray matter is subdivided into *dorsal or posterior horn*, made up of the sensory nuclei where somatosensory information ascends toward the brain stem and thalamus, and in *ventral or motor horn* containing motor neurons innervating skeletal muscles. Information flows from dorsal horn

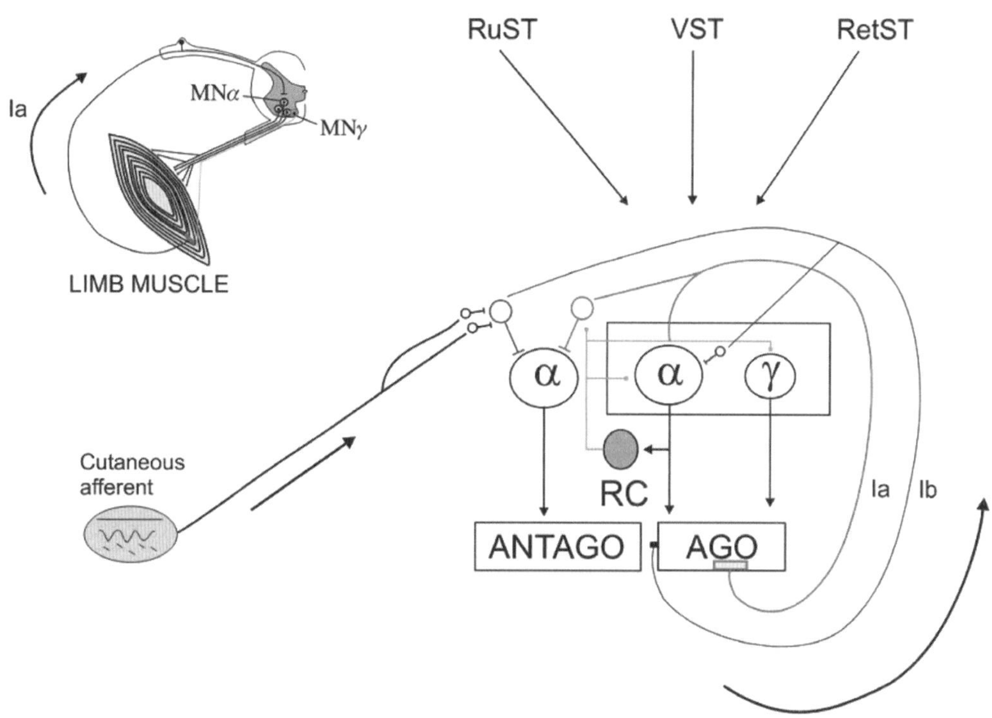

FIGURE 2.5: Ia and Ib loops. Gamma-motor neurons innervate the contractile regions of intrafusal fibers and regulate the sensitivity of muscle spindles; they are coactivated during motion. Ia afferent input (originating from a muscle spindle illustrated by a small red rectangle; spindles are arranged in parallel with muscle fibers) inhibits antagonist muscles via reciprocal inhibitory interneurons. Ib Golgi tendon organ afferents (the main source of autogenic inhibition) are arranged in series with muscle fibers. They are sensitive to changes in tension. Flexor motoneurons may receive excitation not only from antagonist extensor muscles, but also from extensors operating at other joints. Afferent fibers from Golgi tendon organs provide a negative feedback to the homonymous and synergic alpha-motoneurons and excite the antagonist motoneurons via an excitatory interneuron. Ib inhibitory interneurons receive convergent input originating from the skin and joints. α: alpha-motoneuron, γ: gamma-motoneuron; RuST: rubrospinal tract; VST: vestibulospinal tract; RetST: reticulospinal tract. VST stimulates directly alpha-motoneurons. Adapted from Manto et al. (2008).

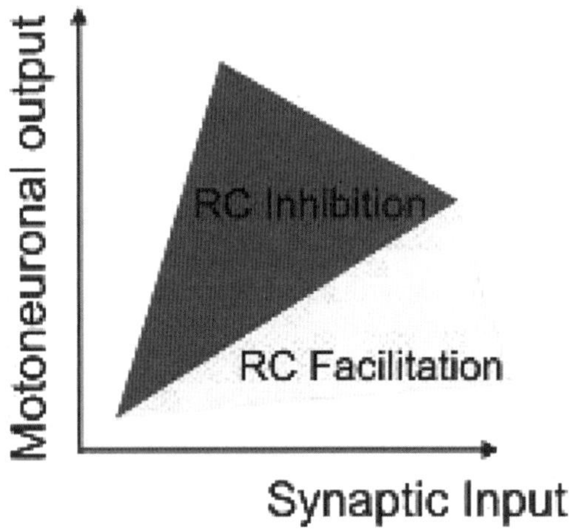

FIGURE 2.6: Renshaw cells (RC) modulate the motoneuronal output from the spinal cord (spinal recurrent inhibition). In case of RC facilitation via motor axon collaterals, the output from motoneuronal output is decreased compared to the state of RC inhibition. Adapted from Manto et al. (2008).

to motoneurons (or between different groups of motoneurons) and is modulated by numerous interneurons. In particular, the inhibitory Renshaw cells allow a regulation of motor neuronal activity (Figures 2.5 and 2.6). Between the two horns, the *intermediate zone* contains neurons whose axons project either in the ventral horn or in the brain stem and cerebellum.

2. The white matter consists mainly of axons grouped into tracts. Three large columns are found:
    - *Dorsal column*: composed by primary sensory fibers afferent to the brain stem
    - *Lateral column*: contains both axons ascending to the higher levels of the central nervous system and axons coming from brain stem nuclei and cortex that project on spinal motoneurons and interneurons
    - *Ventral column*: made up of fibers transmitting to the higher levels of the central nervous system pain and thermal sensations, as well as descending motor axons that control axial muscles and posture. Moreover, ventrally to the central canal, the *ventral commissure* is made up of packed axons that cross from one side to the other side of the spinal cord. Some of them transmit information about sensation of pain; others are involved in the control of posture.

The principal *spinal tracts* are the following:

1. Anterior column
    - *Longitudinal medial tract*: from mesencephalon to spinal cord. It represents the connection between vestibular nuclei and the nuclei controlling the eyes and head muscles

movements, allowing the maintenance of balance by the synergical movements of the eyes and head.

- *Anterior tectospinal tract*: part of the extrapyramidal pathways. Fibers start in the mesencephalic tectum (superior quadrigeminal tuberculi), cross the midline, and synapse with motoneurons at the level of the cervical anterior horn.
- *Vestibular spinal anterior tract*: made up of crossed and uncrossed fibers from the vestibular lateral nucleus of Deiters (located in the bulb). This tract controls postural tone and is responsible of the maintenance of balance by adjusting head, trunk, and limb movements in response to vestibular inputs.
- *Anterior rubrospinal tract*: made up of a crossed group of fibers from the red nucleus of the mesencephalon to the cervical cord where they make a synapse with motoneurons of the ventral horn; it is also part of the extrapyramidal pathways.
- *Anterior reticular spinal tract*: a direct path from the mesencephalic and pontine reticular nuclei to the ventral horn, where motoneurons undergo a facilitatory effect.
- *Anterior (direct) cortical–spinal tract*: part of the pyramidal pathway. Fibers have their origin in the motor cortex. They pass in the internal capsule, the brain stem, and reach the anterior column. Groups of fibers cross the midline through the anterior commissure and make synapses with motoneurons of ventral horn. Although the tract is named "direct," the information carrying voluntary inputs is crossed.
- *Olivospinal tract*: from the inferior olive to the spinal anterior horn.
- *Anterior spinothalamic tract*: has fibers that emerge in the posterior horn, cross the midline, and ascend the brain stem until the lateroventral-posterior thalamic nucleus. This tract transmits information about tact and pressure. Once in the brain stem, it is joined by the lateral spinothalamic tract, both forming the spinal lemniscus.

2. Lateral column:
    - *Vestibular spinal lateral tract*
    - *Lateral rubrospinal tract*
    - *Lateral tectospinal tract*
    - *Lateral reticular spinal tract*
    - *Crossed cortical-spinal tract*: part of the pyramidal pathway and carries conscious motor information.
    - *Lateral spinothalamic tract* (STT): has fibers that start in the posterior horn, cross the midline, and ascend the brain stem up to the lateroventral-posterior thalamic nucleus. It transmits information about pain and thermal sensibility.
    - *Crossed ventral spinal cerebellar tract* (VSCT or Gowers's tract): made up of fibers originating in the Bechterew nucleus. After a first crossing of the midline, fibers ascend contralaterally

in the lateral column, reach the pons, mesencephalon, superior cerebellar peduncle, and finally get to the anterior lobe of the cerebellar cortex (see Section 3). Once in the cerebellum, they cross again. VSCT transmits proprioceptive unconscious information.
- *Direct dorsal spinal cerebellar tract* (DSCT or Flechsig's tract): has fibers that originate in the cells of Clark's nucleus (dorsal horn) and ascend homolaterally. In the bulb, they enter the cerebellum through the inferior cerebellar peduncle and end in the anterior lobe cerebellar cortex. This tract transmits proprioceptive unconscious information concerning the trunk.

3. Dorsal column:
    - *Goll fasciculus gracilis*: located medially in the dorsal column. It conveys fine tactile and conscious proprioceptive information.
    - *Burdach fasciculus cuneatus*: located laterally in the dorsal column. Fibers constituting this tract emerge from the upper thoracic and cervical spinal ganglia. In the bulb, they make synapses with the cells of the Burdach nucleus whose axons constitute the bulbothalamic tract (medial lemniscus). Fine tactile and conscious proprioceptive information is transmitted.

2.4.1 Organization of Motoneurons

Motoneurons innervating the axial musculature (trunk) are located medially. The group of motoneurons innervating the distal limbs muscles are located more laterally. There is also a ventrodorsal organization for motor neurons dedicated to extensor muscles versus those activating flexor muscles. Similarly, the central axons of the dorsal root ganglion neurons are arranged somatotopically in the dorsal column.

2.4.2 Spinal Cord and Reflexes

The spinal cord is a major site of reflex activity for the nervous system, and reflexes are important contributors in tremor genesis and maintenance. The *spinal reflex loop* results in stereotyped responses (contraction or relaxation) executed by somatic muscles in response to inputs from muscles, skin, or joint/tendon receptors. Five elements are involved in the reflex arch: receptor, afferent nerve fibers, nervous station, efferent nerve fiber, target structure (for instance: thermal receptor activated by heat → T ganglion cell branches → synapse on motoneuron → descending axon → somatic muscle resulting in movement of defense). The stretch reflex will be discussed in Section 8.

2.5 PERIPHERAL NERVES

The cranial nerves emerge from the encephalon, the spinal nerves from the spinal cord (see paragraph 2.4). Nerves can be classified as:

- *sensory nerves*: made up by sensory nervous fibers that transmit information from periphery to central structures;
- *motor nerves* (effectors): made up by motor fibers targeting muscles and glands;
- *mixed nerves*: containing both types of fibers. A spinal nerve is a typical example of a mixed nerve, containing a ventral motor root and a dorsal sensory root. Both are compacted to form the nervous trunk. As a mixed nerve, spinal nerves contain all the four types of fibers.

Cranial nerves innervate the head and neck. The vagal nerve (nerve X) reaches the thoracic and abdominal organs.

I	optic
II	olfactory
III	oculomotor
IV	trochlear
V	trigeminal
VI	abducens
VII	facial and intermediate
VIII	vestibulocochlear
IX	glossopharyngeal
X	vagus
XI	spinal accessory
XII	hypoglossal

2.6 NEUROMUSCULAR JUNCTION

The motor neuron axon innervates a specialized zone of the muscle membrane called the end-plate. Terminal axonal branches get close to the muscle fiber membrane where they form multiple grape-like varicosities (synaptic boutons) and lie in correspondence to the junctional folds of the muscle fiber membrane. Synaptic boutons release acetylcholine (Ach) that reaches the Ach receptor of the muscle membrane, opening the sodium channel. As a result, a depolarizing potential (called end-plate potential) occurs (see also Chapter 3). To produce an effective depolarization of the postsynaptic cell, the activation of the postsynaptic voltage-dependent Na^+ channel located outside the end-plate is required (Kandel and Siegelbaum, 1991).

2.7 MUSCLE

Skeletal muscle fibers are made of myofibrils whose functional contractive unit is the sarcomere. This latter is composed of fibrillar proteins called thin and thick filaments (actin, tropomyosin, and troponin). Grouped myosin molecules correspond to thick filaments. During contractions, motion of the thin filaments (relative to the thick ones) occurs. This phenomenon is called the filaments sliding. Contraction is induced by cell depolarization. The strength of contraction is related to the initial muscle length. Indeed, the muscle has elastic properties and acts as a spring in response to lengthening (see Chapter 1). Muscle stiffness increases during contraction. These changes modify filaments sliding speed, thus influencing contraction intensity (Ghez, 1991).

A given motor neuron and all muscle fibers innervated represent the *motor unit*, the smallest functional unit controlled by the motor system. The whole group of muscle fibers belonging to a motor unit has similar physiological and biochemical features.

Motor units can be classified as:

- *fast fatigable*: The unit contracts and relaxes rapidly, produces high force level but, for repeated stimuli, fatigues rapidly as well.
- *slow fatigable*: The unit has a longer contraction time and is highly resistant to fatigue, but is much less powerful than the fast fatigable motor units. These units are active mainly in tasks requiring sustained contractions.

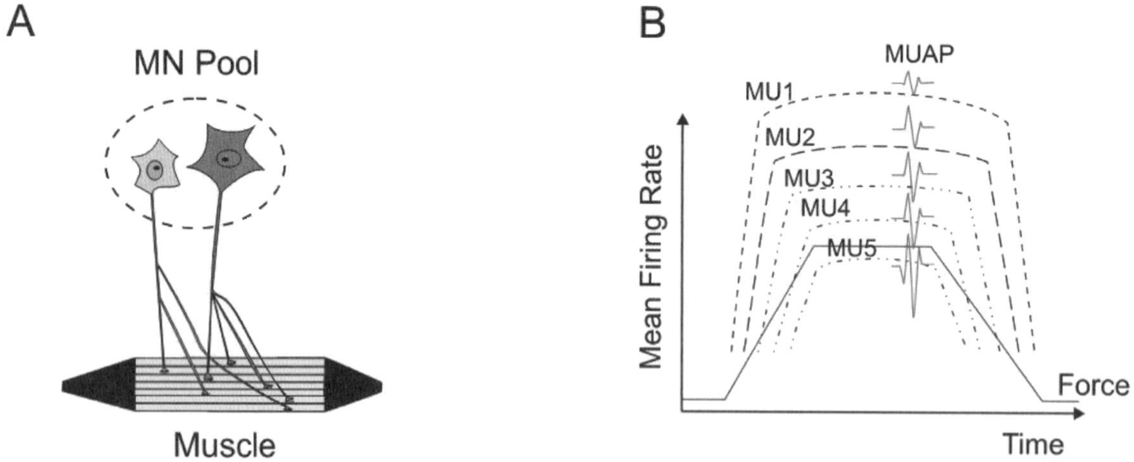

FIGURE 2.7: (a) Illustration of the innervation of a muscle fibers by two motor neurons (MN). (b) Size principle in a voluntary motor task consisting of an isometric contraction (production of a trapezoidal force). The firings of five motor units (MU1 to MU5) are illustrated. The smallest motor units are recruited first, generating the smaller motor unit action potentials (MUAPs illustrated in brown). These smaller motor units have the highest firing rates. Motor units activated subsequently have lower firing rates. Modified from Manto et al. (2008).

- *Fast fatigue resistant*: The unit has a fast fatigable-like time contraction and a slow fatigable-like resistance, but unlike the slow fatigable motor units, they can produce about twice as much force.

2.7.1 Recruitment of Motoneurons

To build up muscle contraction force, the nervous system follows the *size principle* and the *rate modulation.* Weakest inputs activate the smallest cell bodies because of their lowest threshold (Figure 2.7, see also Chapter 6 for the decomposition of the myoelectric signal). As the synaptic inputs increases in strength, larger motor neurons are progressively recruited. Increases in firing rate of motor units produce increasing force output because of the summation of successive twitches. The motor units' frequency of discharge is about 8–11 Hz during standard voluntary contraction, and it can switch up to about 25 Hz for sustained contractions. Motor units are activated asynchronously, thus minimizing jerks and producing smooth movements. According to the type of muscle fibers activated, spectral shifts in the median frequency of the power spectrum may be analyzed (see Chapter 6 about signal processing).

2.8 MUSCLE SPINDLES AND OTHER PERIPHERAL RECEPTORS

These are specialized nervous structures sensitive to a given form of energy (mechanical, thermal, chemical, etc.). The energy is transformed in electrochemical energy to enter in the sensory system to be processed. The sensory receptor is the first neuron in each sensory pathway. Localization of the sensation is elaborated, thanks to the receptive field of receptors and the central sensory neurons. Duration is encoded through the receptor adaptation to sustained stimuli, considering that receptor potential tends to decrease in response to a long duration stimulus.

The modalities of stimuli captured by the somatic sensory system are the following:

- *touch*: elicited by mechanical stimulation of body surface;
- *proprioceptive sensations*: elicited by mechanical displacements of the muscles and joints;
- *vibrations*;
- *pain*, elicited by noxious stimuli;
- *thermal sensations*, elicited by cold and warm stimuli.

Specific receptors are:

- *cutaneous and subcutaneus mechanoceptors*: Meissner's corpuscle, Pacinian corpuscle, Ruffini corpuscle, Merkel's receptor, hair-guard and hair-tylotrich, hair-down;
- *muscle spindles* (primary and secondary), Golgi tendon organs, mechanoreceptors;
- *nociceptors*: mechanical, thermal, and polymodal.

Full proprioceptive sensitivity depends on the combined actions of muscle receptors, joint receptors, and cutaneous mechanoreceptors. It is generally accepted that although cutaneous and joint receptors are involved in kinesthesia (defined as the sense of position and movement of the extremities), the main afferent signals originate from muscle spindles (Fallon and Macefield, 2007). These mechanoceptors that are entwined around specialized muscle fibers are called intrafusal fibers (Gordon and Ghez, 1991). They respond to changes in muscle length. Moreover, they discharge at higher rates in response to vibrations, which cause length variation and give the illusion of limb motion. Indeed, the illusion of joint movement is triggered by vibration (usually around 70–90 Hz) of either the muscle belly or preferably the muscle tendon. The vibration may generate a tonic vibration reflex (TVR) during which the muscle contracts involuntarily. Spindles can detect very small length changes.

Muscle spindles are innervated by groups I and II afferent fibers. Sensory afferents terminate in the central region of intrafusal fibers, while the gamma fibers innervate the polar regions relatively rich in contractile elements. Thereby, activation of a gamma efferent causes shortening of the intrafusal polar regions, which in turn, stretches the noncontractile central region. Such a stretch from both ends leads to an increase in firing rate of the sensory endings. Moreover, to grant a flow of information despite the muscle contraction (thereby fibers are shortening), the central nervous system keeps gamma motor neurons excited during the muscle contraction.

From a physiological point of view, the stretch reflex (or myotatic reflex) is the muscle contraction in response to muscle lengthening. It is linked to the reciprocal innervation that allows relaxation of the antagonist muscles. Stretch reflex consists of two components: a brisk and short-lasting phasic contraction, which is trigged by the dynamic change in muscle length, and a weaker but longer-lasting tonic contraction determined by the static stretch of the muscles. A balance exists between facilitating and inhibiting control on reflex loops. Actually, cerebral cortex and other higher centers adjust stretch reflex intensity according to the motor tasks' performance.

The activity of muscle spindles can be assessed using a technique called microneurography (Burke et al., 1978). Insulated microelectrodes (usually in tungsten) are inserted through the skin, for instance, at the level of the fibula head (knee) to record the electrical activities within the common peroneal nerve.

REFERENCES

Andersen BB, Korbo L, Pakkenberg B. A quantitative study of the human cerebellum with unbiased stereological techniques. *J Comp Neurol.* 1992;326(4):549–560.

Buisseret-Delmas C. Sagittal organization of the olivocerebellonuclear pathway in the rat. I. Connections with the nucleus fastigii and the nucleus vestibularis lateralis. *Neurosci Res.* 1988;5(6):475–493.

Buisseret-Delmas C, Angaut P. The cerebellar nucleocortical projections in the rat. A retrograde labelling study using horseradish peroxidase combined to a lectin. *Neurosci Lett.* 1988;84(3):255–260.

Burke D, Hagbarth KE, Lofstedt L. Muscle spindle activity in man during shortening and lengthening contractions. *J Physiol (Lond).* 1978;277:131–142.

Carpenter MB. *Core Text of Neuroanatomy*, 3rd edn. Baltimore, MD: Williams & Wilkins, 1985.

Clendenin M, Ekerot CF, Oscarsson O. The lateral reticular nucleus in the cat. III. Organization of component activated from ipsilateral forelimb tract. *Exp Brain Res.* 1974;21(5):501–513.

Clendenin M, Ekerot CF, Oscarsson O. The lateral reticular nucleus in the cat. IV. Activation from dorsal funiculus and trigeminal afferents. *Exp Brain Res.* 1975;24(2):131–144.

Colin F, Ris L, Godaux E. Neuroanatomy of the cerebellum. In: *The Cerebellum and Its Disorders.* Manto MU and Pandolfo M (eds). Cambridge, UK: Cambridge University Press, 2002, pp. 6–29.

Dow RS. The evolution and anatomy of the cerebellum. *Biol Rev.* 1942;17:179–220.

Eccles JC, Ito M, Szentagothai J. *The Cerebellum as a Neuronal Machine.* New York: Springer-Verlag, 1967.

Ekerot CF. The lateral reticular nucleus in the cat. VI. Excitatory and inhibitory afferent paths. *Exp Brain Res.* 1990;79(1):109–119.

Ekerot CF. The lateral reticular nucleus in the cat. VII. Excitatory and inhibitory projection from the ipsilateral forelimb tract (iF tract). *Exp Brain Res.* 1990;79(1):120–128.

Ekerot CF. The lateral reticular nucleus in the cat. VIII. Excitatory and inhibitory projection from the bilateral ventral flexor reflex tract (bVFRT). *Exp Brain Res.* 1990;79(1):129–137.

Fallon JB, Macefield VG. Vibration sensitivity of human muscle spindles and Golgi tendon organs. *Muscle Nerve.* 2007;36:21–29.

Ghez C. The control of movement. In: *Principles of Neural Science.* Kandel ER, Schwartz JH, Jessel TM (eds). New York: Elsevier, 1991, pp. 530–547.

Goldman MS, Kelly PJ. The surgical treatment of tremor disorders. In: *Handbook of Tremor Disorders.* Findley LJ, Koller WC (eds). New York: Marcel Dekker, 1995, pp. 521–562.

Gorgon J and Ghez C. Muscle receptor and spinal reflex: the stretch reflex. In: *Principles of Neural Science.* Kandel ER, Schwartz JH, Jessel TM (eds). New York: Elsevier, 1991, pp. 564–580.

Ito M and Yoschida M. The cerebellar-evoked monosynaptic inhibition in Deiters' neurones. *Experientia.* 1964;20:515–516.

Ito M. *The Cerebellum and Neural Control.* New York: Raven Press, 1984.

Kandel ER, Schwartz JH, Jessel TM. *Principles of Neural Science.* New York: Elsevier, 1991.

Kandel ER and Siegelbaum SA. Directly gated transmission of the nerve–muscle synapse. In: *Principles of Neural Science.* fKandel ER, Schwartz JH, Jessel TM (eds). New York: Elsevier, 1991, pp. 135–152.

Lainé J, Axelrad H. Lugaro cells target basket and stellate cells in the cerebellar cortex. *NeuroReport.* 1998;9(10):2399–2403.

Loeb C, Favale E. *Neurologia di Fazio Loeb.* Rome: Società Editirice Universo. 2003.

Manto M, Sauvage C, Roark RM. Unifying hypothesis for the motoneuronal code in neurological disorders. *Biosci Hypoth.* 2008;2:93–99.

Manto MU, Pandolfo M. *The Cerebellum and Its Disorders.* Cambridge, UK: Cambridge University Press, 2002.

Oscarsson O. Functional organization of spinocerebellar paths. In: *Handbook of Sensory Physiology, vol II. Somatosensory System.* Iggo A (ed). Berlin: Springer-Verlag, 1973, pp. 339–380.

Pouzat C, Hestrin S. Developmental regulation of basket/stellate cell → Purkinje cell synapses in the cerebellum. *J Neurosci.* 1997;17(23):9104–9112.

Serapide MF, Cicirata F, Sotelo C, Pantó MR, Parenti R. The pontocerebellar projection: longitudinal zonal distribution of fibers from discrete regions of the pontine nuclei to vermal and parafloccular cortices in the rat. *Brain Res.* 1994;644(1):175–180.

Trott JR, Apps R, Armstrong DM. Zonal organization of cortico-nuclear and nucleo-cortical projections of the paramedian lobule of the cat cerebellum. 1. The C1 zone. *Exp Brain Res.* 1998;118(3):298–315.

• • • •

CHAPTER 3

Physiology of the Nervous System

3.1 MEMBRANE POTENTIAL: BASIC CONCEPTS

Neuronal membrane acts as a barrier for the diffusion of ions, maintaining actively concentration gradients. Clouds of negative and positive ions spread over the inner and outer surfaces of the membrane. Concentration of sodium is higher in the extracellular side, and concentration of potassium is higher within the cell. These gradients are maintained thanks to the sodium–potassium pump (Na^+K^+ATPase). This enzyme pumps three Na^+ ions out of the cell and two K^+ ions into the cell for one ATP molecule. Moreover, neuronal cell membrane presents multiple ion channels (see Figure 3.1 for an example).

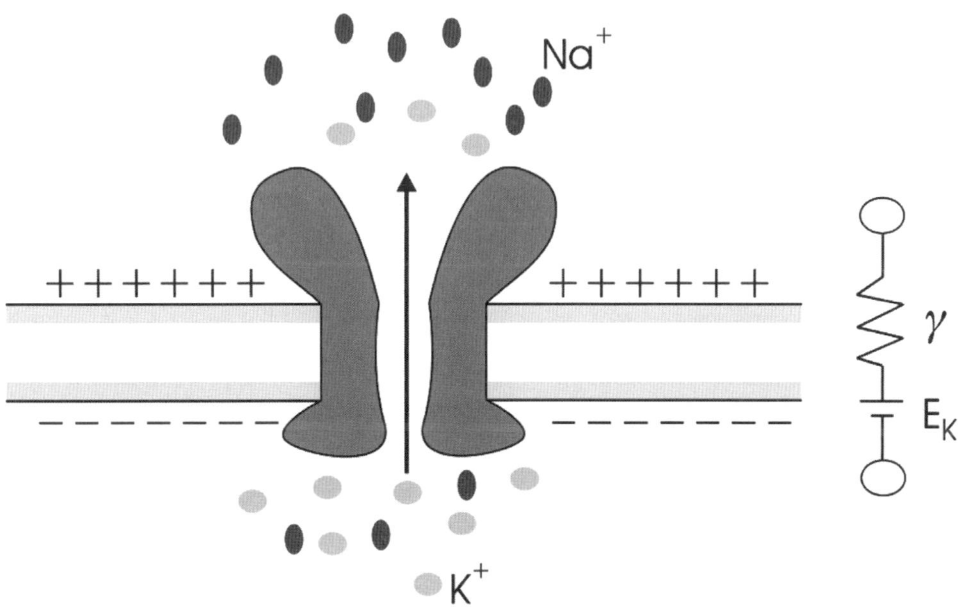

FIGURE 3.1: Representation of a cellular membrane. A specialized protein causes the efflux of potassium from the internal side of the cell (green circles). This protein is selectively permeable to this ion. A conductor γ represents the conductance of the K^+ channel. E_K represents a battery, corresponding to the potassium Nernst potential.

There is discordance among ions' diffusion down their concentration gradient, through non-gated channels, and the immobility of nonpermeable big anions that are left behind the membrane. Such a disposition determines an electrical potential difference called the *resting membrane potential* whose value in neurons is –60 to –70 mV. The electrical potential across the neuron's membrane can be calculated from the ion concentrations using the Nernst equation:

$$E = RT \times [ln(X_o/X_i)]/nF,$$

where E is the potential across the membrane, R is the resistance, T is the absolute temperature (ions move faster when the temperature increases), X_o is the concentration of the ion outside the membrane, X_i is the concentration of the ion inside the membrane, n is the charge of the ion, and F is the Faraday unit.

The equilibrium potentials of four main ions are given in Table 3.1.

The membrane potential (V_m) at any time can be estimated if the relative permeabilities to ions are known, applying the following Goldman–Hodgkin–Katz equation:

$$V_m = 2.303\,(RT/F)\,\log(P_K\,[K^+]_o + P_{Na}\,[Na^+]_o + P_{Cl}\,[Cl^-]_i)/(P_K\,[K^+]_i\,P_{Na}\,[Na^+]_i P_{Cl}\,[Cl^-]_o)$$

Electrical signals result from brief changes of the resting membrane potential. A flux of sodium entering in (followed by potassium out) of the cell changes the potential across the membrane from –70 mV to positive value in the range of +10/+30 mV. This *depolarization* reflects the reduction of the charge separation, while *hyperpolarization* is an increase in charge separation. In response to a depolarization sufficient to reach a given threshold, the *action potential* occurs in a very short period of time (a few ms), according to the all-or-none law. By contrast, a passive electrochemical process leads to hyperpolarization, which is the electrical basis of the inhibitory response. The speed

TABLE 3.1: Equilibrium potentials calculated from the Nernst equation

ION	CONCENTRATION X_O (mM)	CONCENTRATION X_I (mM)	EQUILIBRIUM POTENTIAL (mV)
Na^+	150	15	+61.5
K^+	5.5	100	–88.2
Cl^-	125	9	–70.4
Ca^{2+}	100	10^{-4}	+183

of action potential propagation is usually directly related to the size of the axon and the presence of myelin around the axon. Myelinated axons conduct impulses faster than unmyelinated axons. Impulses typically move along neurons at a velocity from 1 to 120 m/s. The action potential is followed by a refractory period, during which a second stimulus cannot produce a second action potential.

3.2 THE SYNAPSES

The synapse (Figure 3.2) can be defined as a specialized region of membranes through which neurons are connected. There are two elemental events at the level of synapses: the electrical and the chemical transmission. In the central nervous system, chemical synapses are more abundant, and thanks to their plasticity, they represent a crucial element for processes such as perception, motion, feelings, and learning.

The chemical synapse is characterized by

- a presynaptic terminal enriched in synaptic vesicles mainly deriving from the Golgi apparatus (the end of the axon is referred to as a synaptic knob);
- a synaptic cleft of 20–40 nm (more than adjacent extracellular space);
- a postsynaptic membrane with transmembrane proteins acting as receptors and ion channels.

In response to a presynaptic action potential, vesicles release neurotransmitters. Calcium plays a key role in this process. The neurotransmitters bind subsequently to postsynaptic receptors, inducing the sodium channels' opening and generating depolarization with a postsynaptic action potential. The synapse is equipped with transporters to allow recycling of neurotransmitters. By comparison with the electrical synapses, this organization determines a delay varying from 0.3 ms up to several milliseconds. Because of the release of vesicles and neurotransmitters, chemical synapses tend to amplify nervous signals.

3.2.1 Neurotransmitters

Neurotransmitters are usually divided into excitatory and inhibitory. Their function depends on their target receptors. Each neurotransmitter works with a given specific family of receptors, maintaining a given functional role.

Typical neurotransmitters are glutamate, glycine, and GABA, which bind to ionophoric receptors undergoing a conformational change directly opening the channel. Norepinephrine and serotonin bind to a receptor that is separated from the channel and communicates with it through GTP-binding proteins and other intracellular second messengers.

In terms of synaptic delay, ionophoric receptors are faster. Transmission involving receptors activating second-messenger cascade (metabotropic receptors) can modulate the excitability of neurons and connection strength. They are implicated in learning processes, for instance.

Neurons receive both excitatory and inhibitory synapses from a huge number of axons. Therefore, nervous system neurons must *integrate* synaptic information. Processes of temporal and spatial summation occur (see below). To generate a postsynaptic action potential, a sum of several excitatory postsynaptic potentials (EPSP) should occur, and the result should be strong enough to contrast eventually inhibitory postsynaptic potentials (IPSP) generated by inhibitory synapses. The resulting synaptic inhibition is an important mechanism to control spontaneously active nerve cells, thus determining their firing pattern.

Ionic current can also be transferred directly from one neuron to another through highly specialized synapses called gap junctions. In this case, the pre- and postsynaptic cell membranes are very close (~3.5 nm). Because of this peculiar connection between neurons, an action potential in the presynaptic neuron can generate an action potential in the postsynaptic neuron. Therefore, neurons can synchronize their activity.

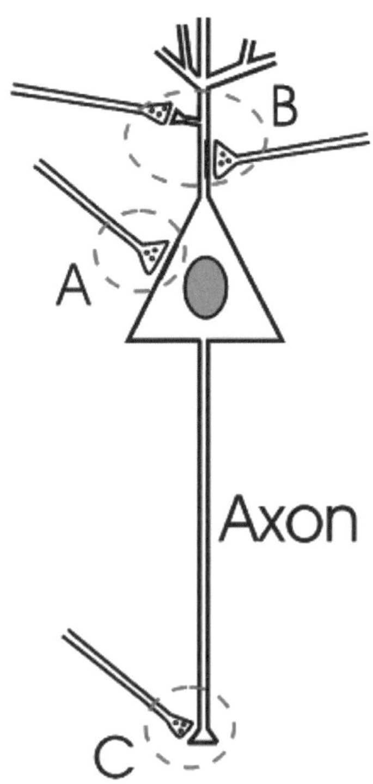

FIGURE 3.2: Categories of synapses. A: Axosomatic; B: axodendritic (left: spine synapse; right: shaft synapse); C: axo-axonic.

TABLE 3.2: Neurotransmitters in the nervous system

MOLECULE	EFFECT	RECEPTORS	ION INVOLVED/ SECOND MESSENGERS	CENTRAL OR PERIPHERAL LOCATION
Glutamate	Excitatory	NMDA AMPA Kainate QuisqualateB	Ca^{2+}, Na^+, K^+ Na^+, K^+ Na^+, K^+ Metabotropic	Central nervous system (CNS)
GABA	Inhibitory	GABA (A) subunits α, β, γ, δ GABA (B) GABA (C)	Cl^- Metabotropic	CNS Retina and CNS
Glycine	Inhibitory	GlyR NMDA	Cl^-	CNS
Amine transmitters · Dopamine · Noradrenaline · Adrenaline · Serotonin · Histamine	Heterogeneous activities	 D(1)/D(5) Adrenoceptors Adrenoceptors 5-HT1/5-HT7 H1/H4		Heterogeneously expressed in different organs and brain
Acetylcholine	Excitatory	Nicotinic AChR Muscarinic AChR	Na^+, K^+, Cl^- Metabotropic	Neuromuscular junction Autonomic nervous system CNS
Substance P and other neuropeptides		Tachykinin receptors		

3.3 EXCITATORY AND INHIBITORY SYNAPSES: NEUROTRANSMITTERS AND THEIR RECEPTORS

As mentioned earlier, postsynaptic potentials are induced by ion channels opening. Neurotransmitters (Tables 3.2 and Figure 3.3) control the permeability of ion channels.

3.3.1 Glutamate

Glutamate is the main excitatory neurotransmitter. Its action is mainly mediated by two kinds of receptors: AMPA receptor and NMDA receptor. With the first one, glutamate acts directly on a low-conductance cation channel, permeable to Na^+ and K^+. The NMDA receptor and its channel

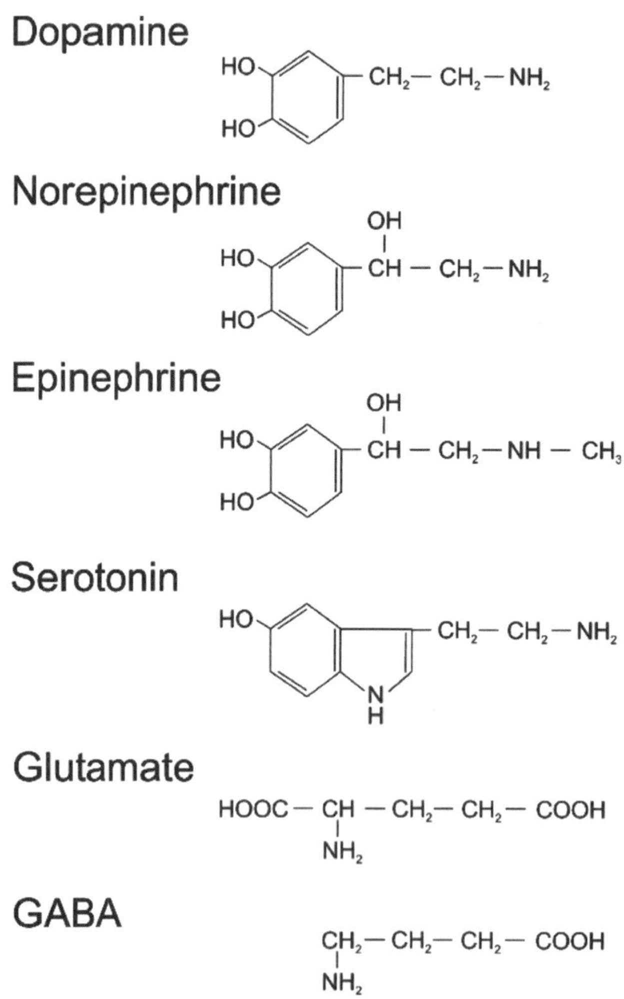

FIGURE 3.3: Main neurotransmitters of the central nervous system involved in tremor pathogenesis.

are characterized by (a) a high-conductance to Na^+, K^+, and Ca^{2+}, (b) the presence of extracellular Mg^{2+}, which delays the ionic conduction unless membrane depolarization is large enough, (c) the fact that glycine is requested for glutamate effectiveness in channel opening. Because of the Mg^{2+}-mediated block of the NMDA receptor at the normal resting membrane potential, non-NMDA receptors contribute largely in excitatory postsynaptic potential, while NMDA receptors are crucial for the late phase of EPSP. The complex NMDA channel receptor could be considered both as a voltage and a transmitter-gated channel. Moreover, NMDA-mediated Ca^{2+} entry in the cell induces a second-messenger cascade. In addition, glutamate is also known for its excitotoxicity because of free radical production in response to increasing concentration of intracellular Ca^{2+} and consecutive Ca^{2+}-dependent proteases.

3.3.2 Gamma-Aminobutyric Acid and Glycine

Gamma-aminobutyric acid (GABA) and glycine are the main inhibitory neurotransmitters. They bind directly to a channel receptor. In addition, GABA receptor has two other binding sites: one for the benzodiazepines and one for the barbiturates. The activation of these sites increases the GABA-induced Cl^- current. Moreover, the presence of any one of the three ligands influences the binding of the other two, with a global potentiation of the effect.

3.4 INTEGRATIVE ACTION OF THE NERVOUS SYSTEM

The sum of neuronal inputs—excitatory and/or inhibitory—leads to effective postsynaptic responses. According to the localization and amplitude, possible synergistic or antagonistic actions, strength of synapses, and others, postsynaptic neurons and networks integrate the inputs received and choose a corresponding behavior through the so-called integrative action, which takes place in the "axon hillock," a region of the membrane located in the initial segment of the axon with a lower threshold. When a postsynaptic potential is generated, it flows along the membrane to the hillock according to the passive properties of the membrane, mainly the time constant (summation of consecutive inputs in the same site) and the length constant (summation of postsynaptic potential that occurred in different sites). Longer time and larger space constantly increase the probability of input summation; thereby, the action potential threshold is more likely reached.

Localization is an important factor in the process of integration. Inputs are stronger when they are closer to the postsynaptic trigger zone. For instance, axodendritic synapses are excitatory, while axosomatic synapses are inhibitory. Such a disposition allows the IPSP-induced depolarization, located between the site of excitatory postsynaptic potential and the trigger zone, to prevent the excitatory input arrival. In terms of synaptic learning, this phenomenon is associated with modifications of strengths of connections between neurons. The efficacy of a synapse can change with experience, providing both memory and learning through long-term potentiation (LTP).

3.5 STRUCTURES INVOLVED IN INFORMATION PROCESSING AND MOVEMENT CONTROL

Some of the anatomical structures described in Chapter 2 have been largely investigated because they are involved in the pathogenesis of physiological and pathological tremor.

3.5.1 Muscle Spindles

Studying the behavior of the vibration-induced illusion of movement, Frima and Grünewald (2007) found out a decrease of this phenomenon with aging in healthy subjects and an impairment of this process in patients affected by essential tremor or idiopathic focal dystonia (see Chapter 5). Besides the hypothesis of abnormal central processing of the afferent information from muscle spindles, the idea that spindles themselves are involved in tremor genesis is emerging. These authors suggest that a sluggish response of the relatively inelastic muscle spindles to dynamic stimuli may result in a slightly delayed and inappropriately prolonged activation of Ia afferent fibers during muscle stretch, *adding a delay in the loop*. Moreover, in the case of essential tremor, inadequate function of the cerebellum (whose function might be to compensate by adjusting timing/gain of the feedback loop to minimise tremor) occurs. Other studies have demonstrated the systematic interaction between imposed joint movement and wrist flexor muscle activity, showing that the imposed movement of large amplitude takes progressive control over the afferent input and markedly increases the entrainment of the muscle activity via the stretch reflex (Cathers et al., 2006).

3.5.2 Spinal Cord

The spinal cord is a major site of reflex activity, and its numerous circuitries have an important role in motor coordination. The majority of reflex pathways are *polysynaptic*, composed of sensory neurons, interneurons, and motoneurons. Interneurons represent an intermediate step between afferences and efferences. Interneurons are under the supervision of supraspinal centers. Moreover, interneuron networks produce complex motor activities continuing after the end of the stimulus, thanks to *reverberating loops*. Interneurons Ia are involved in the coordination between agonist and antagonist muscles, both in reflex and voluntary movement. Renshaw inhibitory interneurons are involved in the recurrent inhibition (see Chapter 2). Inhibitory interneurons are implicated in the control of fine movements such as tactile manipulation of objects.

3.5.3 Brainstem

The brainstem contains important nuclei such as the reticular formation, vestibular nuclei, and the inferior olivary complex (see below). The brainstem nuclei integrate visual and vestibular information and regulate the segmental networks of the spinal cord.

3.5.4 Cerebellum and Inferior Olivary Complex

The cerebellum receives massive information from the periphery (see Chapter 2). In particular, medial and paramedial zones are constantly informed of the status of peripheral receptors. The cerebellum is a key piece for sensory information processing and planning of activities. It governs timing processes, meaning that it holds an internal representation of time, it controls accuracy of movements, coordination, and it is involved in learning. In particular, it tunes dynamic control of learned movement (Sánchez-Campusano et al., 2007). The cerebellum is critically involved in predictions for fast movements and online regulation of slow movements (Diedrichsen et al., 2007).

The mossy fiber–granule cell system carries more temporal information than the climbing fiber system.

3.5.5 Basal Ganglia

The basal ganglia influence movement via thalamocortical projections and via projections to the brainstem nuclei (Pong et al., 2007). The internal segment of the globus pallidus (GPi) receives motor-related cortical signals mainly through the striatum, the external segment of the globus pallidus (GPe), and the subthalamic nucleus (STN). The GPi sends its outputs outside the basal ganglia and plays a key role in motor control (Tachibana et al., 2008).

Disconnecting the basal ganglia motor pathway consistently produces bradykinesia and hypometria, but seldom affects movement initiation time, movement guidance, or the capacity to produce iterative reaches (Desmurget and Turner, 2007). Temporal preparation is disrupted in case of basal ganglia dysfunction (Praamstra and Pöpe, 2007).

3.5.6 Motor Cortex

Neurons in the motor cortices have uniformly distributed preferred directions of activity, consistent with general purpose directional calculations. There is evidence for an orderly mapping of the preferred direction of voluntary motion in the motor cortex (Georgopoulos et al., 2007).

3.5.7 Premotor Cortex

The upper motor neurons in the premotor cortex influence motor behavior through reciprocal connections with the primary motor cortex. One of the core brain areas involved in visually guided reaching is the dorsal aspect of the premotor cortex (PMC; Batista et al., 2007). Lateral premotor cortex is involved in the selection of a movement or a sequence of movements dependent of external cues. Medial parts of the premotor cortex mediate movement based on internal cues. Premotor cortex exerts a control on proximal muscles.

3.5.8 Posterior Parietal Cortex

The posterior parietal cortex (PPC) is responsible for transforming visual information into motor commands. It processes spatial information and the control of eye movements. It contributes to spatial attention. PPC damage after a stroke often leads to the clinical syndrome of neglect, in which patients seem unable to focus attention on events in the contralesional hemifield.

REFERENCES

Batista AP, Santhanam G, Yu BM, Ryu SI, Afshar A, Shenoy KV. Reference frames for reach planning in macaque dorsal premotor cortex. *J Neurophysiol.* 2007;98(2):966–983.

Cathers I, O'Dwyer N, Neilson P. Entrainment to extinction of physiological tremor by spindle afferent input. *Exp Brain Res.* 2006;171(2):194–203.

Desmurget M, Turner RS. Testing basal ganglia motor functions through reversible inactivations in the posterior internal globus pallidus. *J Neurophysiol.* 2008;99(3):1057–1076.

Diedrichsen J, Criscimagna-Hemminger SE, Shadmehr R. Dissociating timing and coordination as functions of the cerebellum. *J. Neurosci.* 2007;27(23):6291–6301.

Frima N, Grünewald RA. Abnormal vibration induced illusion of movement in essential tremor: evidence for abnormal muscle spindle afferent function. *J Neurol Neurosurg Psychiatry.* 2005;76(1):55–57.

Georgopoulos AP, Merchant H, Naselaris T, Amirikian B. Mapping of the preferred direction in the motor cortex. *Proc Natl Acad Sci USA.* 2007;104(26):11068–11072.

Kandel ER, Schwartz JH, Jessel TM. *Principles of Neural Science.* New York: Elsevier 2001.

Marshall FH. The role of GABA(B) receptors in the regulation of excitatory neurotransmission. *Results Probl Cell Differ.* 2008;44:87–98.

Pong M, Horn KM, Gibson AR. Pathways for control of face and neck musculature by the basal ganglia and cerebellum. *Brain Res Rev.* 2007.

Praamstra P, Pope P. Slow brain potential and oscillatory EEG manifestations of impaired temporal preparation in Parkinson's disease. *J Neurophysiol.* 2007;98(5):2848–2857.

Sánchez-Campusano R, Gruart A, Delgado-García JM. The cerebellar interpositus nucleus and the dynamic control of learned motor responses. *J Neurosci.* 2007;27(25):6620–6632.

Schmidt M. GABA(C) receptors in retina and brain. *Results Probl Cell Differ.* 2008;44:49–67.

Tachibana Y, Kita H, Chiken S, Takada M, Nambu A. Motor cortical control of internal pallidal activity through glutamatergic and GABAergic inputs in awake monkeys. *Eur J Neurosci.* 2008;27(1):238–253.

CHAPTER 4

Characterization of Tremor

From the clinical standpoint, tremor is classically classified as rest tremor, postural tremor, and kinetic tremor. Each category of tremor has its own characteristics in terms of:

- frequency
- amplitude
- body segments involved
- distribution (symmetry)
- enhancing/reducing effect of tasks or positions
- associated disabilities.

Neurological examination of a patient complaining and/or presenting tremor includes the following tests or maneuvers (see Chapter 6 for the clinical evaluation of tremor):

- evaluation of oculomotor movements;
- the patient is asked to repeat standard sentences to evaluate dysarthria or other speech deficits;
- the presence of vertical or side-to-side head movements is noted;
- assessment of drawing Archimedes' spiral (see Chapter 6);
- concomitant abnormal movements (included dystonia, tics) are looked for;
- the observation of abnormal movements eventually occurring at rest, the occurrence of the tremor during postural task, or voluntary muscle contraction;
- a possible amyotrophy or muscle hypertrophy is looked for;
- detailed sensory testing (light touch, sense of position, vibratory testing, pin-prick threshold, hot and cold stimuli);
- evaluation of tendon/plantar reflexes;
- the following clinical maneuvers are assessed:
    - holding the upper limbs outstretched parallel to the floor ("Barré test");
    - finger-to-nose test, index-to-index test, knee–tibia test and "Mingazzini test";

TABLE 4.1: Main disorders associated with tremor

TYPE OF TREMOR	DISEASES
Rest tremor	Parkinson's disease Drug-induced Parkinsonism Stroke Posttraumatic tremor
Postural tremor	Essential tremor Enhanced physiological tremor Cerebellar diseases Multiple sclerosis Posttraumatic tremor Metabolic diseases
Kinetic tremor	Cerebellar diseases Essential tremor Multiple sclerosis

- entrainment test (see Section 4.6);
- Romberg position test;
- regular and tandem gait.

Additional clinical tests, metrical scales, laboratory assessment and others tools for quantification of tremor will be described in detail in Chapter 6.

The main disorders associated with postural, rest, and kinetic tremor are given in Table 4.1.

4.1 PHYSIOLOGICAL TREMOR AND ENHANCED PHYSIOLOGICAL TREMOR

Healthy subjects present a physiological tremor (see also Chapter 1), an involuntary rhythmical movement of limb segments typically in the frequency range of 8–12 Hz. The amplitude is small and is barely seen with the naked eye (Cathers et al., 2006). Physiological tremor's frequency is reduced by addition of inertia (Elble, 2003; see Figure 1.1a) and is age-independent (Raethjen et al., 2000). The maintenance of a postural position or action has a typical pronouncing effect on physiological tremor during a standing position, leading to an increase of the amplitude. This is more obvious in the elderly persons (Morrison et al., 2006).

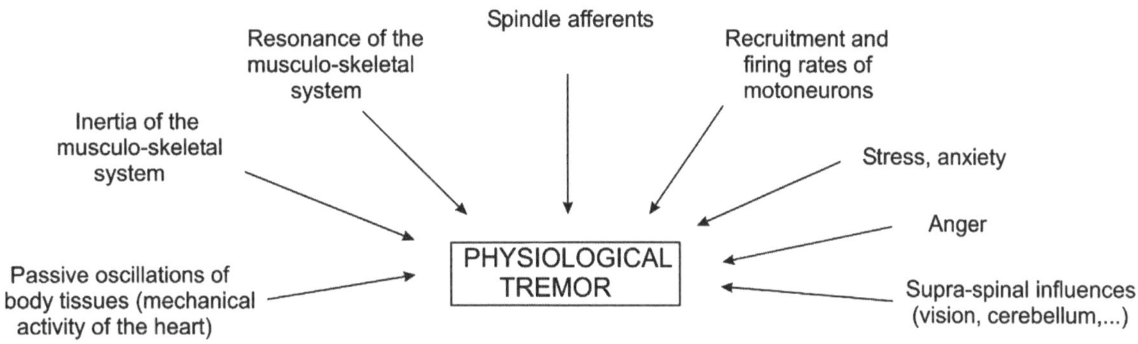

FIGURE 4.1: Factors influencing physiological tremor.

TABLE 4.2: Drugs enhancing physiological tremor

Amiodarone	Neuroleptics
Antihistamine drugs (terfenadine)	Nicotine
β2-Adrenergic agonists (bronchodilators)	Nifedipine
Cinnarizine	Phenelzine
Cyclosporin A	Phenylpropanolamine
Diazepam withdrawal	Pindolol
Donezepil	Prednisone
Flunarizine	Procainamide
Fluoxetine	Pseudoephedrine
Lamotrigine	Theophylline
Lithium	Thyroxine
Methylphenidate	Tricyclic antidepressants
Metoclopramide	Valproic acid
Monoamine oxidase inhibitors	

From Bhidayasiri (2005), Benito-León and Louis (2006), and Song et al. (1993, 2008).

Physiological tremor is augmented by hypoxia (Krause et al., 2000). Mental stress increases its amplitude (this is not specific) but decreases its frequency (Growdon et al., 2000). Factors influencing physiological tremor are summarized in Figure 4.1.

Several metabolic conditions (mainly thyrotoxicosis or hypoglycaemia) and drugs (see Table 4.2) can enhance physiological tremor giving the so-called enhanced physiological tremor, a visible high frequency postural tremor. Muscle fatigue may trigger or exaggerate this tremor. Caffeine significantly increases whole-arm physiological tremor (Miller et al., 1998), while ethanol usually decreases its magnitude (Lakie et al., 1994). Enhancement of postural tremor in subjects treated with amitryptiline (a tricyclic antidepressant drug) is a common phenomenon. Intravenous infusion of adrenaline or noradrenaline increases physiological tremor. This is antagonized by administration of beta-adrenoreceptor blockers (Marsden et al., 1967). Frequency, amplitude, burst duration, and burst pattern of drug-induced tremors can mimic essential tremor (Mehndiratta et al., 2005).

4.2 REST TREMOR

Rest tremor occurs while the body segment is maintained at rest. Rest tremor is typically in the 3- to 6-Hz frequency range. During a clinical evaluation, rest tremor is assessed with the patient sitting with his arms supported against gravity, without any voluntary activities. Tremor may disappear with action. For this reason, rest tremor is often a cause of social embarrassment, rather than disability. Rest tremor is usually asymmetrical, starting distally in the arms. Lips and jaw can be affected. Typically, tremor in the upper limbs reminds the "pill-rolling" movement. In some cases, patients can reduce the tremor by holding one hand with the other or crossing the legs. Rest tremor may increase with mental stress (i.e., counting backwards) or contralateral motion. This is not specific to rest tremor.

The most common cause of rest tremor is idiopathic Parkinson's disease (PD), which is detailed in Chapter 5.

4.3 POSTURAL TREMOR

Postural tremor occurs in body parts during the maintenance of a posture, such as holding a cup. This form of tremor is triggered by maintaining a position against gravity. Some authors consider that physiological tremor and enhanced physiological tremor are a peculiar form of postural tremor. Postural tremor often causes a significant disability. The frequency of postural tremor is usually between 4 and 12 Hz.

Many disorders are associated with postural tremor. Essential tremor (ET) is a typical example (Figure 4.2, see also Chapter 5). Lesions in different anatomical locations of the cerebellar

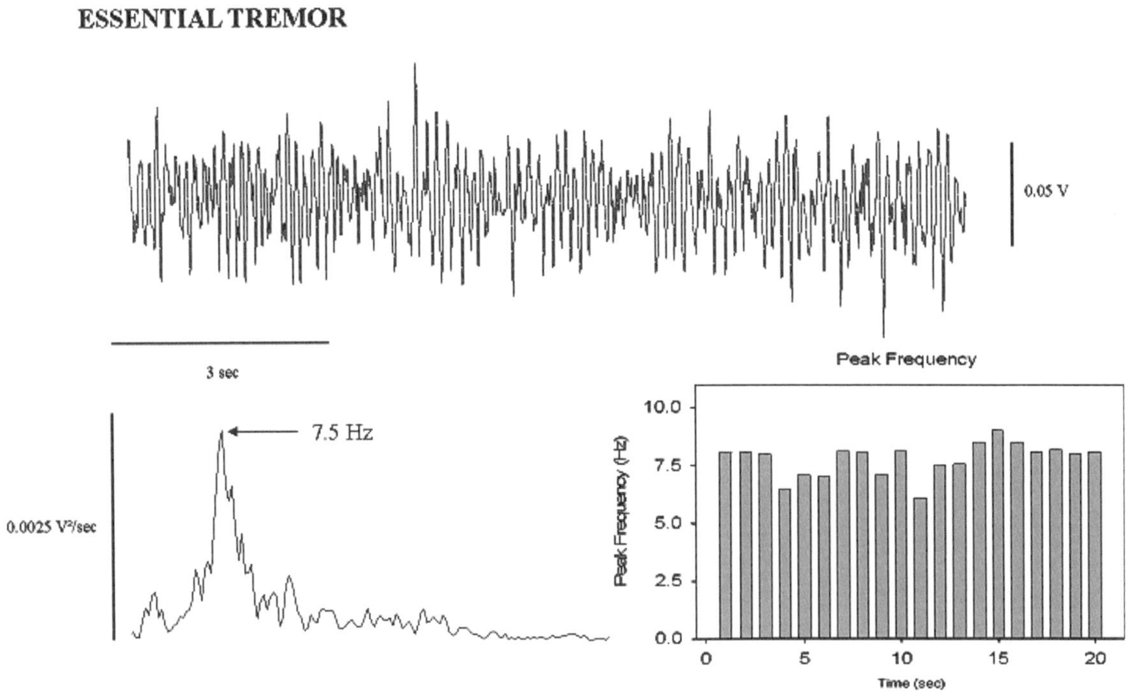

FIGURE 4.2: Monoaxial accelerometry in essential tremor. The patient is asked to maintain the upper limbs horizontally parallel to the floor. A peak frequency of 7.5 Hz is clearly identified on the power spectrum.

pathways result in different kinds of postural tremor (see Table 4.3). An example of postural tremor associated with a cerebellar stroke is illustrated in Figure 4.3. Acute or subacute postural tremor may result from cortical lesions such as mass occupying lesions, ischemic lesions, and arteriovenous malformations (Balci et al., 2007). Postural tremor of the upper limbs is a common manifestation of Wilson's disease, very likely generated within a synchronized cerebellothalamocortical network, comprising the primary sensory–motor cortex, higher cortical motor areas (supplementary motor area, premotor cortex), and posterior parietal cortex (Südmeyer et al., 2006).

4.4 KINETIC TREMOR

Tremor appearing during limb movement and often worsening near the target is defined as a kinetic tremor. Tremorous movements are perpendicular to the main direction of the intended movement and tend to be predominant over proximal musculature. The terms "intention tremor" and "terminal tremor" have also been used in the past. Kinetic tremor is classically tested during finger-to-nose or

TABLE 4.3: Cerebellar postural tremors

TREMOR TYPE	PATHOLOGY	PRECIPITANTS	FREQUENCY (Hz)	DISTRIBUTION
Midbrain tremor	Cerebellar outflow tracts Nigrostriatal lesion	Any posture	2–5	Distal > proximal
Asthenic cerebellar tremor	Cerebellar hemisphere	Fatigue/weakness	Irregular	Proximal + distal
Precision cerebellar tremor	Cerebellar nuclei and brachium conjunctivum	Accurate placements	2–5	Distal
Cerebellar axial postural tremor	Cerebello-olivary system	Any posture	2–10	Proximal > distal
Cerebellar proximal exertional tremor	Cerebellar malformation	Prolonged exercise	3–4	Proximal > distal

Adapted from Brown et al. (1997).

knee–tibia tests. Tremor occurring in cerebellar diseases is the typical example of a kinetic tremor, affecting especially the head and the upper part of the body. The frequency is between 2 and 7 Hz in the large majority of cases. In case of cerebellar injury, kinetic tremor may increase secondarily, involving especially in the proximal joints (shoulder). Addition of inertia tends to improve kinetic tremor. A recent study on the differences between smokers and nonsmokers in terms of kinetic hand tremor showed a greater tremor intensity in smokers, even more apparent in women (Louis, 2007).

FIGURE 4.3: Postural tremor in a patient presenting a cerebellar stroke in the territory of the superior cerebellar artery. (Top) Axial MRI of the posterior fossa: stroke (arrow) in the territory of the superior cerebellar artery. B: brainstem, C.H.: cerebellar hemisphere, V: vermis. The lesion involves the outflow tract of the cerebellar nuclei. (Bottom) Data from a monoaxial accelerometer and from a surface electromyographic (EMG) sensor are shown, as well as the corresponding power spectra. Note the waveform aspect, which is characterized by asymmetry.

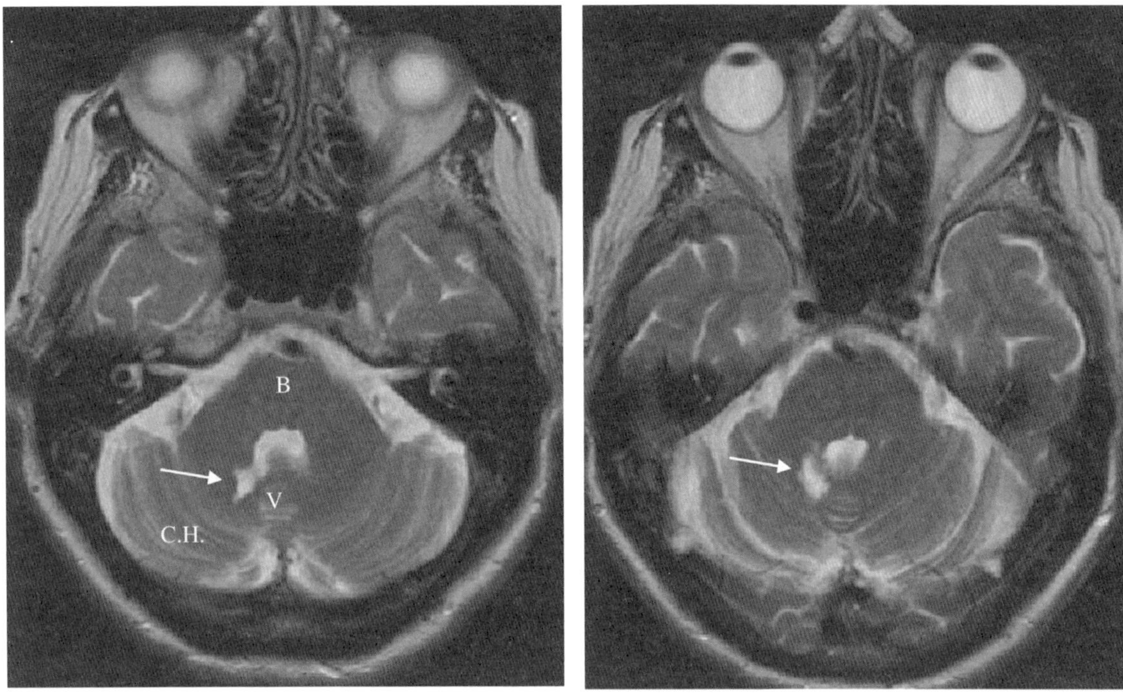
B
V
C.H.

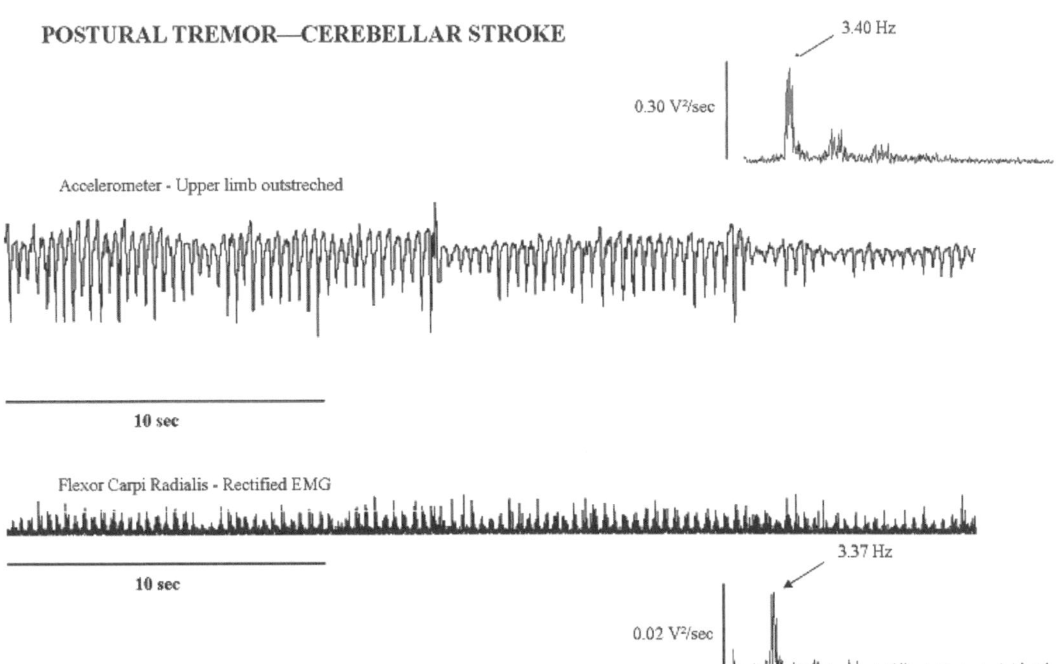
POSTURAL TREMOR—CEREBELLAR STROKE
3.40 Hz
0.30 V²/sec
Accelerometer - Upper limb outstreched
10 sec
Flexor Carpi Radialis - Rectified EMG
3.37 Hz
10 sec
0.02 V²/sec

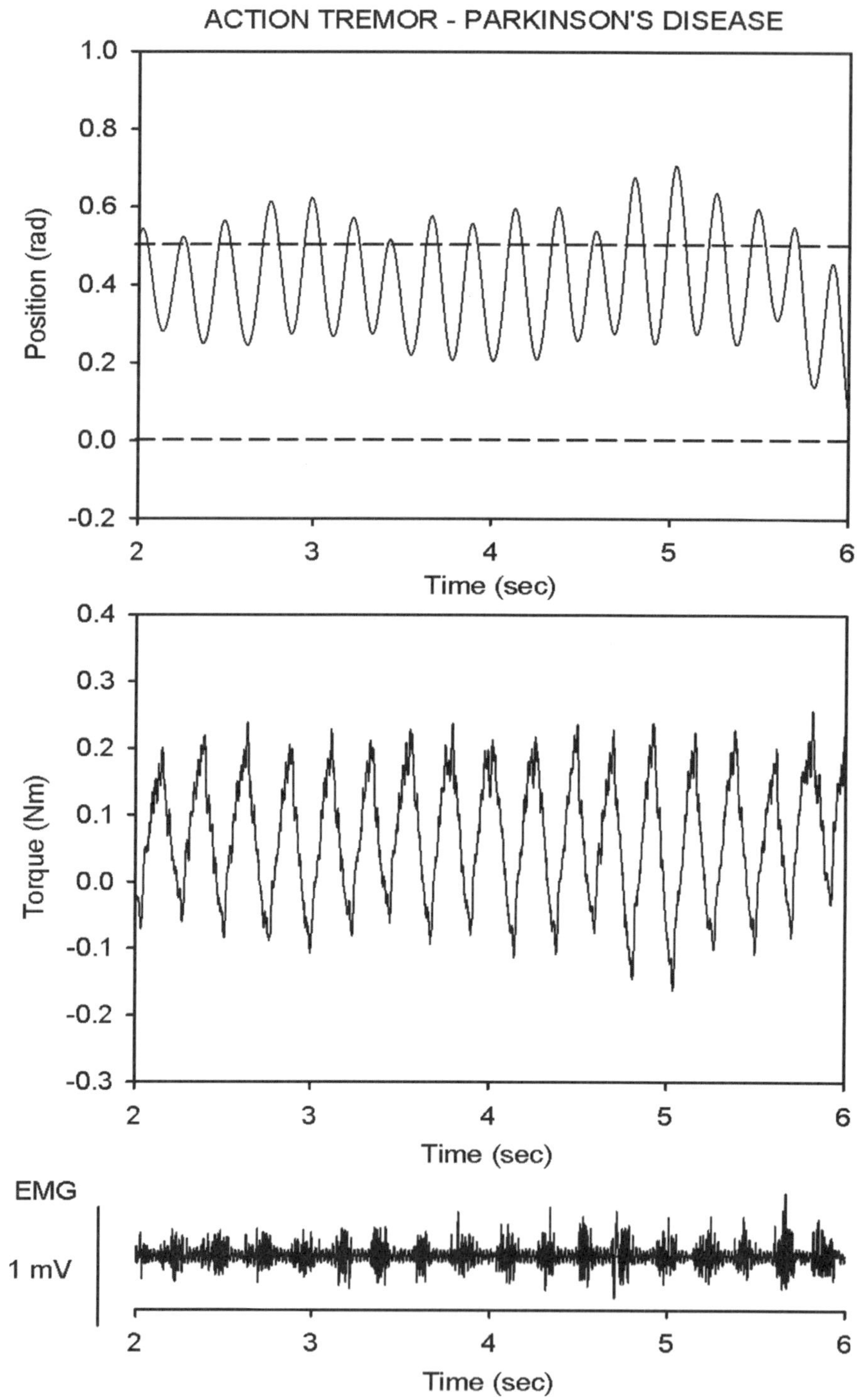

FIGURE 4.4: Action tremor of the right wrist in a patient presenting Parkinson's disease. (Top) Position. The tremor is well discernible. (Middle) Simultaneous torques. (Bottom) Surface EMG activities of the extensor carpi radialis (right side).

4.5 ACTION TREMOR

"Action tremor" refers to any tremor produced by voluntary contraction of muscles. It includes postural, isometric, and kinetic tremor.

Isometric tremor occurs when a voluntary muscle contraction is opposed by a rigid stationary object (Findley and Koller, 1995). Prolonged action-induced clonus, in the case of lesions of the central motor pathways, may mimic action tremor (Fraix et al., 2008).

4.6 OTHER FORMS OF TREMOR

4.6.1 Psychogenic Tremor

Psychogenic tremor is the most common form (55 %) of psychogenic movement disorder (Bhatia and Schneider, 2007). Psychogenic tremor is usually made of a combination of rest, postural, and

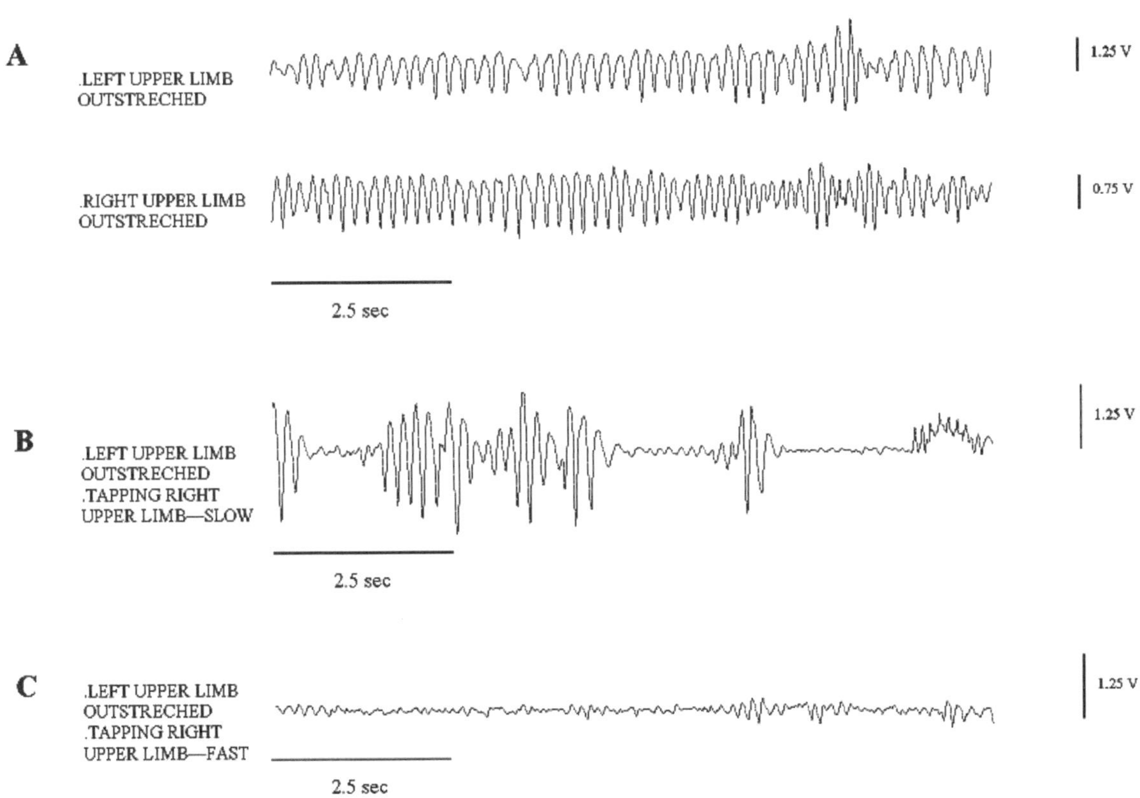

FIGURE 4.5: Psychogenic postural tremor. (a) Accelerometric measurement in both upper limbs during a postural task. Tremor appears regular bilaterally. (b) When the patient performs slow tapping movements with the right upper limb, tremor in the left upper limbs becomes highly variable in intensity and frequency. (c) During fast tapping of the right upper limb, the intensity of tremor decreases markedly on the contralateral upper limb.

kinetic components. It often involves preferentially the upper limbs, followed by the head and legs. It may be continuous or intermittent with typical fluctuations in frequency and amplitude. Psychogenic tremor is often characterized by an abrupt onset, periods of spontaneous remissions, unresponsiveness to antitremor drugs, and absence of concomitant neurological signs (Bhidaysiri, 2005). Preceding events include work-related injuries or other accidents. Psychogenic tremor is influenced by distraction and is entrained at the frequency of a rhythmic movement in the opposite limb (entrainment test), a maneuver which is helpful in daily practice to distinguish a pseudotremor. Measurement with accelerometers of the frequency changes during tapping in a group of psychogenic tremor patients (compared with ET and PD patients) showed large tremor frequency changes and a high intraindividual variability with tapping (Figure 4.5; Zeuner et al., 2003). Comorbidity with psychiatric disorders is common.

4.6.2 Task-Specific Tremor

Task-specific tremor occurs during given actions, for instance, writing representing in this case the primary writing tremor (see also Chapter 5).

4.6.3 Dystonic Tremor

Dystonic tremor is mainly a postural and kinetic tremor in an extremity or in a body part affected by dystonia. It is absent at rest in about half of the cases. Its frequency is typically irregular, varying from 4 to 9 Hz. Amplitude is unsteady. It is usually asymmetrical and remains localized. A typical example is the tremulous spasmodic torticollis. Patients affected may use the "geste antagoniste" or sensory tricks to reduce the tremor (Bhidayasiri, 2005).

Jedynak et al. (1991) studied the clinical and electromyographic characteristics of tremor in patients with idiopathic dystonia. They characterized dystonic tremor as follows: postural, localized, and irregular in amplitude and periodicity, absent during muscle relaxation, exacerbated by smooth muscle contraction, and frequently associated with myoclonus. Although it looks like ET, dystonic tremor is a distinct entity. Dystonic tremor is more irregular with a broader range of frequencies and is usually asymmetrical.

4.6.4 Holmes' Tremor

Holmes' tremor (also called midbrain tremor) is a symptomatic tremor affecting predominantly proximal segments. It has a low frequency (<4.5 Hz). Oscillations are present at rest but worsen during movement and goal-directed tasks. A delay between 4 weeks to 2 years is commonly observed between lesion onset and occurrence of tremor, suggesting a rearrangement of central pathways in the brain or an aberrant plasticity. Lesions are most often located in the upper brain stem, thalamus, and cerebellum, interrupting the pathways in the midbrain tegmentum (Guillain–Mollaret triangle, rubrospinal fibers, nigrostriatal fibers) and the serotoninergic brainstem telencephalic fibers. A unique case of reversible Holmes' tremor associated with hyperglycemia was recently described (Tan et al., 2006).

4.6.5 Cortical Tremor

Cortical tremor is an improper term that actually refers to a specific rhythmic myoclonus. This kind of tremor is characteristic of the familial cortical myoclonic tremor, a rare disorder often misdiagnosed as ET (Bourdain et al., 2006; see Chapter 5).

4.6.6 Orthostatic Tremor

Orthostatic tremor (OT) is a high-frequency tremor (13–18 Hz) predominantly in the legs and trunk (Figure 4.6). The term was first introduced in 1984 as a tremor involving mainly the legs and trunk (Heilman, 1984), although the first description was reported in 1970 (Pazzaglia et al., 1970). OT is absent when the patient walks, is sitting, or lies down. OT is triggered during isometric contraction of the limb muscles (Piboolnurak et al., 2005). OT is divided into primary orthostatic tremor (POT; no lesion detected with imaging techniques) and secondary orthostatic tremor (SOT; see Chapter 5). SOT is different from the paroxysmal high-frequency tremor occurring in up-

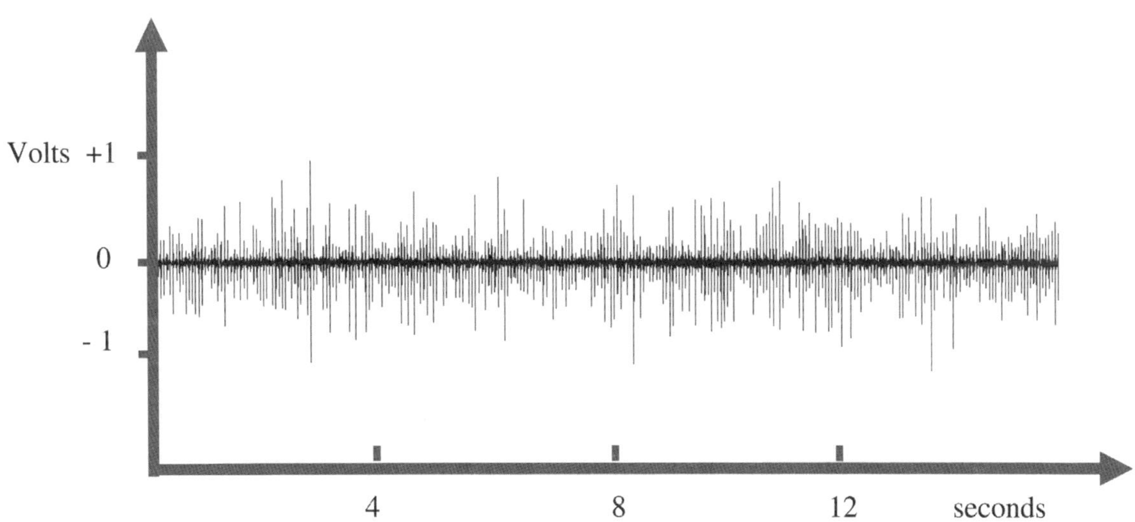

FIGURE 4.6: Orthostatic tremor detected by surface EMG recordings in lower limb muscles during a standing tasks. Amplifier gain: 1000. Frequency of discharges: 17 Hz.

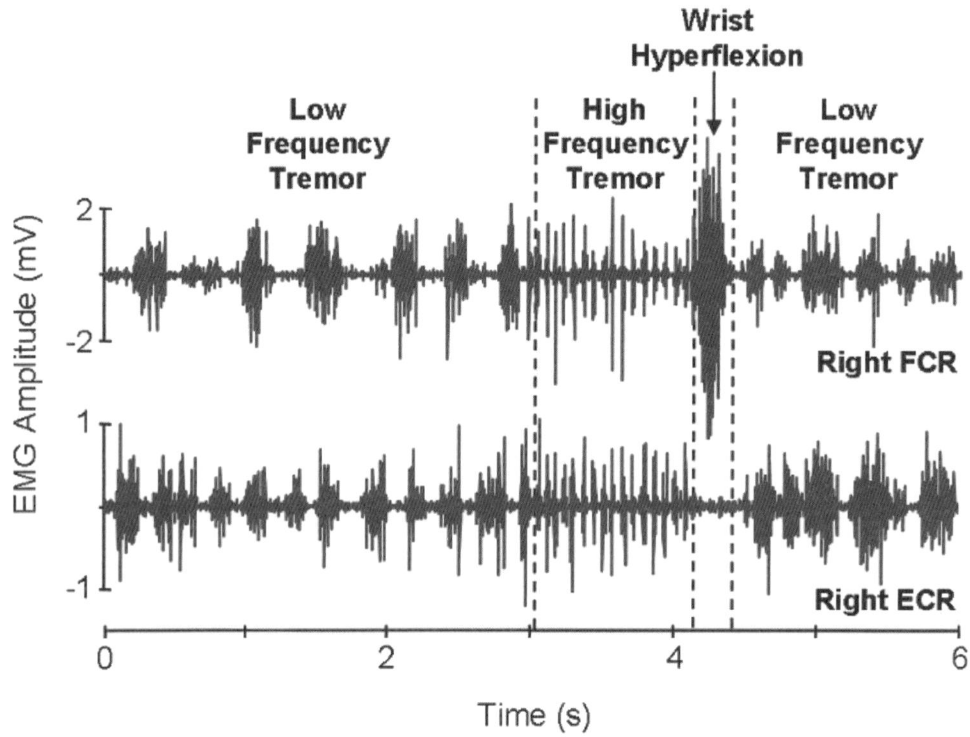

FIGURE 4.7: Episodes of bilateral high-frequency synchronous discharges (BHFSD) occurring in a patient with atrophy of the posterior fossa. The patient exhibits a low frequency tremor followed by paroxysmal synchronous bursts of electromyographic (EMG) activities in the right flexor carpi radialis (FCR) and right extensor carpi radialis (ECR), probably resulting from a sudden activation of cerebellar reverberating circuits. High-frequency bursts are terminated by a forceful wrist hyperflexion. Activities in the FCR and ECR muscles are detected using surface EMG electrodes.

per limbs in patients with posterior fossa lesions (bilateral high-frequency synchronous discharges BHFSD; Figure 4.7). BHFSD affect the upper limbs, occur in a context of low-frequency tremor (3–5 Hz), are intermittent, and are terminated by a forceful flexion of the wrist.

4.6.7 Palatal Tremor

Palatal tremor is a rhythmic involuntary movement of the soft palate (Deuschl et al., 1994) not abolished by sleep, with a range of frequency of 0.3–100 Hz (usually in the range 1.5–3 Hz). Patients often complain of clicking in the ears. Synchronous jerks may affect simultaneously the eyes, the face, the larynx, the neck, the shoulder, and the diaphragm (palatopharyngolaryngo-oculodiaphragmatic myoclonus"; Guillain, 1938). Also known as "palatal myoclonus," palatal tremor is either essential or secondary to a lesion in the brainstem or along cerebellar pathways (Guillain–Mollaret triangle). Ciprofloxacin-induced palatal tremor has been described recently (Cheung et al., 2007).

4.6.8 Tremor After a Peripheral Nerve Injury

Tremor after a peripheral nerve injury appears after a lesion of the peripheral nervous system (Costa et al., 2006). Neurogenic changes may be detected clinically or with neurophysiological investigations in the territory of the damaged nerve (nerve conduction velocities, needle EMG studies, somatosensory-evoked potentials). Concomitant dystonia is present in about 30% of the cases.

REFERENCES

Balci K, Utku U, Cobanoglu S. Two patients with tremor caused by cortical lesions. *Eur Neurol.* 2007;57(1):36–38.

Benito-León J, Louis ED. Essential tremor: emerging views of a common disorder. *Nat Clin Pract Neurol.* 2006;2(12):666–678.

Bhatia KP, Schneider SA. Psychogenic tremor and related disorders. *J Neurol.* 2007;254(5):569–574.

Bhidayasiri R. Differential diagnosis of common tremor syndromes. *Postgrad Med J.* 2005;81:756–762.

Bourdain F, Apartis E, Trocello JM, Vidal JS, Masnou P, Vercueil L, Vidailhet M. Clinical analysis in familial cortical myoclonic tremor allows differential diagnosis with essential tremor. *Mov Disord.* 2006;21(5):599–608.

Brown P, Rothwell JC, Stevens JM, Lees AJ, Marsden CD. Cerebellar axial postural tremor. *Mov Disord.* 1997;12(6):977–984.

Cathers I, O'Dwyer N, Neilson P. Entrainment to extinction of physiological tremor by spindle afferent input. *Exp Brain Res.* 2006;171(2):194–203.

Cheung YF, Wong WW, Tang KW, Chan JH, Li PC. Ciprofloxacin-induced palatal tremor. *Mov Disord.* 2007;22(7):1038–1043.

Costa J, Henriques R, Barroso C, Ferreira J, Atalaia A, de Carvalho M. Upper limb tremor induced by peripheral nerve injury. *Neurology.* 2006;67(10):1884–1886.

Deuschl G, Toro C, Valls-Solé J, Zeffiro T, Zee DS, Hallett M. Symptomatic and essential palatal tremor. 1. Clinical, physiological and MRI analysis. *Brain.* 1994;117(Pt 4):775–788.

Elble RJ. Characteristics of physiologic tremor in young and elderly adults. *Clin Neurophysiol.* 2003;114(4):624–635.

Findley LJ, Koller WC. *Handbook of Tremor Disorders.* New York: Marcel Dekker, 1995.

Fraix V, Delalande I, Parrache M, Derambure P, Cassim F. Action-induced clonus mimicking tremor. *Mov Disord.* 2008;23(2):285–288.

Growdon W, Ghika J, Henderson J, van Melle G, Regli F, Bogousslavsky J, Growdon JH. Effects of proximal and distal muscles' groups contraction and mental stress on the amplitude and frequency of physiological finger tremor. An accelerometric study. *Electromyogr Clin Neurophysiol.* 2000;40(5):295–303.

Guillain G. The syndrome of synchronous and rhythmic palato-pharyngo-laryngo-oculo-diaphragmatic myoclonus. *Proc R Soc Med.* 1938;31:1031–1038.

Heilman KM. Orthostatic tremor. *Arch Neurol.* 1984;41(8):880–881.

Jedynak CP, Bonnet AM, Agid Y. Tremor and idiopathic dystonia. *Mov Disord.* 1991;6(3):230–236.

Krause WL, Leiter JC, Marsh Tenney S, Daubenspeck JA. Acute hypoxia activates human 8–12 Hz physiological tremor. *Respir Physiol.* 2000;123(1–2):131–141.

Lakie M, Frymann K, Villagra F, Jakeman P. The effect of alcohol on physiological tremor. *Exp Physiol.* 1994;79(2):273–276.

Louis ED. Kinetic tremor: differences between smokers and non-smokers. *Neurotoxicology.* 2007;28(3):569–575.

Manto MU, Pandolfo M. *The Cerebellum and Its Disorders.* Cambridge, UK: Cambridge University Press, 2002.

Marsden CD, Foley TH, Owen DAL, McAllister RG. Peripheral beta-adrenergic receptors concerned with tremor. *Clin Sci.* 1967;33:53–65.

Mehndiratta MM, Satyawani M, Gupta S, Khwaja GA. Clinical and surface EMG characteristics of valproate induced tremors. *Electromyogr Clin Neurophysiol.* 2005;45(3):177–182.

Miller LS, Lombardo TW, Fowler SC. Caffeine, but not time of day, increases whole-arm physiological tremor in non-smoking moderate users. *Clin Exp Pharmacol Physiol.* 1998;25(2):131–133.

Morrison S, Mills P, Barrett R. Differences in multiple segment tremor dynamics between young and elderly persons. *J Gerontol A Biol Sci Med Sci.* 2006;61(9):982–990.

Pazzaglia P, Sabattini L, Lugaresi E. Su di un singolare disturbo della stazione eretta (osservazione di tre casi). *Riv Frenatria.* 1970;96:450–457.

Piboolnurak P, Yu QP, Pullman SL. Clinical and neurophysiologic spectrum of orthostatic tremor: case series of 26 subjects. *Mov Disord.* 2005;20(11):1455–1461.

Raethjen J, Pawlas F, Lindemann M, Wenzelburger R, Deuschl G. Determinants of physiologic tremor in a large normal population. *Clin Neurophysiol.* 2000;111(10):1825–1837.

Song IU, Kim JS, Ryu SY, Lee SB, An JY, Lee KS. Donepezil-induced jaw tremor. *Parkinsonism Relat Disord.* 2008.

Soto J, Sacristan JA, Alsar MJ, Sainz C. Terfenadine-induced tremor. *Ann Neurol.* 1993; 33(2):226.

Südmeyer M, Pollok B, Hefter H, Gross J, Butz M, Wojtecki L, Timmermann L, Schnitzler A. Synchronized brain network underlying postural tremor in Wilson's disease. *Mov Disord.* 2006;21(11):1935–1940.

Tan JH, Chan BP, Wilder-Smith EP, Ong BK. A unique case of reversible hyperglycemic Holmes' tremor. *Mov Disord.* 2006;21(5):707–709.

Zeuner KE, Shoge RO, Goldstein SR, Dambrosia JM, Hallett M. Accelerometry to distinguish psychogenic from essential or parkinsonian tremor. *Neurology.* 2003;61(4):548–550.

• • • •

CHAPTER 5

Principal Disorders Associated with Tremor

In this chapter, we describe the diseases which may present with tremor as a main symptom.

5.1 ESSENTIAL TREMOR

Essential tremor (ET) is no more viewed as a monosymptomatic benign disorder, and its clinical spectrum includes both motor and nonmotor features (Benito-Leon and Louis, 2006; Thanvi et al., 2006). ET is one of the most common neurological disorders among adults and is the most common movement disorder in the elderly. Its prevalence is 4% in the population above 65 years of age, with a higher prevalence in White than in African-American populations. No gender differences are reported (Thanvi et al., 2006). A recent prospective, population-based study in Spain of individuals aged 65 years and older revealed an adjusted incidence of ET of 616 per 100,000 person-year (Benito-Leon and Louis, 2006). Incidence and prevalence of ET increase with aging, but it should be kept in mind that the disease can occur at any age. In a screening of 2227 people in Turkey, the prevalence of ET was 3.09% in subjects aged over 18 years (Sur et al., 2008). Several studies have reported an increased risk to develop Parkinson's disease in ET populations. The finding of hyperechogenicity of the substantia nigra—considered as a transcranial sonography risk marker—has been recently proposed as a predictive sign of such a risk (Stockner et al., 2007).

ET is a familial disease in many cases. One possible genetic pattern is a dominant mode of inheritance with an incomplete penetrance by the age of 65 years (Bain et al., 1994). So far, three genetic susceptibility loci (ETM1 on 3q13, ETM2 on 2p24.1, and a locus on 6p23) have been identified in families with the disorder (Deng et al., 2007). In addition, some ET patients have the gly9 susceptibility to the DRD3 gene (dopamine D3 receptor encoding region), which might cause an increased excitability of neuronal circuits. A recent study underlined the possibility of a deficiency of mtDNA multicomplexes in ET (Yoo et al., 2008). Nevertheless, the nonfamilial sporadic form of ET is recognized as well. The familial form of ET is characterized by an earlier age of onset than the sporadic form (Louis and Ottman, 2006).

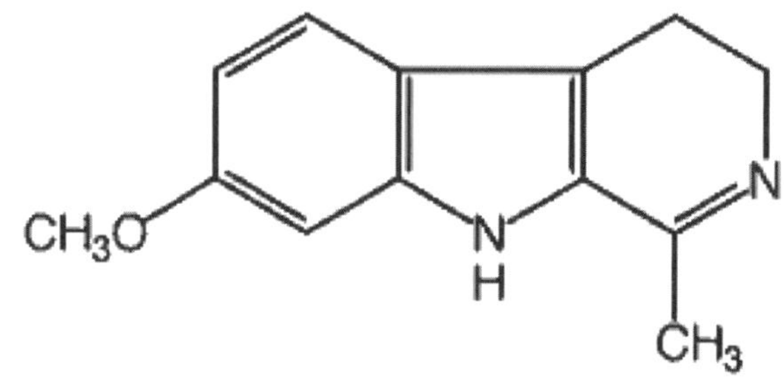

FIGURE 5.1: Structure of a beta-carboline alkaloid (BCA).

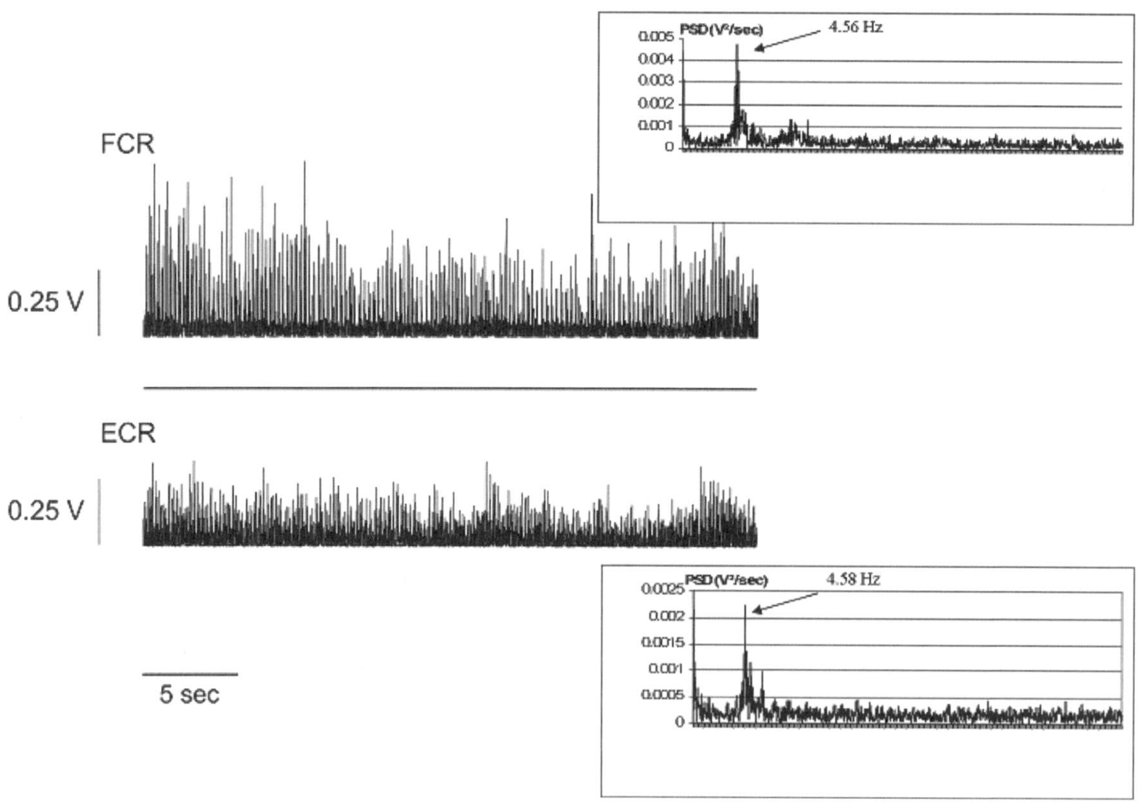

FIGURE 5.2: Postural tremor in the upper limbs during the task of maintaining the upper limbs outstretched parallel to the floor in a patient presenting essential tremor. (Left) Recordings in the right flexor carpi radialis (FCR) and extensor carpi radialis (ECR) using surface electromyography (EMG). Rectified traces are shown. (Right) Power spectra show peaks of frequency, respectively, at 4.56 and 4.58 Hz. Bursts of EMG activity are synchronous in the two antagonist muscles.

ET might be triggered by environmental factors (Jiminez-Jiminez et al., 2007). The β-carboline alkaloids (BCA, Figure 5.1) are the most studied toxic factors. These toxins are normally present in the human diet. Higher concentrations have been reported in blood of ET patients compared to control subjects (Louis et al., 2002).

ET is characterized by a bilateral postural tremor of moderate to severe amplitude at a frequency of 4–12 Hz (Figure 5.2). A kinetic tremor is often observed in association with the postural

TABLE 5.1: Clinical diagnostic criteria for definite ET

Inclusion criteria
Bilateral postural tremor with or without kinetic tremor, involving hands and forearms, that is visible and persistent
Duration >5 years
Exclusion criteria
Other abnormal neurological signs (except Froment's sign*)
Presence of known causes of increased physiological tremor
Concurrent or recent exposure to tremorogenic drugs or the presence of a drug withdrawal state
Direct or indirect trauma to the nervous system within 3 months before the onset of tremor
Historical or clinical evidences of psychogenic origins
Convincing evidence of sudden onset or evidence of stepwise deterioration
Probable ET
Tremor may be confined to one body part other than the hands, and duration is longer than 3 years
Possible ET
Other neurological disorders are present (parkinsonism) or other isolated tremors (task-specific tremor)

*Froment's sign: rhythmical resistance to passive movement of a limb about a joint that can be detected when there is a voluntary action of another body part.

oscillations (Brennan et al., 2002). A rest tremor may also be present, usually in the case of a longer disease duration (Cohen et al., 2003). A bimodal distribution of tremor frequencies is relatively common in ET.

Under the same conditions of observation, there is usually little variation in frequency for postural tremor of the hands (usually <0.6 Hz). With time, tremor amplitude tends to increase, whereas the frequency decreases (Elble et al., 1994). ET appears most frequently in the hands, followed by the head, voice, tongue, legs, and trunk. Typically, a more or less symmetrical tremor affects the extremities and progresses medially. Head tremor is milder than limb tremor and usually is of the side-to-side type. Alcohol intake temporary reduces ET in up to 75% of patients (Paulson, 1892; Lou and Jankovic, 1991). Diagnosis of ET is mainly clinical, based on medical history and physical examination. Clinical diagnostic criteria for definite, probable, or possible ET, according to the Consensus statement of the Movement Disorder Society on Tremor, are reported in Table 7 (Deuschl et al., 1998).

5.2 IDIOPATHIC PARKINSON'S DISEASE

Parkinson's disease (PD) was originally described by James Parkinson in 1817. PD is a progressive neurodegenerative disorder with an estimated prevalence of 0.3% in the U.S. population (Rao et al., 2006). The Rotterdam Study revealed a prevalence increasing to 4 to 5% in subjects older than 85 years (de Lau et al., 2004).

Pesticides (Dick, 2006) and increased body mass index are probably associated with a higher risk of PD (Hu et al., 2006). By contrast, coffee and tea drinking might be associated with a lower risk (Hu et al., 2007).

The majority of PD cases are sporadic. However, a classical Mendelian pattern of inheritance has been demonstrated in several families. Several genes linked to familial forms of disease (encoding alpha-synuclein, parkin, DJ-1, PINK-1, and LRRK2) have been identified. These families show either recessive or dominant inheritance patterns and may exhibit typical and/or atypical symptoms, with an age of onset extending from the second to the sixth decade (Thomas and Beal, 2007; MacInerney-Leo et al., 2005).

5.2.1 Motor Tests

Distal resting tremor of 3 to 6 Hz, rigidity (sustained increase of resistance throughout the range of passive movement about a joint), bradykinesia, and asymmetrical onset are cardinal signs. Other well-known signs include late-onset postural instability and micrographia. Response to an adequate therapeutic challenge of levodopa or a dopamine agonist is one of the key features for the diagnosis (guidelines for the diagnosis of Parkinson's disease, 2003). Parkinsonian patients often present an abnormal posture called camptocormia (ranging from mild to severe), which is characterized by an excessive anterior flexion of the spine (Bonneville et al., 2008). Bradykinesia is characterized by a difficulty to perform rapid and sequential tapping of the fingers or the heels, difficulty twiddling or circling the hands rapidly around each other in front of the body, reduced arm swing on the affected side during deambulation. Small, shuffling steps and troubles in initiating ambulation are common. Involuntary acceleration of gait results in the typical festinating gait. Heel-to-toe ambulation is impaired (Rao et al., 2006). Rigidity of the muscles on passive movement is characteristic of PD. Unlike spasticity (resulting from upper motor neuron lesions), the passive movement of the joints reveals continuous resistance throughout the entire range of motion, the so-called lead-pipe rigidity. In addition, patients with PD may show a cogwheel type of rigidity. Rigidity and cogwheel phenomenon may be triggered or enhanced by contralateral movement (Guttman et al., 2003).

Typically, the parkinsonian tremor is asymmetrical, at least initially, and affects the upper limb before involving the ipsilateral leg after a period of about 2 years. Tremor of the lips, jaw, or tongue may also occur. Head or voice tremor is rare, unlike in ET. A "pill-rolling" resting tremor is characteristic of PD. Postural tremor is also present in most cases. In many patients, two separate peaks of tremor may be distinguished by spectral analysis, and this pattern is considered very suggestive of basal ganglia disease by some authors (Habib-ur-Rehman, 2000). Kinetic tremor has been also described in PD, albeit uncommonly (Kraus et al., 2006). Patients affected by a Parkinson-plus syndrome, in particular, multiple system atrophy (MSA), often exhibit a slight to marked kinetic tremor during the course of the disease.

A study on the relation between center of pressure excursions and postural/resting limb tremor in PD patients, pointed out that the amplified tremor in PD can manifest itself in the center of pressure dynamics and influence postural stability (Kerr et al., 2008).

5.2.2 Nonmotor Tests

Decreased scores in cognitive tests are associated with greater impairment in motor performances (Verbaan et al., 2007). Among the symptoms of autonomic failure, orthostatic dizziness, bladder dysfunction, and constipation are considered to have great impact on daily life (Magerkurth et al., 2005). A decreased olfactory function has been reported.

5.2.3 Neuropathology

PD is due to the degeneration of dopaminergic neurons. Oxidative stress and mitochondrial dysfunction have been implicated as important contributors to neuronal death in substantia nigra (SN) of PD (Fukae et al., 2007). However, other neurotransmitter systems such as the serotonergic

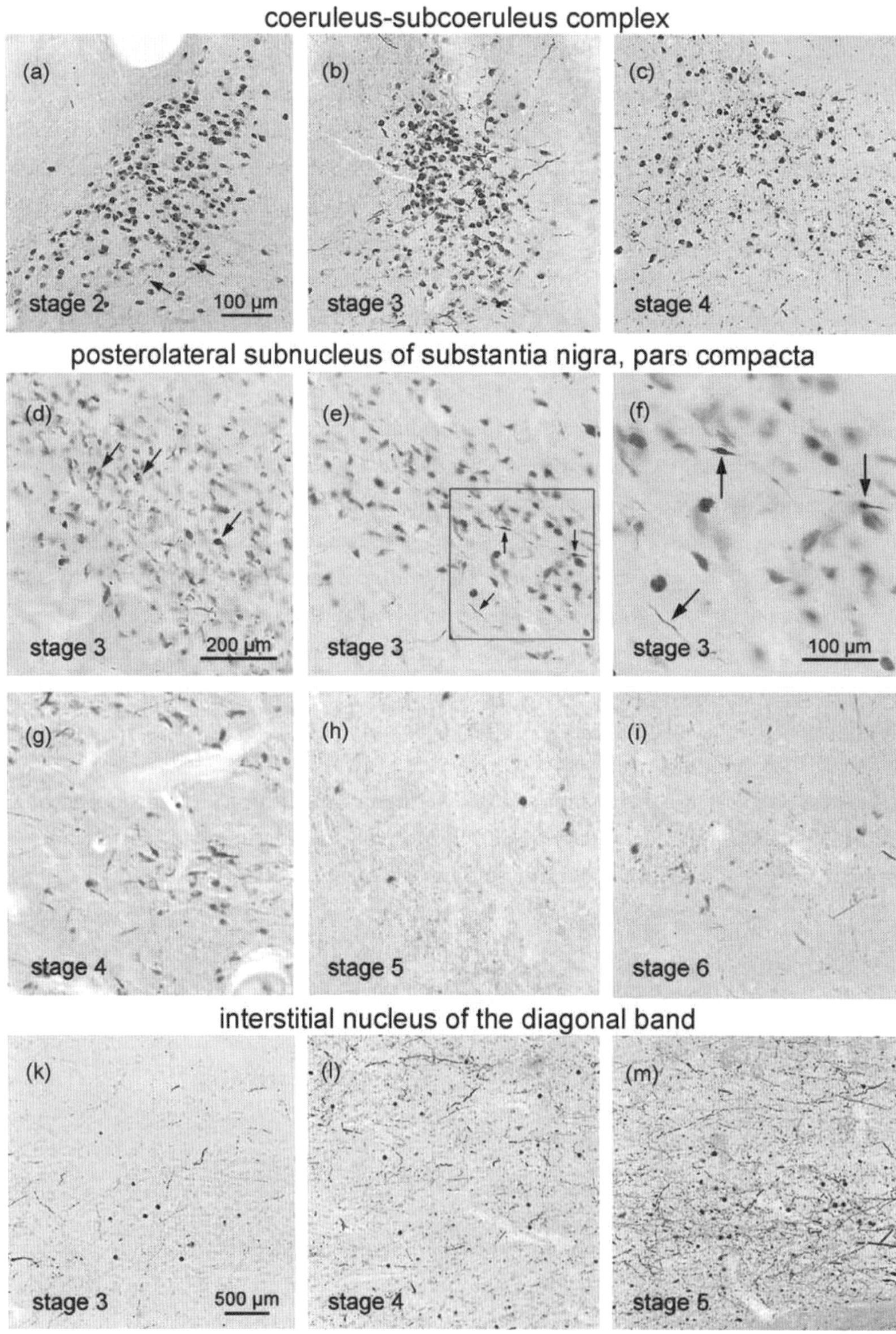
coeruleus-subcoeruleus complex
(a)
stage 2
100 µm
(b)
stage 3
(c)
stage 4
posterolateral subnucleus of substantia nigra, pars compacta
(d)
stage 3
200 µm
(e)
stage 3
(f)
stage 3
100 µm
(g)
stage 4
(h)
stage 5
(i)
stage 6
interstitial nucleus of the diagonal band
(k)
stage 3
500 µm
(l)
stage 4
(m)
stage 5

system also show signs of degeneration (Scholtissen et al., 2006). Progressive loss of SN neurons and presence of Lewy bodies are essential neuropathological features of PD. Recent neuropathological studies suggest that nigral degeneration is only part of a more extended brain degeneration that starts in the medulla oblongata and then spreads to the mesencephalon and cerebral cortex (Figure 5.3). Correspondingly, the clinical symptoms occurring in PD go far beyond parkinsonism. As mentioned earlier, autonomic dysfunction, olfactory disturbances, depression, and even dementia are frequently encountered in PD (Klockgether, 2004).

5.2.4 Differential Diagnosis

A list of extrapyramidal syndromes associated with tremor is given in Table 5.2. Parkinson plus syndromes, mainly progressive supranuclear palsy (PSP), multiple system atrophy (MSA), corticobasal ganglionic degeneration (CBD), Lewy body disease (LBD), vascular Parkinsonism, and Parkinsonism with no clear etiology are distinct from PD (Thanvi et al., 2005; Weiner, 2005). Parkinsonism on one side of the body in association with a contralateral atrophy of brain structures points towards the hemiparkinsonism–hemiatrophy syndrome (HPHA; Klawans, 1981). These patients are relatively young, exhibit a unilateral rest tremor, and may have a history of dystonia.

Data collected in an Italian center (over 2900 parkinsonian patients) showed that PD accounts for no more than 70% of all parkinsonisms (Pezzoli et al., 2004). Clinicopathological studies based on brain bank material from the UK and Canada have shown that clinicians diagnose the disease incorrectly in about 25% of patients (Tolosa et al., 2006).

FIGURE 5.3: Development of PD-related brain stem lesions. (a) The coeruleus–subcoeruleus complex is mildly involved at stage 2 [arrows point to Lewy neurites (LNs)]. (b) Severely affected at stage 3 (note the increase in LNs). (c) Severely depleted of melanin-containing projection neurons at stage 4. Bar in (a) is valid for (b and c). (d) The posterolateral and, thereafter, the posteromedial subnuclei of the substantia nigra, pars compacta, become first involved at stage 3 [arrows point to Lewy bodies (LBs) in melanized neurons]. Elongated LNs with spindle-shaped enlargements at irregular intervals (arrows) occur in the vulnerable subnuclei [the framed portion of (e) appears at higher magnification in (f)]. (g) Loss of melanized nerve cells is severe at stage 4. The vulnerable subnuclei are almost depleted of projection neurons at stages 5–6 (h and i). It is difficult to find surviving LB-containing nerve cells in these portions of the substantia nigra. Bar in (d) is valid for (e) and (g–i). (k) Like the substantia nigra, the interstitial nucleus of the diagonal band begins to develop LNs and LBs at stage 3. The density of both LNs and LBs increases in the following stages 4–5 (l–m). Bar in (k) is valid for (l and m). α-Synuclein immunoreactions, 100-μm thick PEG-sections. From Braak et al. (2003), with permission from Elsevier.

TABLE 5.2: Extrapyramidal syndromes associated with tremor
Parkinson's disease
"Parkinson-plus" (striatonigral degeneration SND, progressive supranuclear palsy PSP, multiple system atrophy MSA, olivopontocerebellar atrophy OPCA)
Neuroleptic-induced parkinsonism
Stroke in basal ganglia
Wilson's disease
Dystonia
Normopressure hydrocephalus
Midbrain lesions
Huntington's disease
Repeated head trauma (boxers, "punch-drunk syndrome")
Intoxication (CO, cyanide, manganese)
Spinocerebellar ataxias
Hallevorden–Spatz disease
Basal ganglia calcifications (hyperparathyroidism)
Alzheimer's disease
Parkinsonism dementia complex
Psychogenic

5.2.5 Neuroimaging

Positron emission tomography (PET) and single-photon emission computed tomography (SPECT) allow the in vivo assessment of the neurochemical, hemodynamic, or metabolic consequences of the degeneration of the nigrostriatal dopaminergic system in PD. The extent of striatal dopaminergic denervation can be quantified with radiotracers. Striatal uptake is markedly decreased in PD, affecting more the putamen than the caudate nucleus. This is inversely correlated with the severity of motor signs and with the duration of the disease (Thobois et al., 2001).

Hyperactivation in the cerebellum and motor cortex and hypoactivation in the basal ganglia in patients with PD have been described by neuroimaging studies. Recent evidence support the hypothesis that hyperactivation in the ipsilateral cerebellum is a compensatory mechanism for the defective basal ganglia. Moreover, hyperactivation in the contralateral primary motor cortex might not be just a compensatory response but seems directly related to upper limb rigidity (Yu et al., 2007).

5.3 CEREBELLAR DISEASES

Pathological conditions usually affect cerebellar function through several mechanisms: decrease in blood flow, edema, mechanical compression, invasion of cerebellar parenchyma, inflammatory and/or immune processes, and direct cytotoxic effect (Manto, 2002). The entities most commonly encountered in daily practice are cerebellar stoke, neurodegenerative ataxias (dominant spinocerebellar ataxias, Friedreich's ataxias, and other recessive forms), paraneoplastic cerebellar syndromes, cerebellitis, multiple sclerosis, multiple systemic atrophy (MSA), posttraumatic, metabolic, or toxic cerebellar ataxias, psychogenic ataxias.

Facing cerebellar disorders, some general principles should to be kept in mind:

1. focal cerebellar lesion located laterally generate signs ipsilaterally (see Chapter 2);
2. diffuse cerebellar disease, such as degenerative ataxias, usually cause relatively symmetric deficit;
3. cerebellar signs due to nonprogressive disease tend to undergo attenuation with time;
4. symptoms suggestive of cerebellar damage may be encountered in patients presenting lesions along the efferent or afferent cerebellar pathways outside the cerebellum (Manto, 2002).

As a rule, the symptoms of cerebellar disorders are influenced more by the location and the rate of progression of the disease than by the underlying pathology. As a consequence, an infection and a tumor, for instance, may produce similar cerebellar syndrome if they share the same location and rate of progression (Manto, 2002). The most common symptoms are gait difficulties, unsteadiness, headache, nausea/vomiting, dizziness, clumsiness in the limbs, speech difficulties, tremor, blurred vision, diplopia, feebleness, and sensory complaints.

Clinical signs may be a clue to the cerebellar area affected. Oculomotor deficits associated with ataxia of stance and gait indicate a *middle zone disease*. *Lateral lesions* are more likely to produce a combination of limb dysmetria (Figure 5.4), kinetic tremor, hypotonia, dysdiadochokinesia, impaired check, and excessive rebound. Isolated dysarthria has been reported in *intermediate zone* damage. Moreover, cerebellar dysarthria—characterized by scanning speech—is also observed in lesions located in the *cerebellar hemispheres* (Manto, 2002).

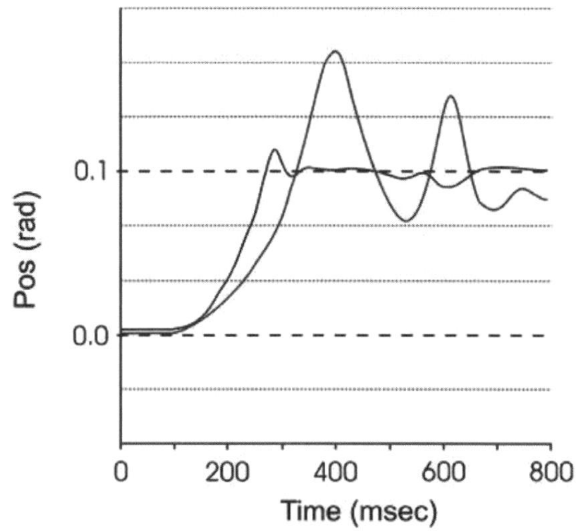

FIGURE 5.4: Overshoot in a cerebellar patient. The figure illustrates a typical cerebellar hypermetria (blue line) of the right hand in a right-handed patient presenting a severe cerebellar atrophy and exhibiting clinically a marked axial and appendicular cerebellar syndrome with oculomotor ataxia, dysmetria of limbs, and gait ataxia. The hypermetric movement is followed by attempts to reach the aimed target. Normometria in a control subject (black line). Aimed target: 0.1 rad.

Tremor associated with cerebellar diseases is mainly a low-frequency kinetic tremor. Many cerebellar patients exhibit a postural tremor. Isometric tremor as well as titubation may occur (Rondot and Bathien, 1995). The tremor may be bilateral or most often ipsilateral to the cerebellar lesion. In case of postural tremor, eye closure and body displacements enhance the oscillations. Atrophy of the anterior lobe of the cerebellum may be associated with a very suggestive 3-Hz leg tremor (Figure 5.5). The majority of these patients have a history of alcohol abuse (see also Section 4 below).

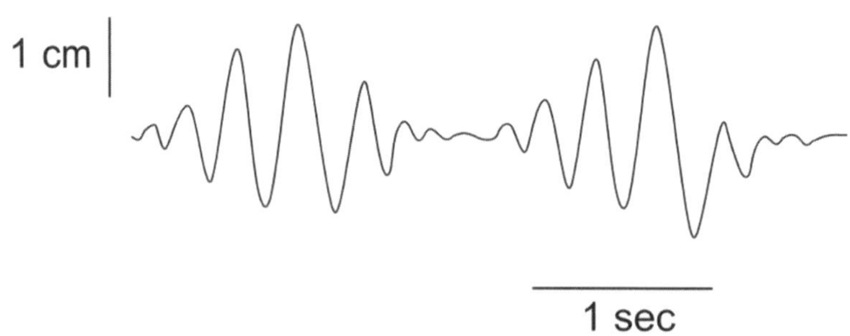

FIGURE 5.5: A 3-Hz leg tremor in a patient presenting atrophy of the anterior lobe of the cerebellum secondary to excessive alcohol consumption. Tremor is recorded during a sustained leg elevation with 90° flexion of the knee and hip joints, while the patient is supine. Tremor has a typical waxing and waning amplitude with a spindle-like aspect and predominates in one leg. Recordings from the right ankle.

TABLE 5.3: Classes of drugs and toxic agents inducing a postural and/or kinetic tremor

A. Anticonvulsants: Phenytoin Carbamazepine overdose	C. Other drugs: Lithium salts Amiodarone Procainamide Cyclosporin and calcineurin inhibitors Bismuth Bromides/ bromvalerylurea Mefloquine Isoniazid Lindane Perhexiline maleate Metronidazole Glycoprotein IIb/IIIa inhibitors Nicotine	D. Drug abuse and addiction: Cocaine Heroin Phencyclidine Ceremonial herbs (kava) Methadone	F. Toluene/ Benzene Derivatives G. Carbon monoxide H. Chemical weapons: Diphenylarsinic acid poisoning	J. Saxitoxin (shellfish poisoning) K. Cyanide L. Complications following scorpion sting
B. Antineoplastics: 5-FU (5-fluorouracil) Ara-C cytosine (arabinoside) Methotrexate Capecitabine Epothilone D		E. Metals: Mercury Lead Manganese Aluminium Thallium Germanium Uranium Vanadium	I. Insecticides/ herbicides Chlodecone: Paraquat Phosphin Carbon disulfide	

5.3.1 Cerebellar Syndrome Induced by Drugs or Toxic Agents

As mentioned in Chapter 4, many drugs may enhance tremor. We summarize in Table 5.3 the classes of drugs and toxic agents which may induce cerebellar syndromes presenting with postural and/or kinetic tremor.

5.4 MISCELLANEOUS

5.4.1 Tremor of the Globe

Tremor of the eyes has frequencies varying from 2 to 70 Hz because of the mechanical properties of the eye.

5.4.2 Jaw Tremor

Jaw tremor can be seen in various neurological disorders such as ET, PD, dystonia, branchial myoclonus, hereditary geniospasm, task-specific tremor, Whipple's disease (showing oculomasticatory myorhythmia: rhythmic contractions in the 1- to 3-Hz frequency range), as well as in normal situations such as shivering (hypothermia), and subclinical physiological jaw tremor (Gonzalez-Alegre et al., 2006). Frequency of jaw tremor is usually lower than 12 Hz.

5.4.3 Voice Tremor

Tremulousness of speech is not disease specific. It is encountered under physiological circumstances ("vibrato"; 4–7 Hz), in ET and PD (frequency 4–7 Hz), in amyotrophic lateral sclerosis (ALS). Speech analysis in ALS patients reveals tremor in about 60% of cases. Vocal cords display flaccid weakness, palatal/labial weakness. Tongue fasciculations may be observed because of denervation and loss of motor units, particularly in the bulbar form of ALS which affects predominantly cranial nerves (see Chapter 2). A rapid vocal flutter (7–10 Hz) may occur also (Aronson et al., 1992). Hypernasal resonation and imprecise articulation are common.

5.4.4 Multiple Sclerosis

Multiple sclerosis (MS) is a demyelinating disease of the central nervous system. Tremor is estimated to occur in about 25% to 60% of patients with MS (Koch et al., 2007). In MS, the most prevalent forms of tremor are postural tremor and kinetic tremor. Tremor of the head, neck, vocal cords, trunk, and limbs have all been described in association with MS. A true rest tremor is unusual (Koch et al., 2007). A case of Holmes tremor has been described recently (Yerdelen et al., 2008). So far, there are no reports of tremor of the palate, tongue, or jaw (Alusi et al., 1999).

5.4.5 Dystonic Tremor

Dystonic tremor occurs in generalized, segmental, and focal dystonias. In primary dystonia, tremor may anticipate the occurrence of clear signs of the disease. Secondary criteria like persistent focal tremor in one extremity, jerky and irregular tremors, gestes antagonistes, or selective responsiveness to antidystonic therapeutic agents are helpful to demarcate this entity. The occurrence of dystonic tremor in symptomatic cases is often observed after thalamic lesions, peripheral injuries, or in dystonia associated with sympathetic reflex dystrophy (Deuschl, 2003; see below).

Torticollis is the most common manifestation of dystonia (Jankovic et al., 1991). Even in the first description by Oppenheim, tremor was mentioned as a symptom associated with dystonic movements (Dubinsky, 1995). Isolated tremor of the trunk or head, especially of slow frequency (2–5 Hz; in the case of the head in a "no–no" direction), may be the initial manifestation of focal dystonia (Rivest and Marsden, 1990). A prospective study on 114 patients of cervical dystonia (CD) revealed that head tremor is common in CD and is more commonly associated with hand tremor (Pal et al., 2000).

The pathophysiology is not fully established (see Chapter 7). Impaired reciprocal inhibition and disturbances in other inhibitory reflex pathways at various levels are incriminated.

5.4.6 Orthostatic Tremor

Orthostatic tremor (OT) is a rare disorder of middle-aged or elderly people. OT is likely underdiagnosed. On average, there is a diagnostic delay of 5.7 years. In most cases, the condition remains relatively unchanged over the years, but symptoms may also spread proximally to involve the trunk and arms (Gerschlager et al., 2004). Cases of OT in brothers and in monozygotic twins were recently reported (Fischer et al., 2007; Contarino et al., 2006).

OT is characterized by unsteadiness on standing, which remits on walking or sitting (Piboolnurak et al., 2005). Lifting the standing patient off the ground abolishes the tremor. When walking, tremor disappears from the nonweight-bearing limb, but may persist in the weight-bearing leg and in truncal muscles (Heilman, 1984). Thereby, patients stand on a wide base but walk normally. Only a fine ripple of muscle activity is visible (Britton et al., 1992). Patients may describe a shaking sensation or cramps of the lower extremities upon standing. Typically, they are not able to stand in place for a long time because of unstability and the sensation of fatigue. A typical 14- to 18-Hz tremor of the lower limbs is pathognomonic (see Figure 4.6). Surface or needle EMG confirms the diagnosis.

Isometric contraction of arm and leg muscles triggers a high-frequency tremor in these patients (Boroojerdi et al., 1999). A single case report of orthostatic jaw tremor has been described (Schrag et al., 1999).

OT is divided into primary orthostatic tremor (POT) characterized by the absence of lesions detectable with imaging techniques and secondary orthostatic tremor (SOT), a symptomatic condition that has been described in hydrocephalus because of nontumoral aqueduct stenosis, in relapsing polyradiculoneuropathy (Gabellini et al., 1990), in pontine lesions (Benito-Leon et al., 1997), after head trauma (Sanitate and Meerschaert, 1993), as a paraneoplastic syndrome associated with small cell lung cancer (Gilhuis et al., 2005), in cerebellar diseases, and in the case of thiamine deficiency (Nasrallah and Mitsias, 2007). The possibility that hyperthyroidism might trigger OT has been raised recently (Tan et al., 2008).

5.4.7 Primary Writing Tremor

This is a task-specific tremor, mainly asymmetrical and focal, that interferes with handwriting. This tremor is often independent of posture or goal-directed motion. A bilateral presentation has been reported (Jimenez-Jimenez et al., 1998). Usually, it occurs alone. However, a concomitant postural tremor can be observed in some cases (Pahwa, 1995). Coexistence of tremor with dystonia has also been described (Elble et al., 1990). Although most of the patients have a negative family history for movements disorders, some relatives may be affected by ET, PD, (Pahwa in Hand book of tremor disorders), or dystonia (Hayashi and Koide, 1997).

Two subgroups are defined, depending on whether tremor appears during writing (type A: *task-induced tremor*) or according to the hand position used in writing (type B: *position-sensitive tremor*). The two groups differ also for the presence (type B) or absence (type A) of co-contracting EMG pattern and tremor induced by tendon taps to the volar aspect of the wrist (Bain et al., 1995).

Patients affected present significantly reduced writing speeds. A 4.1- to 7.3-Hz rhythmical activity predominantly in the intrinsic hand and forearm muscles is detected by polymyography performed during writing. A higher range of median frequency of 5.5 Hz has been detected by accelerometry. Normal subjects write with a median of 4.6 Hz (Bain et al., 1995). Rothwell and colleagues described a patient who complained of jerking of the right forearm on writing (Rothwell et al, 1979). In this patient, the active pronation of the arm produced several beats of pronation/supination tremor. Moreover, tendon taps to the volar surface of the wrist, to the finger extensors, and to the pectoralis major as well as forcible supination of the wrist (delivered by a torque motor) elicited bursts of tremor.

5.4.8 Complex Regional Pain Syndrome

This descriptive term refers to patients complaining of pain and presenting combinations of sensory, motor, and circulatory manifestations. The term algodystrophy is also related to this pathological condition. Complex regional pain syndrome (CRPS) has been classified in type I (formely "reflex sympathetic dystrophy" because of the presumed key role of the sympathetic nervous system) and type II (formerly "causalgia") occurring after partial injury of a peripheral nerve. The peripheral trauma is sometimes very minor, such as a simple foot sprain. CRPS is distinct from the syndrome of "painful legs and moving toes," for which a history of peripheral trauma or peripheral nerve dysfunction may occur (Spillane et al., 1971).

Tremor, spasms, dystonic deformities, and exaggerated tendon reflexes may develop in relation to CRPS. Postural, kinetic tremor, and enhanced postural tremor have been reported, usually with frequencies in the 3- to 6-Hz range (Verdugo and Ochoa, 2000). A prospective study on 145 patients affected by CRPS revealed motor disturbances in 97% of cases (Birklein et al., 2000).

5.4.9 Fragile-X-Associated Tremor/Ataxia Syndrome

Fragile-X-associated tremor/ataxia syndrome (FXTAS) is an adult-onset neurodegenerative disorder that affects carriers (principally males) of premutation alleles (55–200 CGG repeats) of the fragile X mental retardation 1 (FMR1) gene. Some adult male premutation carriers develop a neurological syndrome involving intention tremor, ataxia, dementia, parkinsonism, and autonomic dysfunction (Hagerman and Hagerman, 2004; Greco et al., 2006). Female carriers of the FMR1 premutation who presented with symptoms of tremor and ataxia have been described also. None of the women had dementia (Hagerman et al., 2004).

5.4.10 Familial Cortical Myoclonic Tremor with Epilepsy

The acronym FCMTE (familial cortical myoclonic tremor with epilepsy) has been proposed recently on the basis of both clinical and electrophysiological criteria: irregular postural myoclonic tremor of the distal limbs, familial history of epilepsy, autosomal dominant inheritance, and a rather benign outcome (Regragui et al., 2006). The diagnosis is confirmed by electrophysiological features suggesting a cortical reflex myoclonus (enhanced C-reflex, giant somatosensory-evoked potentials SEPs, premyoclonus cortical spikes detected by the back-averaging method), and response to antiepileptic drugs. The genetic analysis of these families shows a linkage to chromosome 8q24 for Japanese families and a linkage to chromosome 2p for Italian families (Regragui et al., 2006). The similarities of this syndrome with myoclonic epilepsies argue for an abnormality of a gene encoding ion channels.

5.4.11 Posttraumatic Tremor

As mentioned in Chapter 4, tremor may also rise as a consequence of central or peripheral trauma. Holmes tremor is an example of postural symptomatic tremor because of central trauma (Bhidayasiri, 2005), including brainstem hemorrhage (Raina et al., 2007). A 46-year-old man who developed unilateral dystonic tremor 13 years after contralateral combat-induced head trauma has been reported (Newmark and Richards, 1999). Hashimoto et al. (2002) have described a 55-year-old man who developed, 7 days after a neck and spine trauma, a disabling coarse postural tremor in the right upper extremity, along with numbness in the ulnar side of the right forearm. A C7–C8 radiculopathy was detected, while brain and cervical spine MRI were both unremarkable. Loading and ischemic nerve block reduced the tremor frequency, in agreement with a mechanical reflex mechanism underlying the tremor (see Chapter 1). Costa and coworkers have described a patient with proximal right upper limb tremor secondary to direct peripheral nerve lesion caused by prior thoracic surgery (Costa et al., 2006).

Magnetic resonance evidence of lesions in the dentatoe–thalamic pathway was described in patients with severe posttraumatic tremor (Krauss et al., 1995). Lesions may involve the Guillain–Mollaret triangle (see Chapter 2).

A contribution of predisposing factors to the pathogenesis of the posttraumatic tremor has been proposed (Jankovic and Van der Linden, 1988).

5.4.12 Tremor in Postinfectious Syndromes

West Nile virus is considered the most common arboviral infection in North America. Meningoencephalitis, acute flaccid paralysis, and movement disorders including tremor, myoclonus, or parkinsonism are the main clinical presentations. Maculopapular rash and tremor were demonstrated as independently associated with the infection (Tilley et al., 2007). Beside the multiple somatic complaints and abnormalities in motor skills and executive functions, tremor is reported as a common long-term problem among patients who have had West Nile virus infection (Carson et al., 2006).

Postinfectious encephalopathy with tremor may develop depending on possible lesions in the brain in cases of cytomegalovirus infection, toxoplasmosis, HIV infection, and malaria (Coelho et al., 1996; Mizuno et al., 2006; Cardoso, 2002). A midbrain abscess may generate Holmes tremor (Pezzini et al., 2002).

5.4.13 Drug-Induced Tremor

Tremor is the most frequent drug-induced movement disorder. Drug intake (and true or 'hidden' neuroleptic or psychoactive drug intake, see Tables 4.2 and 5.3) should be suspected as a possible cause of movement disorders. Drug-induced acute movement disorders usually disappear spontaneously after withdrawal of the suspected drug (Montastruc and Durrieu, 2004). The delay for disappearance of tremor may vary from days to years.

The most common drug-induced tremor is enhanced physiological tremor after use of sympathomimetics such as pseudoephedrine, bronchodilators, or theophylline, and antidepressants such as tricyclic agents or fluoxetine. Approximately 25% of patients taking valproic acid therapy (taken mainly for treatment of seizures or headaches) exhibit a postural tremor 3 to 12 months after starting therapy. Lowering the dosage decreases the intensity of tremor. Lithium induces a fine postural tremor of the hands (frequency 8–12 Hz). Lithium toxicity may cause permanent damage to the cerebellum that precipitates postural and intention tremors. Amiodarone may cause a dose-dependent reversible neurologic syndrome consisting of postural tremor, ataxia, and peripheral neuropathy. Symptoms develop in the first weeks of treatment and improve slowly after dose reduction or discontinuation. Neuroleptic agents such as haloperidol or dopamine-receptor-blocking drugs like metoclopramide may induce a parkinsonian tremor (Smaga, 2003). Anticalcic agents may also trigger a parkinsonian syndrome in the elderly.

5.4.14 Alcohol Abuse and Alcohol Withdrawal Syndrome

Alcohol is a major toxic agent for the brain, in particular for the cerebellum (Pentney, 2002). Chronic alcohol consumption leads to several neurological complications: cerebellar dysfunction (Gilman et al., 1981), painful small-fiber peripheral neuropathy (whose severity increases during alcohol withdrawal; Dina et al., 2008), impairment of cerebral vasoreactivity, thus eventually resulting in an increase in ischemic brain damage (Sun et al., 2008). During acute alcohol intoxication, patients exhibit slurred speech, gaze-evoked nystagmus, anterior–posterior oscillations in Romberg test, a broad-based ataxic gait. A 3-Hz postural leg tremor and kinetic tremor are frequently found in chronic alcoholic patients (Manto and Jacquy, 2002). One of the most severe consequences of alcoholism is the Wernicke–Korsakoff encephalopathy, historically described with confusion, ataxia, ophtalmoplegia, confabulation, and amnesic disturbances (Pearce, 2008).

A suddenly decrease in alcohol consumption or a sudden abstinence may lead to alcohol withdrawal syndrome's (AWS) comprising mild to moderate postural tremor, irritability, anxiety, or agitation. The most severe manifestations of withdrawal is characterized by delirium tremens, hallucinations, and seizures (Saitz, 1998).

5.4.15 Peripheral Neuropathy

Acute or chronic demyelinating (Guillain–Barre´ syndrome and Charcot–Marie–Tooth syndrome, respectively) and dysgammaglobulinemic neuropathies are the most common causes of tremor secondary to neuropathy. An action tremor resembling ET evolves. Postural tremor has been described in other disorders causing neuropathies including X-linked bulbospinomuscular atrophy, multifocal motor neuropathy, and human T lymphotropic virus 1-associated neuropathy (Habib-ur-Rehman, 2000). Tremor has been reported also in a few cases of peripheral neuropathy associated with diabetes, uremia, alcohol intake, and during treatment with drugs causing a peripheral neuropathy (Smith, 1995).

5.4.16 Psychogenic Tremor

Movement disorders—especially tremor, dystonia, and myoclonus—may be the presentation of a psychogenic disorder. Approximately 3% of patients followed in specialized clinics have a psychogenic disease which represents a conversion disorder or part of somatoform disorder (Reich, 2006). Tremor may affect any region of the body. Cases of psychogenic palatal tremor have been described

(Pirio Richardson et al., 2006). Comorbidity with other movement disorders is not exceptional. A sudden onset and the variability in tremor features are common (see also Chapter 4). The "coactivation sign of psychogenic tremor" and the absence of finger tremor suggest a psychogenic origin. Quantitative analysis of tremor shows decreasing amplitudes in most organic tremors when the extremity is loaded with additional weights. In contrast, tremor amplitude increases with added inertia for most of the cases with psychogenic tremor (Deuschl et al., 1998).

According to a modified Fahn's criteria, the diagnosis of psychogenic tremor is considered in the following conditions:

- exclusion of the major causes of symptomatic tremor
- exclusion of ET and PD
- absence of evidence for any other neurological disorders
- a period without tremor of at least 2 weeks during the observation period (Bhidayasiri, 2005)

5.5 ROLE OF BRAIN IMAGING IN THE DIFFERENTIAL DIAGNOSIS OF TREMOR

5.5.1 Computerized Tomography and Magnetic Resonance Imaging

Computerized tomography (CT scan) and magnetic resonance imaging (MRI) are useful for the diagnosis of tremors associated with to neurodegenerative diseases, vascular diseases, or posttraumatic tremor. *Brain MRI* has become a key tool for the morphological assessment of brain structures, being more sensitive and specific than the CT scan. For instance, brain MRI may detect an enlargement of the inferior olivary nucleus in palatal tremor as a result of a transsynaptic degeneration of the Guillain–Mollaret triangle (Yokota et al., 1989; Barron et al., 1982).

5.5.2 Positron Emission Tomography

Positron emission tomography (PET; see Chapter 7) with ^{18}F-Dopa and specific tracers to measure dopamine transporter (DAT; SPECT) binding provide an accurate measure of degeneration in the dopamine system in PD (Antonini et al., 2001). SPECT ligands provide a marker for presynaptic neuronal degeneration (Marshall et al., 2003). Low striatal uptake of the radioligand differentiates PD from ET (Benamer et al., 2000) and can be used to monitor disease severity (Breit et al., 2006). SPECT seems accurate in differentiating idiopathic PD (PD) from vascular parkinsonism and drug-induced parkinsonism (Vlaar et al., 2008). Moreover, PET might allow the selection of at-risk PD subjects (in early or preclinical stage) for neuroprotective intervention trials as well as for the evaluation of the effects of neurorestorative interventions in patients who are at an advanced stage (Bohnen and Frey, 2003).

5.5.3 Transcranial Sonography

Transcranial sonography assesses echogenicity of the substantia nigra as a characteristic of PD. This procedure might help to differentiate idiopathic PD from atypical parkinsonian syndromes at early stages (Gaenslen et al., 2008). Transcranial sonography is commonly used for stroke assessments.

REFERENCES

Alusi SH, Glickman S, Aziz TZ, Bain PG. Tremor in multiple sclerosis. *J Neurol Neurosurg Psychiatry.* 1999;66(2):131–134.

Antonini A, Moresco RM, Gobbo C, De Notaris R, Panzacchi A, Barone P, Calzetti S, Negrotti A, Pezzoli G, Fazio F. The status of dopamine nerve terminals in Parkinson's disease and essential tremor: a PET study with the tracer [11-C]FE-CIT. *Neurol Sci.* 2001;22(1):47–48.

Aronson AE, Ramig LO, Winholtz WS, Silber SR. Rapid voice tremor, or "flutter," in amyotrophic lateral sclerosis. *Ann Otol Rhinol Laryngol.* 1992;101:511–518.

Bain PG, Findley LJ, Thompson PD, Gresty MA, Rothwell JC, Harding AE, Marsden CD. A study of hereditary essential tremor. *Brain.* 1994;117(Pt 4):805–824.

Bain PG, Findley LJ, Britton TC, Rothwell JC, Gresty MA, Thompson PD, Marsden CD. Primary writing tremor. *Brain.* 1995;118(Pt 6):1461–1472.

Barron KD, Dentinger MP, Koeppen AH. Fine structure of neurons of the hypertrophied human inferior olive. *J Neuropathol Exp Neurol.* 1982;41:186–203.

Benamer TS, Patterson J, Grosset DG, Booij J, de Bruin K, van Royen E, Speelman JD, Horstink MH, Sips HJ, Dierckx RA, Versijpt J, Decoo D, Van Der Linden C, Hadley DM, Doder M, Lees AJ, Costa DC, Gacinovic S, Oertel WH, Pogarell O, Hoeffken H, Joseph K, Tatsch K, Schwarz J, Ries V. Accurate differentiation of parkinsonism and essential tremor using visual assessment of [^{123}I]-FP-CIT SPECT imaging: the [^{123}I]-FP-CIT study group. *Mov Disord.* 2000;15(3):503–510.

Benito-León J, Rodríguez J, Ortí-Pareja M, Ayuso-Peralta L, Jiménez-Jiménez FJ, Molina JA. Symptomatic orthostatic tremor in pontine lesions. *Neurology.* 1997;49(5):1439–1441.

Benito-León J, Louis ED. Essential tremor: emerging views of a common disorder. *Nat Clin Pract Neurol.* 2006;2(12):666–678.

Bhidayasiri R. Differential diagnosis of common tremor syndromes. *Postgrad Med J.* 2005;81:756–762.

Birklein F, Riedl B, Sieweke N, Weber M, Neundörfer B. Neurological findings in complex regional pain syndromes—analysis of 145 cases. *Acta Neurol Scand.* 2000;101(4):262–269.

Bohnen NI, Frey KA. The role of positron emission tomography imaging in movement disorders. *Neuroimaging Clin N Am.* 2003;13(4):791–803.

Bonneville F, Bloch F, Kurys E, du Montcel ST, Welter ML, Bonnet AM, Agid Y, Dormont D, Houeto JL. Camptocormia and Parkinson's disease: MR imaging. *Eur Radiol.* 2008;18:1710–1719.

Boroojerdi B, Ferbert A, Foltys H, Kosinski CM, Noth J, Schwarz M. Evidence for a non-orthostatic origin of orthostatic tremor. *J Neurol Neurosurg Psychiatry.* 1999;66(3):284–288.

Braak H, Del Tredici K, Rüb U, de Vos RA, Jansen Steur EN, Braak E. Staging of brain pathology related to sporadic Parkinson's disease. *Neurobiol Aging.* 2003;24(2):197–211.

Breit S, Reimold M, Reischl G, Klockgether T, Wüllner U. [(11)C]D-Threo-methylphenidate PET in patients with Parkinson's disease and essential tremor. *J Neural Transm.* 2006;113(2): 187–193.

Brennan KC, Jurewicz EC, Ford B, Pullman SL, Louis ED. Is essential tremor predominantly a kinetic or a postural tremor? A clinical and electrophysiological study. *Mov Disord.* 2002;17(2):313–316.

Britton TC, Thompson PD, van der Kamp W, Rothwell JC, Day BL, Findley LJ, Marsden CD. Primary orthostatic tremor: further observations in six cases. *J Neurol.* 1992;239(4):20–217.

Cardoso F. HIV-related movement disorders: epidemiology, pathogenesis and management. *CNS Drugs.* 2002;16(10):663–668.

Carson PJ, Konewko P, Wold KS, Mariani P, Goli S, Bergloff P, Crosby RD. Long-term clinical and neuropsychological outcomes of West Nile virus infection. *Clin Infect Dis.* 2006;43(6): 723–730.

Coelho JC, Wiederkehr JC, Cat R, Carrero JE, de Oliveira ED, Campos AC, Cat I. Extrapyramidal disorder secondary to cytomegalovirus infection and toxoplasmosis after liver transplantation. *Eur J Pediatr Surg.* 1996;6(2):110–111.

Cohen O, Pullman S, Jurewicz E, Watner D, Louis ED. Rest tremor in patients with essential tremor: prevalence, clinical correlates, and electrophysiologic characteristics. *Arch Neurol.* 2003;60(3):405–410.

Contarino MF, Welter ML, Agid Y, Hartmann A. Orthostatic tremor in monozygotic twins. *Neurology.* 2006;66(10):1600–1601.

Costa J, Henriques R, Barroso C, Ferreira J, Atalaia A, de Carvalho M. Upper limb tremor induced by peripheral nerve injury. *Neurology.* 2006;67(10):1884–1886.

de Lau LM, Giesbergen PC, de Rijk MC, Hofman A, Koudstaal PJ, Breteler MM. Incidence of parkinsonism and Parkinson disease in a general population: the Rotterdam Study. *Neurology.* 2004;63(7):1240–1244.

Deng H, Le W, Jankovic J. Genetics of essential tremor. *Brain.* 2007;130(Pt 6):1456–1464.

Deuschl G, Bain P, Brin M. Consensus statement of the Movement Disorder Society on Tremor. Ad Hoc Scientific Committee. *Mov Disord.* 1998;13(3):2–23.

Deuschl G, Köster B, Lücking CH, Scheidt C. Diagnostic and pathophysiological aspects of psychogenic tremors. *Mov Disord.* 1998;13(2):294–302.

Deuschl G. Dystonic tremor. *Rev Neurol (Paris).* 2003;159(10, Pt 1):900–905.

Dick FD. Parkinson's disease and pesticide exposures. *Br Med Bull.* 2006;79–80:219–31.

Dina OA, Khasar SG, Alessandri-Haber N, Green PG, Messing RO, Levine JD. Alcohol-induced stress in painful alcoholic neuropathy. *Eur J Neurosci.* 2008;27(1):83–92.

Dubinsky RM. Tremor and dystonia. In: *Handbook of Tremor Disorders.* Findley LJ, Koller WC (eds). New York: Marcel Dekker, 1995.

Elble RJ, Higgins C, Leffler K, Hughes L. Factors influencing the amplitude and frequency of essential tremor. *Mov Disord.* 1994;9(6):589–596.

Elble RJ, Moody C, Higgins C. Primary writing tremor. A form of focal dystonia? *Mov Disord.* 1990;5(2):118–126.

Findley LJ, Koller WC. *Handbook of Tremor Disorders.* New York: Marcel Dekker, 1995.

Fischer M, Kreß W, Reiners K, Rieckmann P. Orthostatic tremor in three brothers. *J Neurol.* 2007;254(12):1759–1760.

Fukae J, Mizuno Y, Hattori N. Mitochondrial dysfunction in Parkinson's disease. *Mitochondrion.* 2007;7(1–2):58–62.

Gabellini AS, Martinelli P, Gullì MR, Ambrosetto G, Ciucci G, Lugaresi E. Orthostatic tremor: essential and symptomatic cases. *Acta Neurol Scand.* 1990;81(2):113–117.

Gaenslen A, Unmuth B, Godau J, Liepelt I, Di Santo A, Schweitzer KJ, Gasser T, Machulla HJ, Reimold M, Marek K, Berg D. The specificity and sensitivity of transcranial ultrasound in the differential diagnosis of Parkinson's disease: a prospective blinded study. *Lancet Neurol.* 2008;7(5):417–424.

Gerschlager W, Münchau A, Katzenschlager R, Brown P, Rothwell JC, Quinn N, Lees AJ, Bhatia KP. Natural history and syndromic associations of orthostatic tremor: a review of 41 patients. *Mov Disord.* 2004;19(7):788–795.

Gilhuis HJ, van Ommen HJ, Pannekoek BJ, Sillevis Smitt PA. Paraneoplastic orthostatic tremor associated with small cell lung cancer. *Eur Neurol.* 2005;54(4):225–226.

Gilman S, Bloedel JR, Lechtenberg R. Disorders of the cerebellum. In: *Contemporary Neurology Series.* Davis FA (Ed). Philadelphia. 1981.

Gonzalez-Alegre P, Kelkar P, Rodnitzky RL. Isolated high-frequency jaw tremor relieved by botulinum toxin injections. *Mov Disord.* 2006;21(7):1049–1050.

Greco CM, Berman RF, Martin RM, Tassone F, Schwartz PH, Chang A, Trapp BD, Iwahashi C, Brunberg J, Grigsby J, Hessl D, Becker EJ, Papazian J, Leehey MA, Hagerman RJ, Hagerman PJ. Neuropathology of fragile X-associated tremor/ataxia syndrome (FXTAS). *Brain.* 2006;129(Pt 1):243–255.

Guttman M, Kish SJ, Furukawa Y. Current concepts in the diagnosis and management of Parkinson's disease. *CMAJ.* 2003;168(3):293–301.

Habib-ur-Rehman. Diagnosis and management of tremor. *Arch Intern Med.* 2000;160(16):2438–2444.

Hagerman PJ, Hagerman RJ. Fragile X-associated tremor/ataxia syndrome (FXTAS). *Ment Retard Dev Disabil Res Rev.* 2004;10(1):25–30.

Hagerman RJ, Leavitt BR, Farzin F, Jacquemont S, Greco CM, Brunberg JA, Tassone F, Hessl D, Harris SW, Zhang L, Jardini T, Gane LW, Ferranti J, Ruiz L, Leehey MA, Grigsby J, Hagerman PJ. Fragile-X-associated tremor/ataxia syndrome (FXTAS) in females with the FMR1 premutation. *Am J Hum Genet.* 2004;74(5):1051–1056.

Hashimoto T, Sato H, Shindo M, Hayashi R, Ikeda S. Peripheral mechanisms in tremor after traumatic neck injury. *J Neurol Neurosurg Psychiatry.* 2002;73:585–587.

Hayashi M, Koide H. Idiopathic torsion dystonia and writing tremor within a family. *Brain Dev.* 1997;19(8):556–558.

Heilman KM. Orthostatic tremor. *Arch Neurol.* 1984;41(8):880–881.

Hu G, Jousilahti P, Nissinen A, Antikainen R, Kivipelto M, Tuomilehto J. Body mass index and the risk of Parkinson disease. *Neurology.* 2006;67(11):1955–1959.

Hu G, Bidel S, Jousilahti P, Antikainen R, Tuomilehto J. Coffee and tea consumption and the risk of Parkinson's disease. *Mov Disord.* 2007;22(15):2242–2248.

Italian Neurological Society; Italian Society of Clin Neurophysiol; Guidelines for the Treatment of Parkinson's Disease 2002. The diagnosis of Parkinson's disease. *Neurol Sci.* 2003;24(3):S157–S164.

Jankovic J, Van der Linden C. Dystonia and tremor induced by peripheral trauma: predisposing factors. *J Neurol Neurosurg Psychiatry.* 1988;51(12):1512–1519.

Jankovic J, Leder S, Warner D, Schwartz K. Cervical dystonia: clinical findings and associated movement disorders. *Neurology.* 1991;41(7):1088–1091.

Jimenez-Jimenez FJ, Cabrera-Valdivia F, Orti-Pareja M, Gasalla T, Tallon-Barranco A, Zurdo M. Bilateral primary writing tremor. *Eur J Neurol.* 1998;5(6):613–614.

Jiménez-Jiménez FJ, de Toledo-Heras M, Alonso-Navarro H, Ayuso-Peralta L, Arévalo-Serrano J, Ballesteros-Barranco A, Puertas I, Jabbour-Wadih T, Barcenilla B. Environmental risk factors for essential tremor. *Eur Neurol.* 2007;58(2):106–113.

Kerr G, Morrison S, Silburn P. Coupling between limb tremor and postural sway in Parkinson's disease. *Mov Disord.* 2008;23:386–394.

Klawans HL. Hemiparkinsonism as a late complication of hemiatrophy: a new syndrome. *Neurology.* 1981;31:625–628.

Klockgether T. Parkinson's disease: clinical aspects. *Cell Tissue Res.* 2004;318(1):115–120.

Koch M, Mostert J, Heersema D, De Keyser J. Tremor in multiple sclerosis. *J Neurol.* 2007;254(2): 133–145.

Kraus PH, Lemke MR, Reichmann H. Kinetic tremor in Parkinson's disease—an underrated symptom. *J Neural Transm.* 2006;113(7):845–853.

Krauss JK, Wakhloo AK, Nobbe F, Tränkle R, Mundinger F, Seeger W. Lesion of dentatothalamic pathways in severe post-traumatic tremor. *Neurol Res.* 1995;17(6):409–416.

Lou JS, Jankovic J. Essential tremor: clinical correlates in 350 patients. *Neurology.* 1991;41(2, Pt 1): 234–238

Louis ED, Zheng W, Jurewicz EC, Watner D, Chen J, Factor-Litvak P, Parides M. Elevation of blood beta-carboline alkaloids in essential tremor. *Neurology.* 2002;59(12):1940–1944.

Louis ED, Ottman R. Study of possible factors associated with age of onset in essential tremor. *Mov Disord.* 2006;21(11):1980–1986.

McInerney-Leo A, Hadley DW, Gwinn-Hardy K, Hardy J. Genetic testing in Parkinson's disease. *Mov Disord.* 2005;20(1):1–10.

Magerkurth C, Schnitzer R, Braune S. Symptoms of autonomic failure in Parkinson's disease: prevalence and impact on daily life. *Clin Auton Res.* 2005;15(2):76–82.

Manto M. Clinical signs of cerebellar disorders. In: *The Cerebellum and Its Disorders.* Manto MU, Pandolfo M (eds). Cambridge, UK: Cambridge University Press, 2002.

Manto M, Jacquy J. Alcohol toxicity in the cerebellum. In: *Clinical Aspects.* Manto M.U. and Pandolfo M (eds). Cambridge, UK: Cambridge University Press, 2002, pp. 336–341.

Marshall V, Grosset D. Role of dopamine transporter imaging in routine clinical practice. *Mov Disord.* 2003;18(12):1415–1423.

Mizuno Y, Kato Y, Kanagawa S, Kudo K, Hashimoto M, Kunimoto M, Kano S. A case of postmalaria neurological syndrome in Japan. *J Infect Chemother.* 2006;12(6):399–401.

Montastruc JL, Durrieu G. Drug-induced tremor and acute movement disorders. *Therapie.* 2004;59(1):97–103.

Nasrallah KM, Mitsias PD. Orthostatic tremor due to thiamine deficiency. *Mov Disord.* 2007; 22(3):440–441.

Newmark J, Richards TL. Delayed unilateral post-traumatic tremor: localization studies using single-proton computed tomographic and magnetic resonance spectroscopy techniques. *Mil Med.* 1999;164(1):59–64.

Pahwa R. Primary writing tremor. In: *Handbook of Tremor Disorders.* Findley LJ, Koller WC (eds). New York: Marcel Dekker, 1995.

Pal PK, Samii A, Schulzer M, Mak E, Tsui JK. Head tremor in cervical dystonia. *Can J Neurol Sci.* 2000;27(2):137–142.

Paulson GW. Benign essential tremor: features that aid in diagnosis. *Postgrad Med J.* 1982;71(1):105–107.

Pearce JM. Wernicke–Korsakoff encephalopathy. *Eur Neurol.* 2008;59(1–2):101–104.

Pentney R. Alcohol toxicity in the cerebellum: fundamental aspects. In: *The Cerebellum and Its Disorders.* Manto MU and Pandolfo M (eds). Cambridge, UK: Cambridge University Press, 2002, pp. 327–339.

Pezzini A, Zavarise P, Palvarini L, Viale P, Oladeji O, Padovani A. Holmes' tremor following midbrain Toxoplasma abscess: clinical features and treatment of a case. *Parkinsonism Relat Disord.* 2002;8(3):177–180.

Pezzoli G, Canesi M, Galli C. An overview of parkinsonian syndromes: data from the literature and from an Italian data-base. *Sleep Med.* 2004;5(2):181–187.

Piboolnurak P, Yu QP, Pullman SL. Clinical and neurophysiologic spectrum of orthostatic tremor: case series of 26 subjects. *Mov Disord.* 2005;20(11):1455–1461.

Pirio Richardson S, Mari Z, Matsuhashi M, Hallett M. Psychogenic palatal tremor. *Mov Disord.* 2006;21(2):274–276.

Raina GB, Velez M, Pardal MF, Micheli F. Holmes tremor secondary to brainstem hemorrhage responsive to levodopa: report of 2 cases. *Clin Neuropharmacol.* 2007;30(2):95–100.

Regragui W, Gerdelat-Mas A, Simonetta-Moreau M. Cortical tremor (FCMTE: familial cortical myoclonic tremor with epilepsy). *Neurophysiol Clin.* 2006;36(5–6):345–349.

Rao SS, Hofmann LA, Shakil A. Parkinson's disease: diagnosis and treatment. *Am Fam Physician.* 2006;74(12):2046–2054.

Reich SG. Psychogenic movement disorders. *Semin Neurol.* 2006;26(3):289–296.

Rivest J, Marsden CD. Trunk and head tremor as isolated manifestations of dystonia. *Mov Disord.* 1990;5(1):60–65.

Rondot P, Bathien N. Cerebellar tremors: physiological basis and treatment. In: *Handbook of Tremor Disorders.* Findley LJ, Koller WC (eds). New York: Marcel Dekker, 1995.

Rothwell JC, Traub MM, Marsden CD. Primary writing tremor. *J Neurol Neurosurg Psychiatry.* 1979;42(12):1106–1114.

Saitz R. Introduction to alcohol withdrawal. *Alcohol Health Res World.* 1998;22(1):5–12.

Sanitate SS, Meerschaert JR. Orthostatic tremor: delayed onset following head trauma. *Arch Phys Med Rehabil.* 1993;74(8):886–889.

Scholtissen B, Verhey FR, Steinbusch HW, Leentjens AF. Serotonergic mechanisms in Parkinson's disease: opposing results from preclinical and clinical data. *J Neural Transm.* 2006;113(1):59–73.

Schrag A, Bhatia K, Brown P, Marsden CD. An unusual jaw tremor with characteristics of primary orthostatic tremor. *Mov Disord.* 1999;14(3):528–530.

Smaga S. Tremor. *Am Fam Physician.* 2003;68(8):1545–1552.

Smith IS. Tremor in peripheral neuropathy. In: *Handbook of Tremor Disorders.* Findley LJ, Koller WC (eds). New York: Marcel Dekker, 1995.

Spillane JD, Nathan PW, Kelly RE, Marsden CD. Painful legs and moving toes. *Brain.* 1971;94: 541–556.

Stockner H, Sojer M, K KS, Mueller J, Wenning GK, Schmidauer C, Poewe W. Midbrain sonography in patients with essential tremor. *Mov Disord.* 2007;22(3):414–417.

Sun H, Zhao H, Sharpe GM, Arrick DM, Mayhan WG. Effect of chronic alcohol consumption on brain damage following transient focal ischemia. *Brain Res.* 2008;1194:73–80.

Sur H, Ilhan S, Erdo an H, Oztürk E, Ta demir M, Börü UT. Prevalence of essential tremor: a door-to-door survey in Sile, Istanbul, Turkey. *Parkinsonism Relat Disord.* 2008.

Tan EK, Lo YL, Chan LL. Graves disease and isolated orthostatic tremor. *Neurology.* 2008;70(16, Pt 2):1497–1498.

Thanvi B, Lo N, Robinson T. Vascular parkinsonism—an important cause of parkinsonism in older people. *Age Ageing.* 2005;34(2):114–119.

Thanvi B, Lo N, Robinson T. Essential tremor-the most common movement disorder in older people. *Age Ageing.* 2006;35(4):344–349.

Thobois S, Guillouet S, Broussolle E. Contributions of PET and SPECT to the understanding of the pathophysiology of Parkinson's disease. *Neurophysiol Clin.* 2001;31(5):321–340.

Thomas B, Beal MF. Parkinson's disease. *Hum Mol Genet.* 2007;16(Spec No. 2):R183–R194.

Tilley PA, Fox JD, Jayaraman GC, Preiksaitis JK. Maculopapular rash and tremor are associated with West Nile fever and neurological syndromes. *J Neurol Neurosurg Psychiatry.* 2007; 78(5):529–531.

Tolosa E, Wenning G, Poewe W. The diagnosis of Parkinson's disease. *Lancet Neurol.* 2006;5(1):75–86.

Verbaan D, Marinus J, Visser M, van Rooden SM, Stiggelbout AM, Middelkoop HA, van Hilten JJ. Cognitive impairment in Parkinson's disease. *J Neurol Neurosurg Psychiatry.* 2007;78(11):1182–1187.

Verdugo RJ, Ochoa JL. Abnormal movements in complex regional pain syndrome: assessment of their nature. *Muscle Nerve.* 2000;23(2):198–205.

Vlaar AM, de Nijs T, Kessels AG, Vreeling FW, Winogrodzka A, Mess WH, Tromp SC, van Kroonenburgh MJ, Weber WE. Diagnostic value of ^{123}I-ioflupane and ^{123}I-iodobenzamide SPECT scans in 248 patients with parkinsonian syndromes. *Eur Neurol.* 2008;59(5): 258–266.

Weiner WJ. A differential diagnosis of Parkinsonism. *Rev Neurol Dis.* 2005;2(3):124–131.

Yerdelen D, Karatas M, Goksel B, Yildirim T. A patient with multiple sclerosis presenting with Holmes' tremor. *Eur J Neurol.* 2008;15(1):e2–e3.

Yokota T, Hirashima F, Furukawa T, Tsukagoshi H, Yoshikawa H. MRI findings of inferior olives in palatal myoclonus. *J Neurol.* 1989;236:115–116.

Yoo YM, Lee CJ, Lee U, Kim YJ. Mitochondrial DNA in patients with essential tremor. *Neurosci Lett.* 2008;434(1):29–34.

Yu H, Sternad D, Corcos DM, Vaillancourt DE. Role of hyperactive cerebellum and motor cortex in Parkinson's disease. *Neuroimage.* 2007;35(1):222–233.

• • • •

CHAPTER 6

Quantification of Tremor

Several techniques for monitoring and analysis of tremor have been developed (Table 6.1). A quantification of tremor can be achieved with kinematic tools using various miniaturized sensors applied in robotic, industrial, and aerospace fields such as accelerometers or gyroscopes, using videos or force sensors. Wireless techniques offer new perspectives for remote tracking of tremor and become part of the emerging field of body sensor networks (BSN). The choice of relevant sensors, sensor

TABLE 6.1: Available tools for quantification and monitoring of tremor

Clinical scales and scores	The most common scales and procedures are: —Activity of Daily Living (ADL) scales, among which Extended Nottingham ADL Scale —Clinical Tremor Rating Scale (TRS) —Unified Parkinson Disease Rating Scale (UPDRS) —Nine Hole Peg Test (9HPT) —Scoring of spirals
Electromyography	—Surface EMG electrodes (SEMG) —Needle EMG electrodes —Single motor units studies —Long-term recordings (SEMG)
Kinematics	—Accelerometers —Gyroscopes —Electromagnetic tracking systems —Flexible angular sensors —Videos —Systems combining various kinematic sensors
Haptic devices	—Force-feedback devices

packaging, type of communication, power consumption, autonomy, integration, and ergonomic aspects are important factors.

Needless to say, the appropriate placement of the sensor on the body is a key factor also. It is important to take into account that recordings need to be performed in a quiet room, free of vibrations or electrical interferences. Regular calibration of the instrumentation system should not be underestimated. Using standardized recording procedures is essential for intra- and interpatients comparisons.

6.1 CLINICAL EXAMINATION AND CLINICAL SCALES

6.1.1 Clinical Examination

Clinical examination remains a key step, although presenting the following three weaknesses: the assessment is subjective, it provides short-term information, and clinical examination is not correlated with mean tremor amplitude.

The first step in evaluating any patient with tremor is to characterize the tremor on inspection and to look for associated movement disorders. Activation tests are performed, and topographical distribution is noted. Clinical examination has to be careful and extensive, including assessment of cranial nerves, voluntary force in each limb segment, tendon/plantar reflexes, detailed sensory testing, and evaluation of posture/gait.

Neurological examination of a patient complaining and/or presenting tremor includes:

- the observation of the presence of vertical or side-to-side head movement (see Table 6.2 for differential diagnosis);
- speech evaluation: search for a concomitant dysarthria which is evaluated by repetition of standard sentences, for instance: "A mischievous spectacle in Czechoslovakia" (Trouillas et al., 1997);
- testing for oculomotor disturbances (ocular fixation, saccades, pursuit);
- the observation of abnormal movements occurring at rest;
- checking for the occurrence of tremor triggered by postural tasks or voluntary muscle contraction;
- looking for the presence of other abnormal movements (dystonia, tics);
- assessing amyotrophy or muscle hypertrophy;
- evaluating abnormalities of stance and gait.

Rest tremor is detectable while the patient is seated with the upper limbs resting in the lap (showing the dorsal side of the hands) or while the patient is lying down. For lower limbs, rest conditions can be achieved crossing one leg on the other or in supine position. Cognitive tasks, for

TABLE 6.2: Differential diagnosis of head tremor
Essential tremor
Dystonia
Cerebellar ataxias
Parkinson's disease
Myoclonus
Congenital nystagmus
Psychogenic

instance, counting down and tapping of the contalateral foot, can trigger or enhance tremor (Raethjen et al., 2008).

Postural tremor is analyzed by the following maneuvers:

- holding the upper limbs outstretched with the hands in supination, parallel to the floor;
- index-to-index test: the patients has to maintain the two index fingers medially, pointing at each other at a distance of about 1 cm. Forearms are maintained horizontally.

The followings tests are used to evaluate *kinetic tremor* (Manto, 2002):

- finger-to-nose test: patient is asked to make movements of one upper limb with the hand first resting on the thigh and then touching the nose with the index.
- finger-to-finger test: patient touches the examiner's finger, which is moved and stopped in different locations in space.
- knee–tibia test: this maneuver is executed in supine position. The patient is asked to raise one leg and place the heel on the contralateral knee, which is kept motionless. The patient slides down the tibial surface in a regular way up to the ankle. The heel is then raised again up to the resting knee.

Assessment of *writing* includes drawing Archimedes' spiral: the subject is comfortably settled in front of a table, the sheet of the paper being fixed on the table with a tape to avoid motion artefacts. The subject is asked to perform the task without timing requirements. Dominant hand is examined (Figure 6.1). The patient is also asked to write standard sentences.

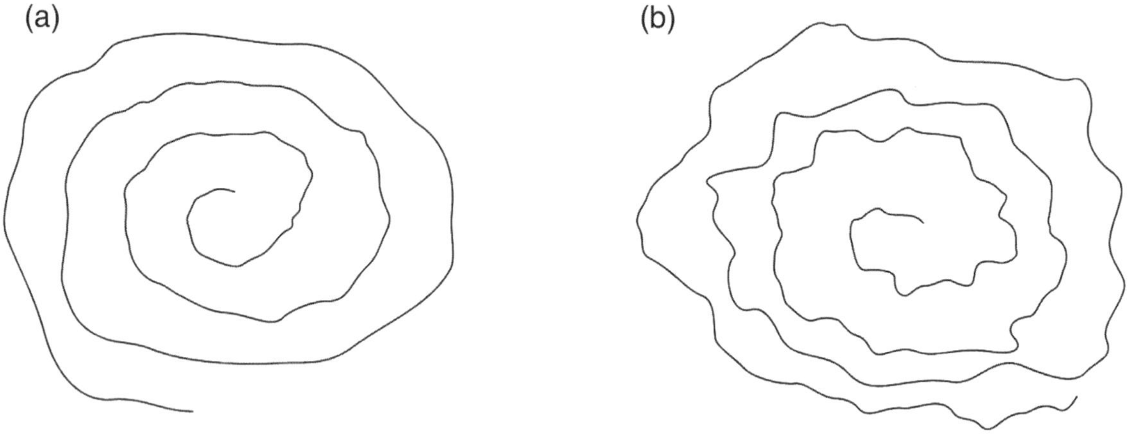

FIGURE 6.1: Drawing of a spiral in a patient with a mild essential tremor (a) and in a second patient with a more advanced disease (b).

6.1.2 Clinical Scales

Clinical scales are used to quantify the degree of severity of a clinical deficit. They systematize the information by assigning certain numbers to certain conditions.

A diagnostic measure aiming to produce quantifiable results and bound to psychometric accuracy standards should meet the following main criteria for quality:

- objectivity: independence of results from examiner;
- reliability: repeating a test application on the same person after a certain amount of time should give the same results;
- validity: a diagnostic method is valid if it actually measures the appropriate variable;
- standardization;
- comparability, economy, and usefulness.

Sensitivity (higher sensitivity means fewer false-negative cases), specificity (higher specificity means more false-positive cases), and predictive values should be taken into account (Masur, 2004).

According to *Weber's law* of psychophysics, the smallest discernible change in tremor ΔT is

$$\Delta T = K \cdot T_1,$$

where T_1 is the initial amplitude of tremor and K is Weber's constant (Gescheider, 1997). Thus, Weber's law states that the smallest discernible change in tremor will be proportional to the initial tremor amplitude. This relationship predicts that any tremor rating procedure will be a nonlinear measure of tremor amplitude. The knowledge of the relationship between tremor rating and precise measures of tremor is useful in interpreting the clinical significance of changes in ratings produced by disease or therapy. Analysis of data from 928 patients revealed a logarithmic relationship between a five-point (0–4) tremor rating procedure and tremor amplitude (Elble et al., 2006). A one-point change in TRS represents a substantial change in tremor amplitude.

The following scales are commonly used to evaluate and quantify tremor:

- *Activity of Daily Living (ADL) scale* is a questionnaire for assessing disability related to everyday functions. It is addressed to the patients themselves. Dressing, mobility, personal hygiene, eating are investigated. The measure of dependence is indicated on a defined scale. Two examples are given in the section Annexes: the *Extended ADL Scale according to Nourie and Lincoln* (Masur et al., 2004) and the *Schwab and England Activities of Daily Living.*
- *Clinical Tremor Rating Scale (TRS)* has been proposed by Fahn, Tolosa, and Marin. Tremor is evaluated through 22 items (see Annex 2).
    - The first 10 items evaluate the clinical manifestation of the tremor and its distribution in various regions of the body.
    - The other items characterize the patients' disability in daily life. Performance tests, such as handwriting and drawing, are included. The test is performed during 30 to 90 min. The range of results is 0–88 with a higher score meaning a more debilitating disease (Masur et al., 2004).
- *The Unified Parkinson Disease Rating Scale (UPDRS)* has been published in 1987 by Fahn, Elton, and the Members of the UPDRS Development Committee. It is composed of four sections (see Annex 3).
    - The first one assesses the psychological effects of the disease and the drugs used.
    - The second section features 13 subitems covering ADL.
    - In the third section (motor examination), the clinical symptoms of parkinsonism are assessed.
    - The fourth section describes the complications and side effects of drug therapies.
    - The test has a duration of 40–60 min. The range of results is 0–154 with impairment increasing from minimum to maximal points (0–5 = no impairment; 154 = maximum clinical impairment) (Masur et al., 2004).
- *Nine_Hole Peg Test* (*9HPT)* is an upper limb motor function test requiring a set of wooden equipment. The patient sits near a table and is asked to place pegs in holes. The examiner records the number of pegs placed in 50 s. The procedure has a duration of about 30 s in healthy persons (Mathiowetz et al., 1985).

6.1.2.1 Evaluation of Drawings. Recently, a new procedure of quantification of drawing of an Archimedes' spiral was developed (Figures 6.2 and 6.3), based on the off-line analysis of its digitized picture obtained with a commercial scanner (Miralles et al., 2006). The following assessment is performed: analysis by means of the cross-correlation coefficient with the spiral template, determination of the mean and the standard deviation of the distance between each point of the spiral drawing and the corresponding point of the spiral model, analysis of the reconstructed spiral using the Fourier transform. The experimental variables were found to be greater in the patients' group (subjects presenting action tremor, the majority of them affected with ET) compared to age-matched controls. A high linear correlation between them and the clinical score given by three

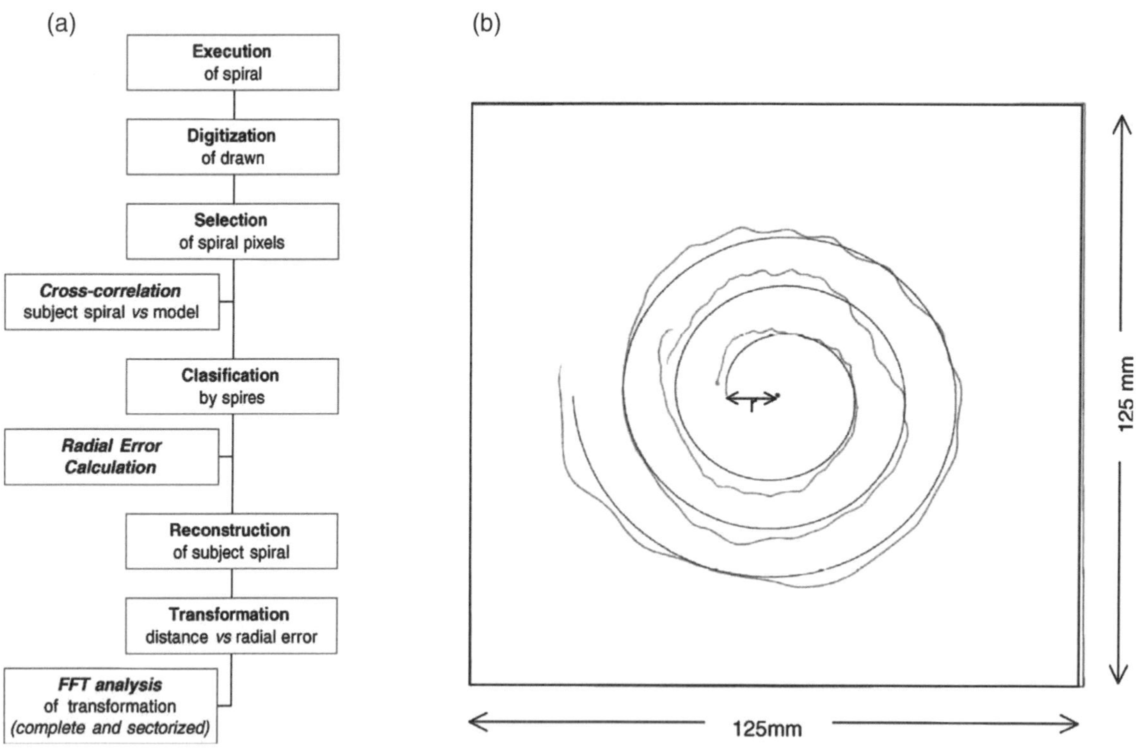

FIGURE 6.2: (a) Diagram of the different steps of the method to quantify spiral drawing. The figure shows the two parts of the procedure: the three types of numerical analysis performed on the data (left) and the progressive reconstruction of the spiral (right). (b) Digitized picture of a spiral specimen obtained from a patient with tremor showing the form and dimension of the spiral template. The initial radius (r) was 10.385 mm (i.e. 54 pixels) with an incremental change of 10.385 mm in the radius between turns. The 125 × 125 mm square that contained the spiral drawing was digitized at a resolution of 650 × 650 pixels. From Miralles et al. (2005), with permission from Elsevier.

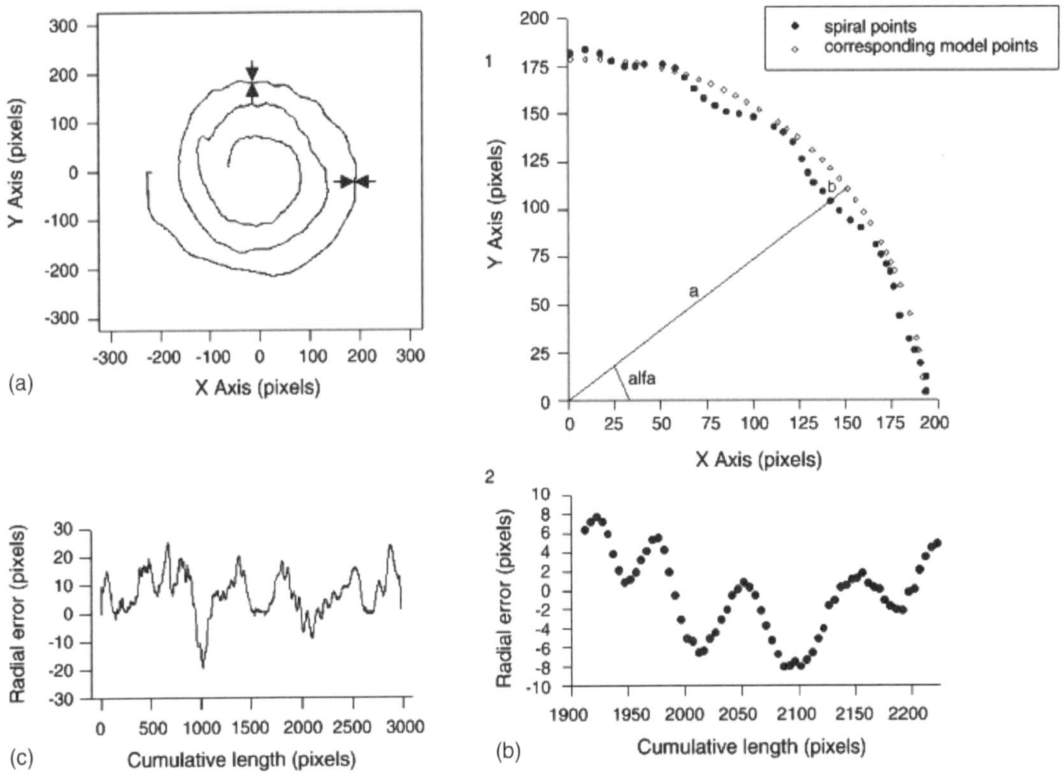

FIGURE 6.3: (a) Spiral reconstituted from the digitized picture shown in Figure 6.2b. Note the good similitude with the original image. (b, 1) Enlargement of the spiral sector located between the arrows in part (a). For the sake of clarity, only a 10% of the spiral points represented in (a) are shown. Knowing the spire to which it pertains and the angle that it forms with the x-axis (alfa), it is possible to calculate the corresponding model spiral point (white dots) for each spiral point (black dots). The radial error is signalled with the small letter b. (b, 2) Transformation of the spiral sector shown in part (b, 1). The spatial coordinates of each spiral points are substituted for another pair in which the x-value is the spiral length from the origin, and the y-value is the radial error. (c) Complete transformation of the spiral represented in part (A). From Miralles et al. (2006), with permission from Elsevier.

neurologists was also found (Spearman's rho coefficients greater than 0.7). The method is recognized to classify the spirals better than human raters.

6.1.2.2 Correlation of Postural Tremor with Quantified Tests. Alusi and colleagues found out that right arm postural tremor scores correlate with right arm Finger-Tapping Test and 9HPT scores ($p < 0.005$). A good correlation of postural tremor scores and patient perceived disability—as quantified by the ADL questionnaire—was also found. Tremor scores from spiral drawings of both dominant and nondominant hands and dominant handwriting had a high correlation with the

9HPT. Tremor scores from the nondominant hand spirals correlated less with the tremor ADL because most of the items on the scale are usually performed by the dominant hand (Alusi et al., 2000).

6.1.2.3 Vocal Tremor Scoring System. The Vocal Tremor Scoring System (VTSS) has been developed recently on the basis of a list of flexible fiber optic findings from patients with essential tremor of voice (ETV) and a review of multiple patient examinations of ETV. The final items include tremor of the palate, the base of tongue, the pharyngeal walls, the larynx (global), supraglottis, and true vocal folds. Such a standardized flexible nasolaryngoscopic examination protocol evaluates vocal tremor in a consistent fashion and could serve to predict which subpopulation of patients with ETV would benefit most from intralaryngeal muscle botulin toxin therapy (Bovè et al., 2006; see also Chapter 8).

6.2 ELECTROMYOGRAPHY AND MOTOR UNIT STUDIES

Surface EMG (SEMG) and needle EMG studies are commonly used to assess muscle discharges. Surface electrodes are arranged approximately 2 cm apart over the muscle. They record electrical potentials generated by the muscle fibers (see Chapter 3). Multiple muscles can be recorded simultaneously. Single motor unit discharges can be also studied to determine the firing rate of individual motor units, the synchronization of units within one muscle, the synchrony and coherence of tremor in homologous muscle groups (Bain, 1995). In healthy subjects, a moderate contraction is associated with motor unit discharges of about 9–16 Hz. Different diseases are associated with distinct durations of EMG bursts detected with surface sensors (Figure 6.4). In a given sensorimotor context, the patterns of motoneuronal discharges might constitute a "motoneuronal signature" which takes into account the size of the recruited motor units, the timing of their respective firings, their firing rates, and their modulation (Manto et al., 2008).

For intramuscular studies, pairs of intramuscular signals are usually recorded from a very small multichannel sterile electrode implanted in the muscle of interest. Decomposition of the myoelectric signals identifies individual motor unit action potentials (MUAPs). Firing statistics are extracted, based upon the template-matching approach (Calder et al., 2008), resolving the spatio-temporal superposition of motor unit potentials, tracking and identifying action potentials from different motor units. Motor unit potentials are identified on the basis of shape and firing time characteristics. Baseline needs to be stable enough. Sampling frequency is usually >5 kHz. The following parameters are extracted for each given task:

- the firing times of each motor unit
- the firing rates

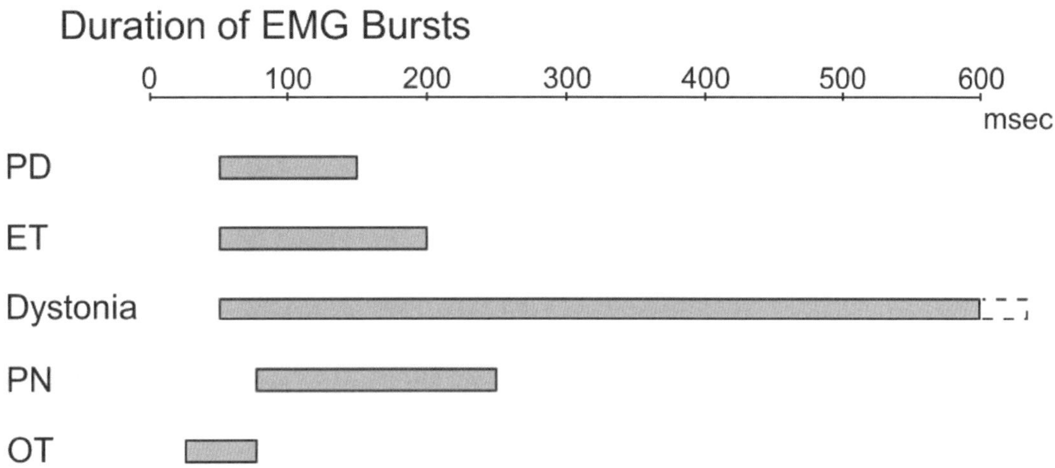

FIGURE 6.4: Duration of electromyographic (EMG) bursts in forearm muscles in various forms of tremor. PD: Parkinson's disease, ET: essential tremor, PN: peripheral neuropathy (mainly paraproteinemic neuropathy, chronic inflammatory demyelinating polyneuropathy CIDP, and hereditary neuropathies such as Charcot–Marie–Tooth disease), OT: orthostatic tremor. In some cases of dystonic tremor, the EMG bursts may last up to 800 ms.

- the recruitment threshold and the time of derecruitment
- the time of peak discharge
- the cross-correlation of the firing rates of concomitant motor units
- the synchronization histogram
- the presence of fibrillations/positive sharp waves (these are markers of denervation; fibrillation potentials have a duration <5 ms, an amplitude below 1 mV, and a discharge frequency between 1 and 50 Hz)
- the presence of fasciculations (polyphasic electrical potentials related to large motor units, see also Chapter 5: amyotrophic lateral sclerosis ALS)
- an index of instability of the motor unit potentials
- the presence of satellite potentials (markers of reinnervation)
- an index of the jitter and blocks of satellite components.

The *Tremor Coherence Analyzer* (TCA) is an electronic device developed for wireless monitoring of physiological variables such as EMG potentials or EEG traces (Figure 6.5). This technique uses surface EMG signals for tremor analysis. The portable system calculates instantaneously the coherence function between these signals, allowing the determination of linear dependencies between two signals (Brunetti et al., 2004; see Section 5).

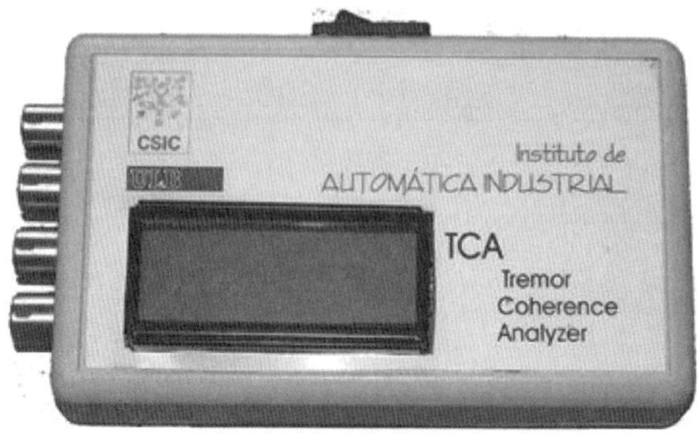

FIGURE 6.5: The Tremor Coherence Analyzer (TCA) is a wireless portable tool calculating instantaneously the coherence between 2 and 4 EMG channels to assess the behavior of the tremor generator and to detect sudden oscillators.

Long-term EMG has proven to be a valid and reliable method for the quantification of pathological tremors. Recent works indicate that these recordings allow a rater-independent classification of parkinsonian versus ET (Breit et al., 2008). A comparison of the technique of EMG with accelerometers and gyroscopes is given in Table 6.3.

The *entrainment test*—a quantified electrophysiological entrainment test performed using accelerometer or surface EMG tremor signals—has been proposed as a sensitive and specific procedure to distinguish psychogenic tremor from dystonic and other organic tremors (McAuley and Rothwell, 2004). The subject is asked to tap at a given frequency with one upper limb, while keeping the contralateral upper limb outstreched. The basic concept is the inability of patients with hysterical tremor to generate voluntary tapping oscillations independently of ongoing tremor oscillation (see Chapter 4). Normal subjects are not able to "mimic" organic tremor: a given tapping frequency will influence strongly the contralateral limb.

Simultaneous use of EMG recorded, during functional magnetic resonance imaging (fMRI) in patients with movement disorders, is complementary to conventional fMRI and may facilitate studies of hyperkinetic movement disorders, including tremor (van Rootselaar et al., 2007).

Surface and needle EMG studies are better interpreted when results of nerve conduction studies, somatosensory-evoked potentials, transcranial magnetic stimulation are available. Additional tests such as resetting of tremor, reciprocal inhibition (H reflex), stretch reflexes, ballistic movements may provide useful information in selected cases.

6.3 KINEMATIC STUDIES

Electronic positional sensors have found various clinical applications including the analysis of general physical activity, gait, posture, trunk, and upper limb movement. Tables 6.4 and 6.5 summarize

TABLE 6.3: Comparison of three sensors

	TYPE OF SENSOR		
	GYROSCOPE	ACCELEROMETER	ELECTROMYOGRAM
Measure	Measures rotational motion	Measures linear motion and gravity concurrently	Measures electrical potentials associated with muscle activity
Influence of gravity	Not influenced	Signal magnitude corrupted by gravity	Not influenced by gravity
Accuration of frequency information	Down to DC (zero frequency)		Frequency information appropriate
Accuration of magnitude information	Down to DC (zero frequency)		Magnitude less reproducible Contact resistance is a significant variable
Integration required to obtain position signal	Only a single integration is needed to obtain angular displacement	Second integration required to obtain displacement	
Signal-to-noise ratio	High signal-to-noise ratio	Low signal-to-noise ratio Gravity component	Low signal-to-noise ratio
Dynamic range	High	High	Low
Electrical contact with subject	No intimate electrical contact with subject	No intimate electrical contact with subject\	Intimate electrical contact with subject required
Size and influ- ence on motion	Solid state gyros (with no spinning wheels) do not influence motion of subject being measured	Piezoelectric accelerometers can be very small with no influence on motion; servo accelerometers much larger	Small sensing elements available, with little influence on motion

TABLE 6.4: Summary of applications of different sensors

MAIN APPLICATIONS	SENSOR				
	ACCELEROMETER	GYROSCOPE	ELECTROMAGNETIC	FLEXIBLE ANGULAR	SENSING FABRICS
Gait analysis	√	√	X	√	X
Posture and trunk movement analysis	√	√	√	√	√
Physical activity analysis	√	X	X	X	X

Adapted from Wong et al. (2007).

TABLE 6.5: Summary of different parameters usually measured using the positional sensors

TYPE OF SENSORS	PARAMETERS MEASURED
Accelerometer	Orientation of body segment Acceleration of lower limbs Velocity and translations of lower limbs Angle of lower limbs Acceleration and angle of upper limb movements Frequency of upper limb movements Acceleration of trunk Step and cycle time of walking Metabolic energy expenditure Tilting angle of trunk
Gyroscope	Velocity and stride length Joint angle of lower limbs Angular velocity of trunk rotation Angular displacement of trunk motions

the sensors' applications and measured parameters in different studies. For historical and practical reasons, accelerometers tend to be the most commonly used (Wong et al., 2007).

New advances in microelectromechanical and wireless technologies enable inertial sensing as an alternative for motion caption. For instance, Brunetti et al. (2006) built up a wireless inertial sensor including three linear accelerometers, three gyroscopes, and three magnetometers. The use of low-mass sensors is important to decrease the low-pass filtering effect due to mass addition.

6.3.1 Accelerometry

Accelerometer (Figure 6.6) measures acceleration along the sensitive axis of the sensor based on Newton's second law (force = mass × acceleration). There are three main types of accelerometers: piezoelectric, piezoresistive, and capacitive types. The characteristics are shown in Table 6.6 (Wong et al., 2007).

Accelerometry is simple, relatively reliable, and remains a convenient technique to measure frequency and amplitude of oscillatory signals (Bain, 1995). Sensors are fixed on the skin at given

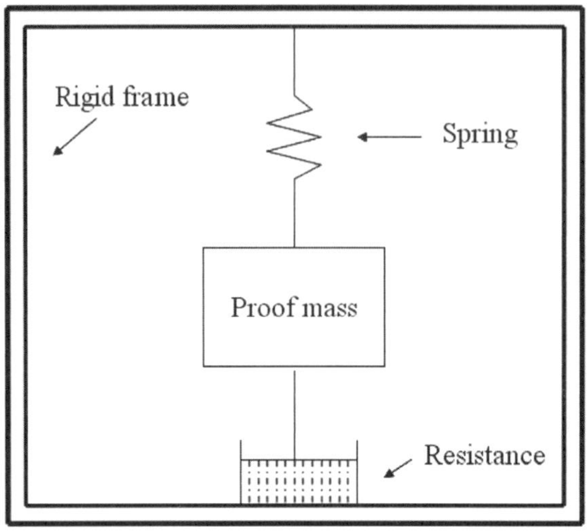

FIGURE 6.6: Basic layout of an accelerometer. In the mass–spring–damper system, the loading force drives a second-order damped harmonic oscillator, where the displacement of the proof mass relative to the rigid frame is considered. Under a constant acceleration condition, the displacement is directly proportional to the given acceleration. Modified from Wong et al. (2007).

anatomical landmarks. Various levels of support of the limbs (Figure 6.7) have been proposed (Morrison and Newell, 2000).

Accelerometers are interesting for intraoperative assessment of disabling motor symptoms such as tremor and side effects during surgery, to optimize the target position electrodes for deep brain stimulation (DBS). The intraoperative neurophysiological monitoring improves sensitivity and adds objective neurophysiological data to the clinical tests for improvement of the quality of stereotactic intervention of movement disorders (Journee et al., 2007). The reader is also referred to Chapter 8 about stereotactic surgery for tremor.

Regular calibration of accelerometers should not be underestimated.

6.3.2 Gyroscopes

A gyroscope is an angular velocity sensor. The angular orientation can then be obtained from integration of the velocity signal (Wong et al., 2007). A gyroscope is based on the measurement of the Coriolis force of vibrating devices. Coriolis force is an apparent force arising in a rotating reference frame and proportional to the angular rate of rotation. A simple model of gyroscope is shown in Figure 6.8.

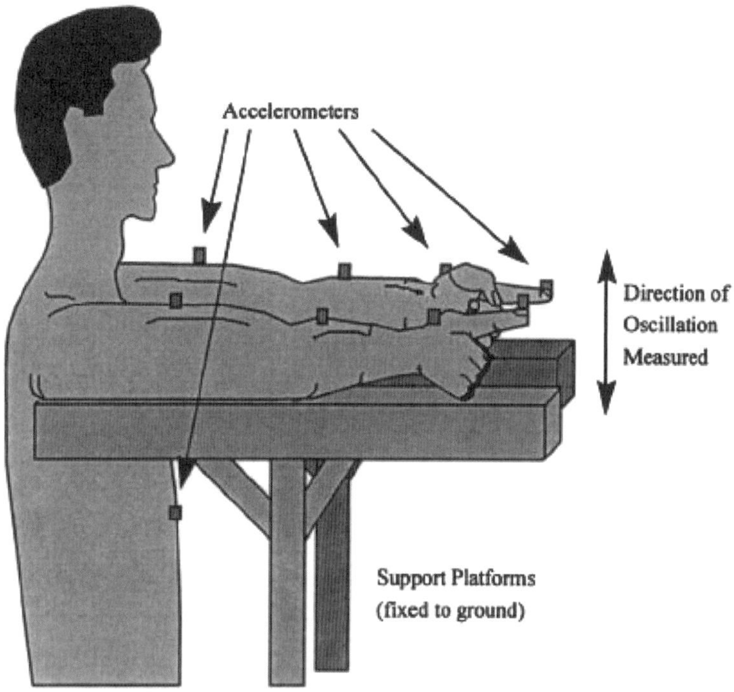

FIGURE 6.7: Schematic diagram of an experimental setup showing the position of the support and body during a condition where all segments of the upper limb—except the index fingers—were supported. From Morrison and Newell (2000), with permission from Elsevier.

There are three basic types: spinning rotor gyroscope, ring laser gyroscope, and vibrating mass gyroscope. The last one presents many advantages for portable applications because of its size, weight, lower power consumption, and cost. An internal mass is constantly vibrating inside the sensor.

6.3.3 Electromagnetic Tracking Systems

The system consists of a transmitter generating a low-frequency magnetic field detected by receivers. Signal processing difficulties may be encountered with this option in case of tremor evaluation.

6.3.4 Flexible Angular Sensor

The change of electrical output resulting from an angular change is measured. Strain gauges in goniometers produce an electrical output which is proportional to the angle. This option is clinically viable.

TABLE 6.6: Comparison of different accelerometers

PARAMETERS	PIEZOELECTRIC	PIEZORESISTIVE		CAPACITIVE
		SILICON	THICK FILM	
Gravitational component	No	Yes	Yes	Yes
Bandwidth	Wide	Moderate	Low	Wide
Self-generating	Yes	No	No	No
Impedance	High	Low	Low	Very high
Signal level	High	Low	Low	Moderate
Linearity	Good	Moderate	Moderate	Excellent
Static calibration (turnover)	No	Yes	Yes	Yes
Cost	High	Low	Low	High
Ruggedness	Good	Moderate	Moderate	Good
Suitable for shock	Yes	No	No	No

Adapted from Wong et al. (2007).

6.3.5 Videos

Videotaping of patients and computerized video motion detecting systems are qualitative and quantitative methods useful to analyze motion disorders (Zhang et al., 1999). Measuring systems based on video imaging are effective in quantifying motor impairments in clinical settings (Swider, 1998). The videotaped UPDRS motor examination (see Section 6.1 above) is widely used for diagnosing PD, with some limitations when patients have mild symptoms (Louis et al., 2002). Several experimental procedures for assessment of tremor validation of new methods benefit from videotaping (Louis et al., 1998, 2000, 2001; Stacy et al., 2007). Videotaping may also be used as a teaching tool to improve the uniform application of tremor rating scales by raters with various levels of experience

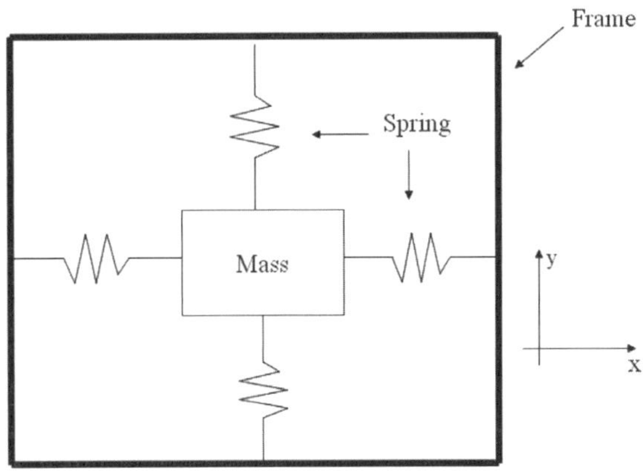

FIGURE 6.8: A simple model for analysing gyroscope behavior. A single mass is suspended by four springs in x and y directions. The frame is presumed to rotate about the z-axis. Assuming small-amplitude motions which will be true in resonant gyroscopes, the x-motion and y-motion can be coupled only through the Coriolis-force term and the term involving angular acceleration. The Coriolis force then induces motion in the third direction which is perpendicular both to the direction of rotation and to the driven motion. Adapted from Wong et al. (2007).

in movement disorders (Louis at al., 2001). However, analysis of videos may be time-consuming, which may become an obstacle.

6.3.6 Optoelectronic Devices

Several optoelectronic devices are commercially available. Most of them use reflective or active (light-emitting diodes LEDs) markers fixed on the skin. A calibration procedure with a reference structure may be required. Although rather expensive, their accuracy and the opportunity to perform simultaneous recordings in multiple sites are advantages.

6.4 HAPTIC DEVICES

Haptic robots and brain–computer interfaces (BCIs) are being used more and more in medicine and virtual reality. We have designed and built a new tool to investigate and monitor sensitively the functions of the wrist. This portable robotic device combines haptic technology with EMG assessment (a mechatronic myohaptic device). The system, called *wristalyzer* (Figure 6.9), allows assessing wrist motion in physiological and pathological conditions by applying loads and mechanical oscillations, taking into account the ergonomy and the angular positioning of the joints, thanks to

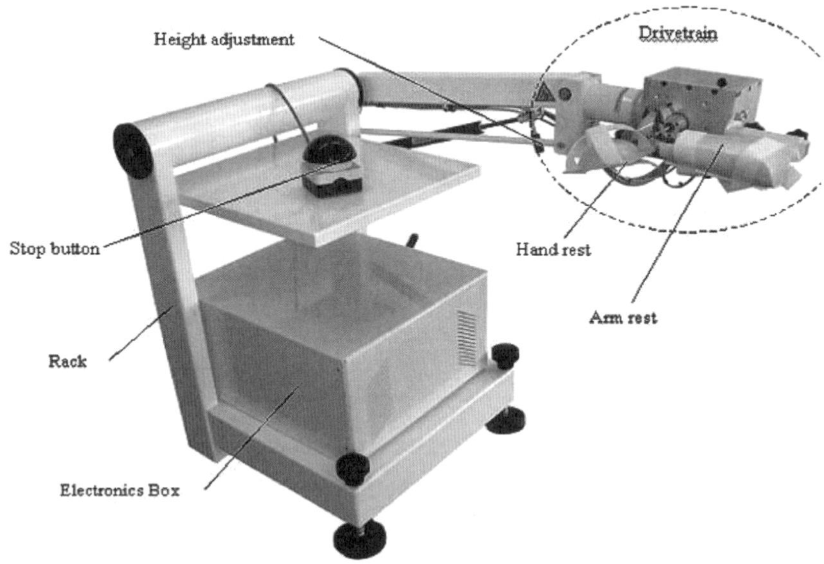

FIGURE 6.9: Illustration of the wristalyzer with its main components: rack with wheels, main unit, drivetrain. The mechatronic device includes a position encoder embedded in the motor casing and a torque sensor on the moving part attaching the hand flap (manipulandum) to the motor.

an adaptable manipulandum. The device characterizes the effects of damping on voluntary motion in neurological patients (Grimaldi et al., 2008).

The wristalyzer's features are the following:

- It works in a free or loaded mode for assessment of metrics of motion (normometria, dysmetria) and tremor. Angular motion, torques, and corresponding EMG activities are recorded. Spectral analysis of position, torque, and EMG trace is performed in real time.
- The device analyses the behavior of the wrist joints and the associated muscle activities during delivery of mechanical oscillations at frequencies from 0.1 to 50 Hz. Therefore, EMG activities in response to stretches can be assessed (Figure 6.10). The stretch responses are abnormal in neurological patients with a paraneoplastic cerebellar syndrome exhibiting dysmetria and kinetic tremor. The reliability of mechanically delivered oscillations has been investigated. Comparison with a conventional accelerometer has demonstrated a high accuracy in terms of reproducibility of frequency and amplitudes of oscillations (Grimaldi et al., 2007).
- The maximal voluntary contraction (MVC) can be measured.
- The device assesses automatically the impedance of the wrist (for assessment of rigidity or spasticity).

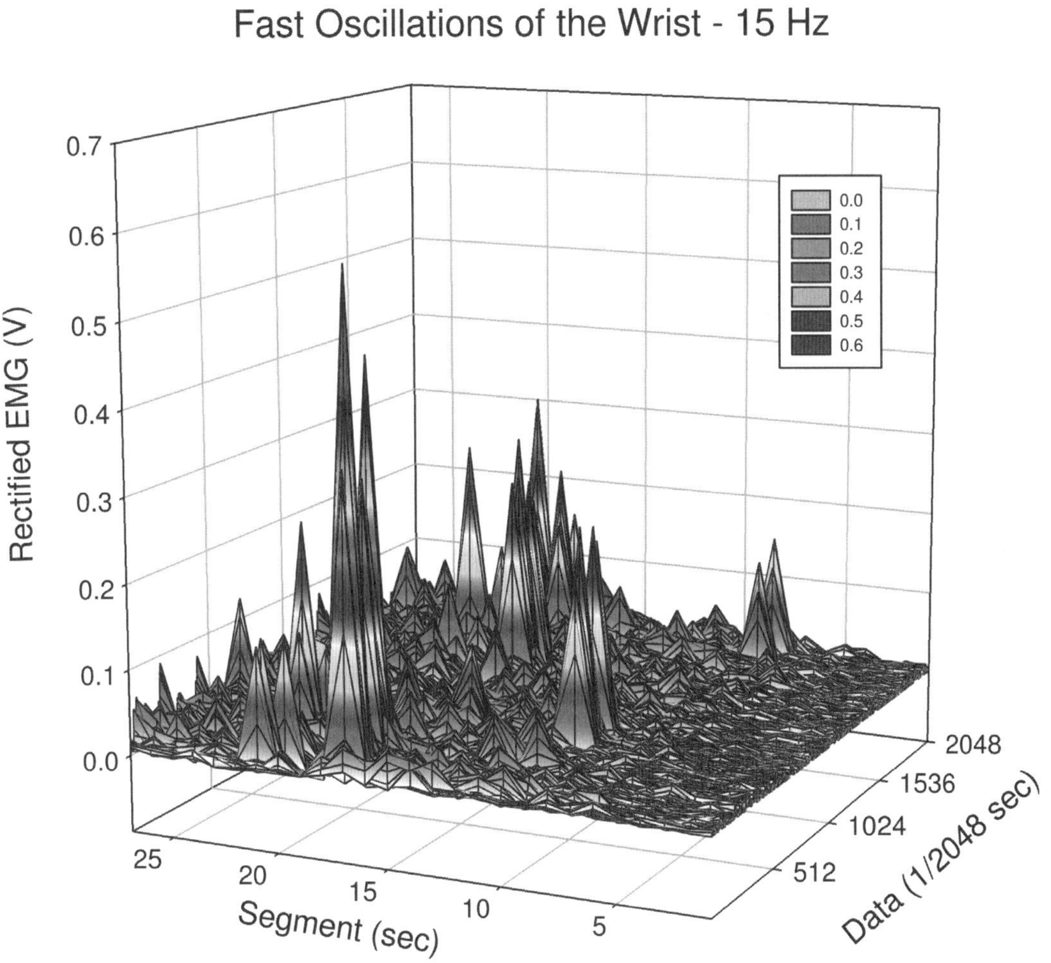

FIGURE 6.10: Effects of high-frequency oscillations (15 Hz) on EMG activity (forearm muscle) in a healthy subject. Note the bursts of EMG activities elicited by rhythmical stretches.

The wristalyzer has typical applications in acute and chronic neurological patients and can be used for rehabilitation or training purposes.

6.5 SIGNAL PROCESSING

6.5.1 Editing

Editing remains a critical step in digital processing. Data editing can be compared to a pre-analysis procedure allowing the detection and removal of degraded signals (DC offset). For tremor recordings, noise may result, for instance, from a problem which occurred during acquisition (instrumentation error, recording in a noisy environment, low frequency noise because of movement artifacts).

Many experts edit through visual inspection of recordings presented graphically. Signal-to-noise ratio (ratio of peak-to-peak signal to root mean square—RMS or quadratic mean—noise) is estimated.

We will not detail here the general principles and techniques of data filtering. Readers are referred to specialized books. Noise minimization is generally required for tremor signals. Frequency-selective filters (lowpass, highpass, bandpass, bandstop filters,or "notch" with a lower and upper cutoff) or adaptative filters may be used. Wavelet (time-scale distribution) denoising may be applied also to bandpass filter a given signal.

In case of triaxial accelerometry, the three-dimensional acceleration vector corresponds to the square root of the sum of the acceleration squared on all three axes. Removal of the 1 G (9.81 m/s^2) gravity vector can be performed.

Tremorous activity is composed of deterministic (nonrandom) and stochastic components. Signal processing is required to interpret time-series data of nonlinear systems and instances in which the frequency content of a signal provides more information than the original waveform. Most spectral estimation algorithms are devised for complete data sets. However, missing samples may occur. Nonparametric adaptative filtering-based tools are now available to deal with missing data samples. The "Gibbs phenomenon" relates to errors at points of discontinuities (a value of about 9% occurs in any region of discontinuity).

6.5.2 Fourier Transforms

Fourier transforms (FT) break down time-domain signals into constituent sinusoids of different frequencies. For periodic functions, the original waveform may be reconstructed from the sinusoidal components by application of the FT. The most used forms of FT are:

- the Fourier trigonometric series;
- the discrete FT: it is considered as a subset of the LaPlace transform;
- the fast FT (FFT): is the most popular method to perform transformation in the frequency domain. The FFT takes advantage of the symmetrical properties of sinusoidal periodic waveforms.

The spectral estimation methods allowing the calculation of the power spectra are the autocorrelation function, the FFT, and the autoregression. Units for power spectra are in power per frequency band.

Physiological waveform data generally represents truncated data segments of continuous signals. A discrete Fourier transform is often used for signals composed of data sampled at given spaced intervals, and a continuous signal may be reconstructed without information loss if the sampling frequency is greater than twice the highest frequency component in the signal (Nyquist critical frequency), to avoid aliasing. Direct application of the FT is typically modified by windowing (data windows such as Blackman/Hamming/Hanning window are applied to reduce the spectral leakage; truncated ends produce artificial side lobes or peaks in the autospectrum, see the Gibbs phenomenon above) and averaging to reduce artifacts in the power spectrum resulting from the analysis of truncated data. Indeed, spectra are averaged to reduce the variance of spectral estimates). Results of spectral densities may be gathered in a time-frequency representation (Figure 6.11).

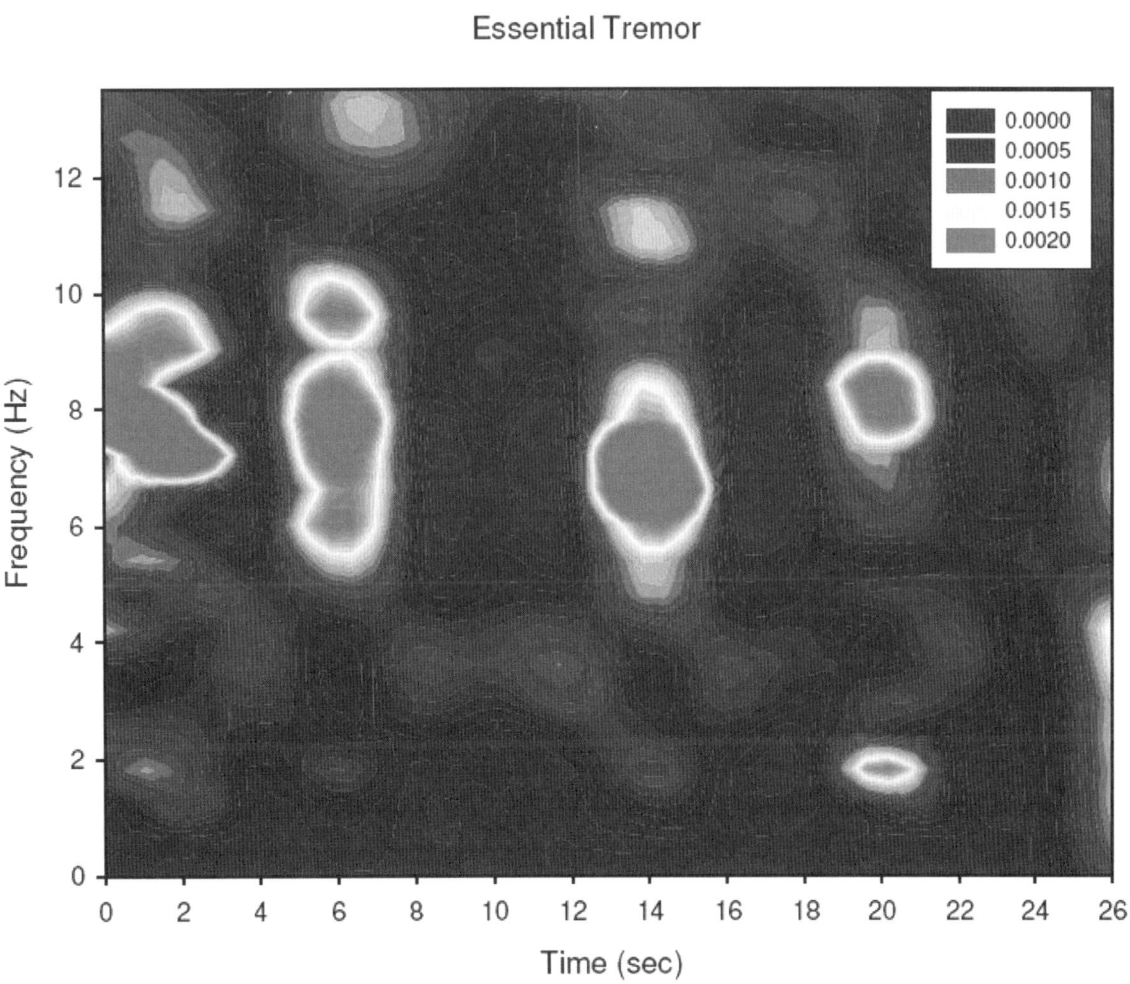

FIGURE 6.11: Time–frequency representation in a patient presenting essential tremor. Spectral densities are obtained using a Hamming window.

6.5.3 Drawbacks of FFT Analysis

There are several drawbacks of this technique. The signal is linearly decomposed as combination of sines and cosines, but most tremors are nonperiodic. The compromise between time and frequency resolution of these methods may not underline the presence of local oscillations in the signal, which might bring key information for the understanding of tremor.

The analysis of short data segments can be done using an autoregressive method of spectral analysis. The time series are first fitted to an autoregressive equation of order k. This can be very useful, for instance, in the analysis of the finger-to-nose test or the heel-to-knee test.

The Hilber analysis for tremor is a recently developed signal processing technique comprising the empirical mode decomposition (EMD) and the Hilbert spectrum (HS; Rocon et al., 2006; Figure 6.12). EMD can automatically detect and isolate tremulous movements from nonshaking movements. EMD decomposes arbitrary time-series into sets of components called intrinsic mode functions (IMFs). Subsequently, the energy of IMFs is assessed as a function of time and frequency on the HS. By contrast with the spectrogram, HS provides a windowing independent time–frequency representation allowing for the detection of events not detected by the conventional analysis.

6.5.4 Spectral Analysis Commonly Applied to Physiological Signals in Tremor

In the field of tremor, data from EMG and accelerometer recordings are treated with classical spectral analysis. Determination of spectrum shape, frequency of spectral peaks, *center frequency* (median value of the area below the power spectrum), harmonic index, as well as changes in frequency over time are some of the parameters resulting from tremor analysis. *Frequency dispersion* is the frequency width of an interval around the center of frequency that contains 66% of the total power spectrum. Frequency dispersion is used to estimate the harmonicity of oscillations. The power within a frequency range may be calculated by the integral of over the relevant frequency range (8–12 and 20–40 Hz).

These parameters allow to distinguish pathological and physiological tremor and to characterize the various types of pathological tremor. In the majority of patients with pathological tremor (PD tremor, ET, and cerebellar tremor), a single spectral peak is found (Machowska-Majchrzak et al., 2007). This peak is persistent for the whole registration period. Control subjects typically show variability in time in terms of frequency of tremor (Farkas et al., 2006). Moreover, pathological tremors show:

- significantly lower frequencies of the largest peak in the power spectrum,
- lower central frequency,

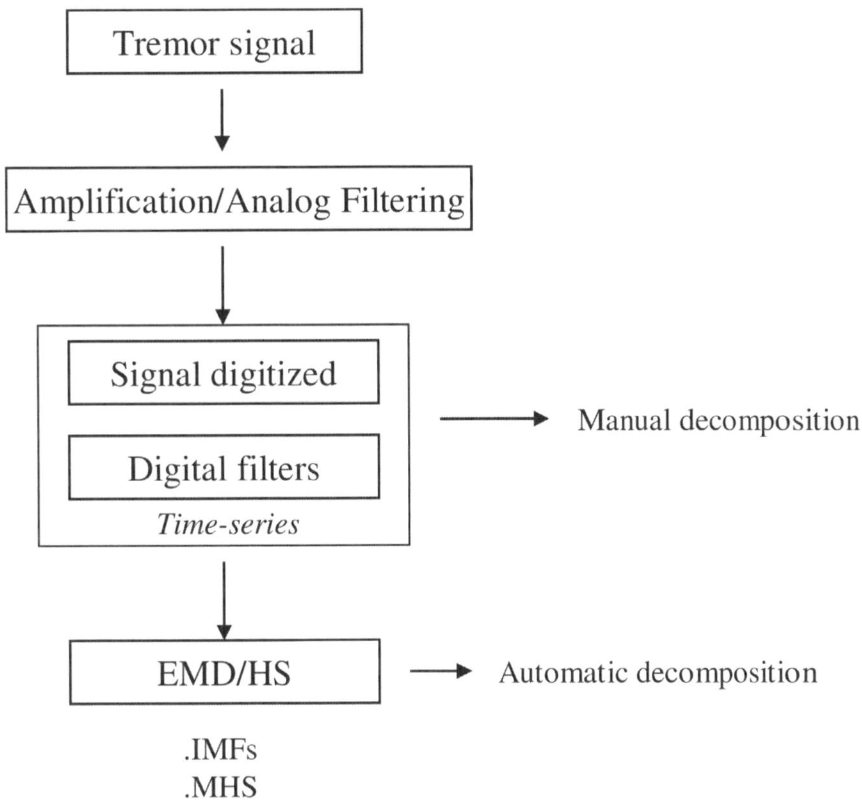

FIGURE 6.12: Block diagram showing the sequence of analysis (adapted from Rocon et al. 2006). Time-series are extracted and decomposed either manually or automatically. Tremor component and voluntary movement are demarcated. The empirical mode decomposition (EMD) provides the intrinsic mode functions (IMFs). Marginal Hilbert spectrum (MHS) $h(\omega)$ is estimated according to the following equation:

$$h(\omega) = \int_0^T H(\varpi,t)\mathrm{d}t \text{ where } \varpi(t) \text{ is the instantaneous frequency.}$$

- a significantly smaller standard deviation of the central frequency,
- a significantly higher harmonic index.

6.5.5 Wavelets

Wavelet transforms colocalize in both frequency domain and time domain and may be used effectively for nonstationary signals such as tremor. The basic idea behind wavelets is to express a signal as a linear combination of given sets of functions (wavelet transform). These are obtained by shifting and dilating a function called mother wavelet. A set of coefficients are computed. The signal is reconstructed as a linear combination of the basic functions weighted by these coefficients. Several mother wavelets are available (Gaussian, Haar, and Daubechies wavelets may be particularly useful for the assessment of EMG bursts).

Wavelet transforms are classified into discrete wavelet transforms (DWT) and continuous wavelet transforms (CWT; Figure 6.13). DWT is based on a specific subset of scale and translation values or representation grid. CWT runs over every possible scale and translation.

Because the signal in tremor has often many transient components which are interesting to isolate and analyze and because of noisy components, wavelet-based denoising can efficiently isolate activities of interest (myoelectric bursts) in many cases. Wavelet denoising has some advantages compared to conventional filtering such as smoothing, allowing a separation of the signal from noise.

Wavelet analysis can be combined with neural network analysis. This is out of the scope of this book.

6.5.6 Fuzzy Logic

In the field of the fuzzy logic (see Chapter 8), a comprehensive methodology has been developed for tremor analysis, including classic and chaotic parameters, complementing the time–frequency analysis. Teodorescu and coworkers have shown that tremor includes a significant chaotic component. The technique appears useful in diagnosis and rehabilitation. Their system includes nonlinear

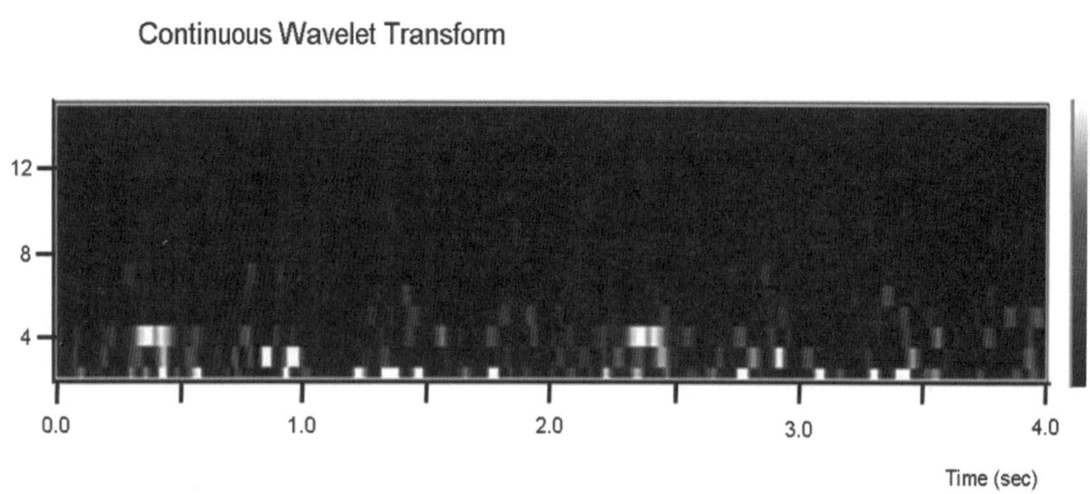

FIGURE 6.13: Continuous wavelet transform (CWT) in a postural tremor of upper limbs. Mother wavelet used: Haar filter. White corresponds to a strong positive correlation between the wavelet shape and the tremor oscillations. Black corresponds to the largest negative value.

data analysis and a fuzzy method to process the signal and to assess the type of tremor (Teodorescu et al., 2001).

6.5.7 Cross-Spectral Analysis and Coherence

Cross-spectral methods provide a powerful tool to investigate the relation between simultaneously recorded signals. These methods have been used in tremor research to study the relation between

- muscle activity (EMG) and magnetoencephalogram (MEG),
- EMG signal and electroencephalogram (EEG),
- EMG signal and other medical measurements,
- EMG signal from different muscles,
- EMG from single motor units and surface EMG.

Moreover, cross-spectral methods might be helpful in daily clinical routine as a diagnostic tool (Timmer et al., 2000).

The estimation of the *power spectrum* of a zero mean process $X_{(t)}$ is defined as the Fourier transform (FT) of the autocovariance function and is performed by a direct spectral estimation based on the FT of the measured data. The *cross-spectrum* $CS(\omega)$ of two zero mean process $X_{(t)}$ and $Y_{(t)}$, similar to the univariate case, is defined as the FT of the cross-correlation function:

$$CCF_{(t')} = (X_{(t)} Y_{(t-t')})$$

The modulus of the cross-spectrum $CS(\omega)$ normalized by the respective autospectra $Sx(\omega)$ and $Sy(\omega)$ gives the *coherence spectrum*:

$$\mathrm{Coh}(\omega) = |CS(\omega| / \sqrt{Sx(\omega)\ Sy(\omega)}$$

The *coherence* can be understood as a measure of linear predictability. The coherence value is 1 whenever $X_{(t)}$ is obtained from $Y_{(t)}$ by a linear operator. The coherence value is equal to zero whether there is no relation or the relation between the processes is a quadratic one; interpretation of coherence does not rely on the linearity of the processes $X_{(t)}$ and $Y_{(t)}$. Simultaneously recorded signals ($X_{(t)}$ and $Y_{(t)}$) may be uncorrelated or present a coherence unequal to 1 because of:

- a nonlinear relationship
- additional influences on $X_{(t)}$ apart from $Y_{(t)}$ (other inputs to the system have not been accounted for)
- estimation bias because of misalignment
- observational noise (Timmer et al., 2000).

REFERENCES

Alusi SH, Worthington J, Glickman S, Findley LJ, Bain PG. Evaluation of three different ways of assessing tremor in multiple sclerosis. *J Neurol Neurosurg Psychiatry.* 2000;68(6):756–760.

Bain P. Are neurophysiological techniques useful? In: *Handbook of Tremor Disorders.* Findley LJ, Koller WC (eds). New York: Marcel Dekker, 1995.

Bové M, Daamen N, Rosen C, Wang CC, Sulica L, Gartner-Schmidt J. Development and validation of the vocal tremor scoring system. *Laryngoscope.* 2006;116(9):1662–1667.

Breit S, Spieker S, Schulz JB, Gasser T. Long-term EMG recordings differentiate between parkinsonian and essential tremor. *J Neurol.* 2008;255(1):103–111.

Brunetti FJ, Rocon E, Pons JL, Manto M. The tremor coherence analyzer (TCA): a portable tool to assess instantaneous inter-muscle coupling in tremor. *Conf Proc IEEE Eng Med Biol Soc.* 2004;1:61–64.

Brunetti F, Moreno JC, Ruiz AF, Rocon E, Pons JL. A new platform based on IEEE802.15.4 wireless inertial sensors for motion caption and assessment. *Conf Proc IEEE Eng Med Biol Soc.* 2006 ;1 Suppl:6497–6500.

Calder KM, Stashuk DW, McLean L. Physiological characteristics of motor units in the brachioradialis muscle across fatiguing low-level isometric contractions. *J Electromyogr Kines.* 2008;18:2–15.

Elble RJ, Pullman SL, Matsumoto JY, Raethjen J, Deuschl G, Tintner R; Tremor Research Group. Tremor amplitude is logarithmically related to 4- and 5-point tremor rating scales. *Brain.* 2006;129(Pt 10):2660–2666.

Farkas Z, Csillik A, Szirmai I, Kamondi A. Asymmetry of tremor intensity and frequency in Parkinson's disease and essential tremor. *Parkinsonism Relat Disord.* 2006;12:49–55.

Gescheider GA. *Psychophysics: The Fundamentals.* Mahwah, NJ: Lawrence Erlbaum Associates, 1997.

Grimaldi G, Lammertse P, Manto M. Effects of wrist oscillations on contralateral neurological postural tremor using a new myohaptic device ('wristalyzer'). *Proceedings of the 4th IEEE-EMBS*, Cambridge, UK, 2007, pp. 44–48.

Grimaldi G, Lammertse P, Van Den Braber N, Meuleman J, Manto M. Effects of inertia and wrist oscillations on contralateral neurological postural tremor using the Wristalyzer, a new myohaptic device. *IEEE Trans Biomed Circuits Syst.* 2008 (in press).

Journee HL, Postma AA, Staal MJ. Intraoperative neurophysiological assessment of disabling symptoms in DBS surgery. *Neurophysiol Clin.* 2007;37(6):467–475.

Louis ED, Ford B, Bismuth B. Reliability between two observers using a protocol for diagnosing essential tremor. *Mov Disord.* 1998;13(2):287–293.

Louis ED, Barnes LF, Wendt KJ, Albert SM, Pullman SL, Yu Q, Schneier FR. Validity and test–retest reliability of a disability questionnaire for essential tremor. *Mov Disord.* 2000;15(3):516–523.

Louis ED, Barnes L, Wendt KJ, Ford B, Sangiorgio M, Tabbal S, Lewis L, Kaufmann P, Moskowitz C, Comella CL, Goetz CC, Lang AE. A teaching videotape for the assessment of essential tremor. *Mov Disord.* 2001;16(1):89–93.

Louis ED, Levy G, Côte LJ, Mejia H, Fahn S, Marder K. Clinical correlates of action tremor in Parkinson disease. *Arch Neurol.* 2001;58(10):1630–1634.

Louis ED, Levy G, Côte LJ, Mejia H, Fahn S, Marder K. Diagnosing Parkinson's disease using videotaped neurological examinations: validity and factors that contribute to incorrect diagnoses. *Mov Disord.* 2002;17(3):513–517.

Machowska-Majchrzak A, Pierzchała K, Pietraszek S. Analysis of selected parameters of tremor recorded by a biaxial accelerometer in patients with parkinsonian tremor, essential tremor and cerebellar tremor. *Neurol Neurochir Pol.* 2007;41(3):241–250.

Manto M. Clinical signs of cerebellar disorders. In: *The Cerebellum and Its Disorders.* Manto MU and Pandolfo M (eds). Cambridge, UK: Cambridge University Press, 2002.

Manto M, Sauvage C, Roark RM. Unifying hypothesis for the motoneuronal code in neurological disorders. *Biosci Hypoth.* 2008;2:93–99.

Masur H, Papke K. *Scale and Scores in Neurology. Quantification of Neurological deficits In Research and Practice.* Stuttgart: Thieme, 2004.

Mathiowetz V, Weber K, Kashman N, Volland G. Adult norms for the Nine Hole Peg Test of finger dexterity. *Occup Ther J Res.* 1985;5:1:24–37.

McAuley J, Rothwell J. Identification of psychogenic, dystonic, and other organic tremors by a coherence entrainment test. *Mov Disord.* 2004;19(3):253–267.

Miralles F, Tarongí S, Espino A. Quantification of the drawing of an Archimedes spiral through the analysis of its digitized picture. *J Neurosci Methods.* 2006;152(1–2):18–31.

Morrison S, Newell KM. Postural and resting tremor in the upper limb. *Clin Neurophysiol.* 2000; 111(4):651–663.

Raethjen J, Austermann K, Witt K, Zeuner KE, Papengut F, Deuschl G. Provocation of Parkinsonian tremor. *Mov Disord.* 2008;15:1019–1023.

Rocon E, Pons JL, Andrade AO, Nasuto SJ. Application of EMD as a novel technique for the study of tremor time series. *Conf Proc IEEE Eng Med Biol Soc.* 2006;(Suppl):6533–6536.

Stacy MA, Elble RJ, Ondo WG, Wu SC, Hulihan J; TRS study group. Assessment of interrater and intrarater reliability of the Fahn–Tolosa–Marin Tremor Rating Scale in essential tremor. *Mov Disord.* 2007;22(6):833–838.

Swider M. The application of video image processing to quantitative analysis of extremity tremor in humans. *J Neurosci Methods.* 1998;84(1–2):167–172.

Teodorescu HL, Kandel A, Hall LO. Report of research activities in fuzzy AI and medicine at USF CSE. *Artif Intell Med.* 2001;21(1–3):177–183.

Timmer J, Lauk M, Köster B, Hellwig B, Häußler S, Guschlbauer B, Radt V, Eichler M, Deuschl G, Lücking CH. Cross spectral analysis of tremor time series. *Int J Bifurc Chaos.* 2000;10: 2595–2610.

Trouillas P, Takayanagi T, Hallett M, Currier RD, Subramony SH, Wessel K, Bryer A, Diener HC, Massaquoi S, Gomez CM, Coutinho P, Ben Hamida M, Campanella G, Filla A, Schut L, Timann D, Honnorat J, Nighoghossian N, Manyam B. International Cooperative Ataxia Rating Scale for pharmacological assessment of the cerebellar syndrome. The Ataxia Neuropharmacology Committee of the World Federation of Neurology. *J Neurol Sci.* 1997;145(2):205–211.

van Rootselaar AF, Maurits NM, Renken R, Koelman JH, Hoogduin JM, Leenders KL, Tijssen MA. Simultaneous EMG-functional MRI recordings can directly relate hyperkinetic movements to brain activity. *Hum Brain Mapp.* 2007.

Wong WY, Wong MS, Lo KH. Clinical applications of sensors for human posture and movement analysis: a review. *Prosthet Orthot Int.* 2007;31(1):62–75.

Zhang T, Wei G, Yan Z, Ding M, Li C, Ding H, Xu S. Quantitative assessment of Parkinson's disease deficits. *Chin Med J (Engl).* 1999;112(9):812–815.

• • • •

CHAPTER 7

Mechanisms of Tremor

In this chapter, we review the pathophysiology of the various forms of tremor. Principal mechanisms and presumed generators are summarized in Table 7.1. Tremors are rarely because of a single factor.

7.1 BLOOD STUDIES

Secondary causes of tremor may be identified by laboratory workup. Tremor is a classic sign of hypoglycemia. Hyperthyroidism may induce tremor, as well as high levels of toxic agents, such as medications (see Chapter 4), metals (manganese, lead, methylmercury, copper, iron overload), etc. (Sadek et al., 2003). Copper overload is associated with Wilson's disease (characterized by corneal deposits, overload of copper in liver, low ceruplasmine level in blood, and increased 24-h cupruria) (Benito-León and Louis, 2006).

A high level of monoclonal IgM in the blood may be found in cases exhibiting a postural tremor. Tremor is associated with monoclonal gammopathy of undetermined significance (MGUS) or with a chronic inflammatory demyelinating neuropathy CIDP (Bain et al., 1996). It is worthy to note that tremor can be the presenting symptom in paraproteinemic neuropathies, but some patients will exhibit tremor several years after disease onset. Peripheral neuropathy may be associated with benign IgG and IgA paraproteinemia, in the absence of antimyelin glycoprotein activity (anti-MAG).

Cerebellar syndrome and tremor may also have an autoimmune origin. Anti-Yo, anti-Hu, anti-Tr and other antineuronal antibodies (Shams'ili et al., 2003), antiglutamic acid decarboxylase antibodies (Vianello et al., 2003), and anti-GM1 ganglioside antibodies (Zappia et al., 2002) should be looked for in case of cerebellar syndrome. Patients presenting anti-Tr antibodies-associated Hodgkin's lymphoma may present myorhythmia (Wiener et al., 2003)

Genetic diseases should not be overlooked.

7.2 BRAIN IMAGING

7.2.1 Computerized Tomography

Atrophy detected by computerized tomography (CT scan) and changes in signal intensity observed on *magnetic resonance* (*MRI*) are well-known in patients with kinetic tremor (Fukuhara et al., 1994). Lesions may involve the brainstem or the cerebellothalamocortical projections. Lesions in the white

TABLE 7.1: Principal mechanisms and generators of tremor

TYPE OF TREMOR	MECHANISMS/GENERATORS
Physiological tremor	• related to the mechanical properties of the oscillating limb • central component attributed to spinal interneuronal systems, subcortical oscillators, or to cortical rhythms
Enhanced physiological tremor	• enhancement of the central component of physiological tremor • a single brainstem source or bilateral oscillators closely linked at or below this level
Essential tremor	• abnormal GABAergic system (see Chapter 3) • abnormal cerebellar afferents from the inferior olive/abnormal bilateral overactivity of cerebellar connections • involvement of the thalamic nuclei • subcortical and subcorticocortical circuits, i.e., cerebellar–thalamic–cortical pathway
Rest tremor and PD	• multiple oscillatory circuits operating on similar frequencies • corticosubthalamopallidothalamic loop • external pallidum and subthalamic nucleus acting as a pacemaker • abnormal synchronization within the striatopallidothalamic pathway • abnormal coupling in a cerebellodiencephalic–cortical loop • functional disequilibrium between GABAergic and DAergic influences (>DA in the caudoventral parts of the GPi)
Kinetic tremor	• cerebellar dysfunction • abnormality of the adaptive cerebellar afferent inflow to motor cortex
Orthostatic tremor	• bilateral overactivity of cerebellar connections • oscillator located in the brainstem (inferior olive) and influenced by cerebral cortex, basal ganglia, and cerebellum

matter with or without basal ganglia involvement are observed in vascular parkinsonism (Rao et al., 2006).

7.2.2 Positron Emission Tomography

Positron emission tomography (PET) imaging investigates in vivo (by measuring regional cerebral flow rCBF and glucose/oxygen/amino acids consumption) neurochemical, hemodynamic, or metabolic processes that underlie movement disorders. PET and specific tracers to measure *dopamine transporter* (DAT) binding can provide an accurate and highly sensitive measure of degeneration in the dopamine system in PD (Antonini et al., 2001; see also Chapter 5).

7.2.3 Magnetoencephalography

Oscillations are a prominent feature of macroscopic human sensorimotor cortical activity as recorded noninvasively with electroencephalography (EEG) and magnetoencephalography (MEG). The advent of whole-scalp MEG systems allows a rapid noninvasive recording from the entire cortex and an accurate localization of neural sources (Schnitzler and al., 2000).

7.2.4 Transcranial Magnetic Stimulation

Transcranial magnetic stimulation (TMS) studies in combination with MRI have been performed mainly for the unraveling of mechanisms of sensorimotor plasticity. Many TMS abnormalities are seen in the different diseases, concurring to show that motor cortical areas and their projections are the main target of the basal ganglia dysfunction in movement disorders (Cantello, 2002).

7.3 CONTRIBUTION OF ANIMAL MODELS IN THE IDENTIFICATION OF THE CAUSE OF TREMOR

Animal models, especially in rodents and monkeys, are widely used in the field of tremor pathogenesis. Drugs under development are first assessed in these models. Figure 7.1 summarizes the sites of action of pharmacological agents used to generate animal models of tremor.

7.4 PHYSIOLOGICAL TREMOR

Although the main determinant of PT is related to the mechanical properties of the oscillating limb, an additional central component coexists. Measurement of PT in the hand and finger in a group of 117 normal subjects aged between 20 and 94 years showed a significant reduction of the tremor frequency by adding inertia (Raethjen et al., 2000). A negative correlation with hand volume was found. The significant EMG peak (50–80% of the recordings) was unaffected by mechanical

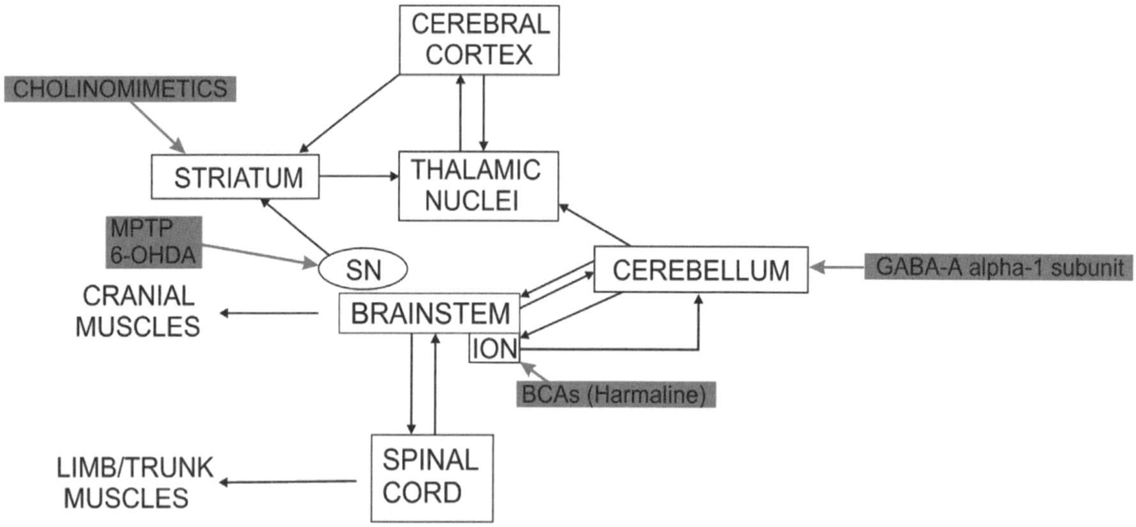

FIGURE 7.1: Sites of action of pharmacological agents used to generate animal models of tremor, especially in rodents and monkeys. Systemic administration of cholinomimetics induces a general tremor by acting on the muscarinic receptors of striatal neurons. MPTP (1-methyl-4-phenyl-1,2,3,6-tetrahydropyridine) is highly toxic for dopaminergic neurons in the substantia nigra (SN). 6-OHDA (6-hydroxydopamine) destroys cathecholaminergic fibers, including nigrostriatal projections. Beta-carboline alkaloids (BCAs) such as harmaline, harmine, and ibogaine produce an action tremor via a facilitatory effect on rhythmic activity (enhancement of electrotonic coupling) of the inferior olivary nucleus (ION) projecting to the cerebellum. Genetic mutants may exhibit a phenotype with tremor. GABA-A receptor alpha-1 subunit knockout mice exhibit a high-frequency postural and kinetic tremor. Adapted from Miwa (2007).

changes in the periphery (Figure 7.2). A significant EMG-EMG coherence was observed in about one-third of the subjects.

The independent central component in the 6- to 15-Hz band has been attributed to spinal interneuronal systems or subcortical oscillators. Recently, the role of cortical rhythms has been underlined. The findings of a 6- to 15-Hz coherence between cortex and EMG have been interpreted as a corticomuscular transmission of the oscillation rather than a peripheral feedback to the cortex (Raethjen et al., 2002).

Recent works highlight that PT occurring while the subject extends the middle finger against a spring has two load-independent components and a load-dependent component (Takanokura and Sakamoto, 2005). The stretch–reflex system causes the load-dependent component, whose frequency is determined by the mechanical property of the elastic load. The load-independent components are likely produced by the supraspinal system.

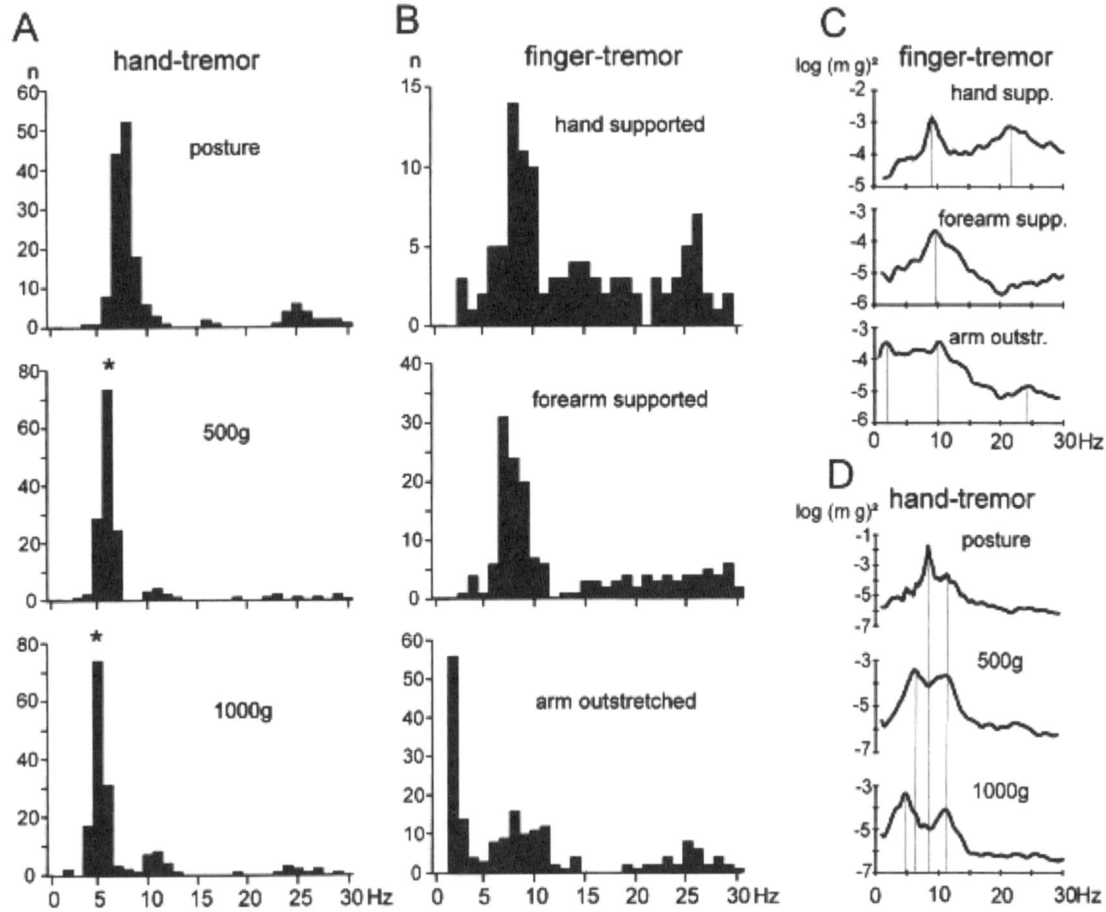

FIGURE 7.2: Distributions of accelerometric peak frequencies under different recording conditions in physiological tremor. The frequencies are displayed representatively for the right side, but the results of the other side were essentially the same. (a) The main frequency band of the hand in the 6- to 12-Hz band clearly dropped under added weight (Wilcoxon test: posture 500 g, $P < 0.001$; 500–1,000 g, $P < 0.001$). There was a small amount of remaining peaks in the 10-Hz range under weight load of the hand. (b) Finger tremor showed three distinct frequency bands (2–4, 6–12, and 15–30 Hz), which strongly depend on the arm position. (c) Examples of coincident occurrence of the different peaks within the same spectra. (d) Example of an additional hand tremor peak around 10 Hz which is not influenced by weight. From Raethjen et al. (2000), with permission from Elsevier.

The main difference in PT magnitudes between the dominant and nondominant side of right-handed individuals consists of an approximately 30% greater amplitude of fluctuations in acceleration for the nondominant hand. Central modulation of neural activity has been proposed to explain this phenomenon (Bilodeau et al., 2008).

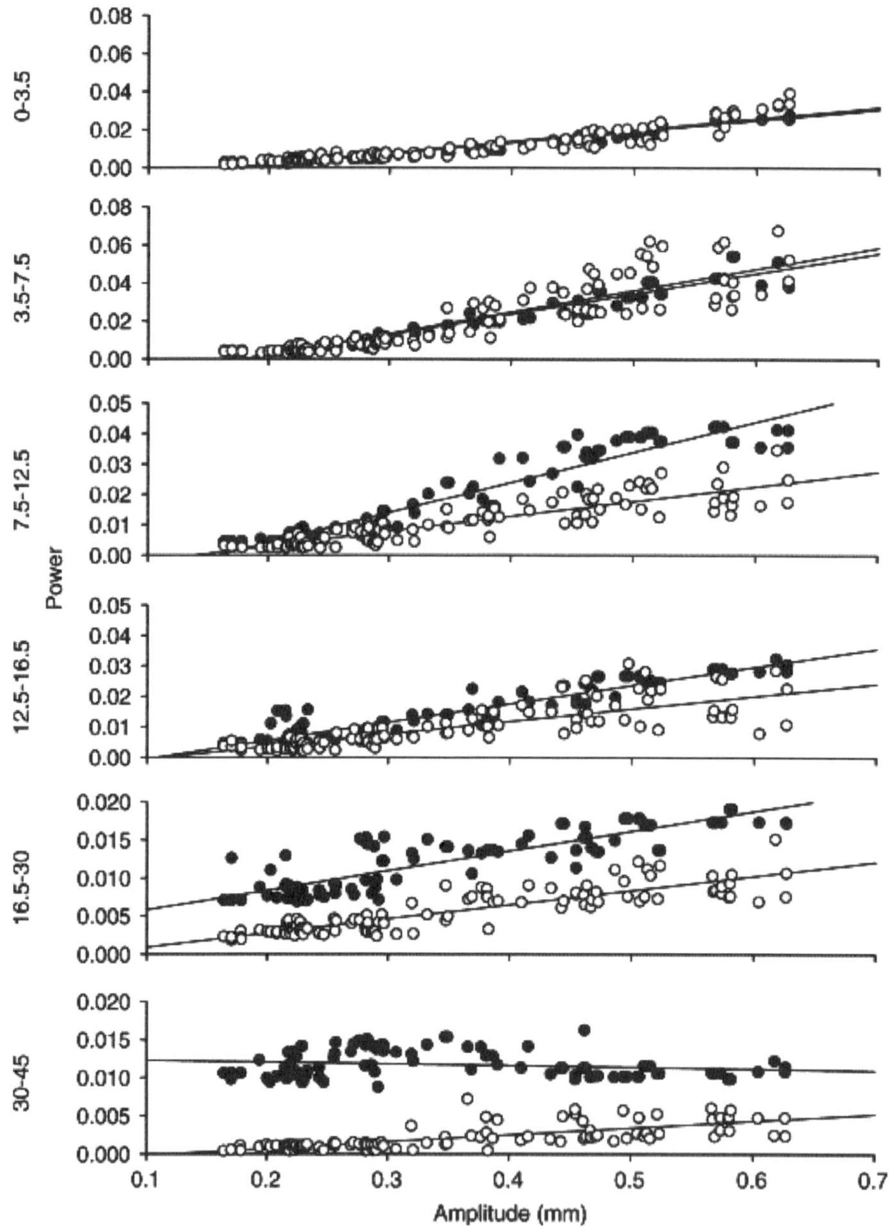

FIGURE 7.3: From top to bottom: in the ordinate axis of each plot, the sum of power found in postsurgical tremor (white dots) and matched controls for each frequency bands. Frequency bands are indicated on the far left. Abscisses represent the tremor amplitude placed in increasing order. Solid lines represent the second-order regression. Note that within the 0- to 3.5-, 3.5- to 7.5-, and 12.5- to 16.5-Hz bands, the power increased concomitantly in both groups as higher 5-s epochs amplitudes were compared. For the 7.5- to 12.5-Hz band, the power increased mostly for controls. As for the 16.5- to 30- and 30- to 45-Hz bands, there was also clearly more power for the controls. From Duval et al. (2005), with permission from Elsevier.

The cortical correlates of PT may be involved in linking different cortical motor centers and might therefore play a role in cortical motor planning (Raethjen et al., 2004).

In a comparison between post-thalatomy parkinsonian tremor and physiological tremor, Duval and colleagues found that the high-frequency component of physiological tremor failed to emerge in patients which underwent thalatomy (Figure 7.3). Therefore, they suggested that the thalamus should be considered as an important component of the generation and/or propagation of high-frequency components of physiological tremor (Duval et al., 2005).

7.5 ENHANCED PHYSIOLOGICAL TREMOR

A typical example of enhancement of postural tremor occurs in subjects treated with amitryptiline (a tricyclic antidepressant drug, see Chapter 4). The proposed mechanism is an enhancement of the central component of physiological tremor (Raethjen et al., 2001).

Transcranial magnetic stimulation (TMS) studies of long-latency reflexes (LLR) and cross-spectral analysis were performed by Köster et al. (1998) in subjects with the syndrome of persistent mirror movements (PMM), characterized by abnormal bilateral corticospinal projections (Köster et al., 1998). The authors made the hypothesis that a supraspinal mechanism of EPT would result in coherent activity between both sides in PMM subjects. The authors concluded that the 8- to 12-Hz component of EPT is transmitted transcortically.

It has been proposed that a single brainstem source or bilateral oscillators closely linked at or below this level may participate in tremor genesis of EPT (O'Sullivan et al., 2002).

7.6 POSTURAL TREMOR AND ESSENTIAL TREMOR

7.6.1 Postural Tremor

Intralimb coordination analysis in healthy subjects showed that motion is influenced by a compensatory synergy organized around the wrist and shoulder joints. This compensatory synergy reduces the coordination of the four degrees of freedom (arm links) to a single degree of freedom arm control task (Morrison and Newell, 1996). Coherence analysis of intersegment tremor relation has confirmed that the upper limbs maintain a degree of independence in both postural and rest tremor, although they exhibit the same intralimb pattern of compensatory organization.

7.6.2 Essential Tremor

Slight increases in the concentrations of glutamate and reduced levels of GABA, glycine, and serine in cerebrospinal fluid (CSF) provide a neurochemical basis for the central oscillations observed in ET and also raise the possibility that a genetically determined metabolic disorder is involved in its etiology (Mally et al., 1996). Harmaline, a beta-alkaloid derivative, induces a temporary essential-like tremor in animals via its action at the level of the olivocerebellar pathway (Wilms et al., 1999. See also Figure 7.1). The inhibition of GABA-A receptors results in enhanced electrical coupling of cerebellar afferents in the inferior olive (Stratton and Lorden, 1991). Harmaline might also act directly upon other brain structures involved in tremor genesis.

Deletion of GABA-A receptor α1 subunits determines the loss of 50% of all GABA-A receptors in the brain (Kralic et al., 2002), including the motor pathways in the brainstem, cerebellum, thalamus, and basal ganglia. Mice carrying this deletion exhibit motor incoordination that mimics human ET, although tremor frequency is higher in rodents (Kralic et al., 2005).

Another argument for the involvement of the inferior olivary nuclei and thalamus in the generation of ET comes from the significant glucose hypermetabolism of the medulla and thalami in ET patients (Hallet and Dubinsky, 1993).

The excitability of the cerebellothalamocortical pathway in ET seems normal, although recent data indicate that this might not be true at a cellular level. Resetting of tremor is feasible with motor cortex stimulation but not with cerebellar stimulation, suggesting that the abnormal oscillatory activity in ET does not originate in the cerebellar cortex itself, but that the abnormal activity is likely transmitted through the cerebellothalamocortical pathway and driven by abnormal cerebellar afferent input from areas such as the inferior olive (Pinto et al., 2003). The involvement of the thalamic neurons could be because of the fact that they generate intrinsic membrane oscillations (see also the inferior olive), behaving as oscillators or even resonators (Llinas and Paré, 1995). Thalamocortical firing patterns vary with their membrane potential. The interaction between their time-dependent cation current, the low-threshold calcium conductance, and the calcium-dependent conductance can generate oscillations from 0.5 to 4 Hz.

The case of a patient with ET who improved after a sensorimotor stroke (a small cortical infarct near the left precentral region) supports the presence of cortical or transcortical motor loops mediating ET (Kim et al., 2006). The functional changes in the sensorimotor cortex of patients with ET after stereotactic ventralis intermedius (Vim) thalamotomy (see also Chapter 8) indicate that the fundus of the central sulcus (Brodmann area 3a, see Chapter 2) is one of the key relays in the tremor circuit (Miyagishima et al., 2007). Indeed, deep brain stimulation (DBS) of the thalamic region is effective on tremor in ET, and recent studies reveal an increased efficacy of stimulation within the subthalamic area compared to the thalamus (Vim), suggesting a role of this structure (Herzog et al., 2007).

It seems plausible that tremor oscillations build up in different subcortical and subcorticocortical circuits, which temporarily entrain each other (Raethjen et al., 2007). The absence of significant coherence between the magnetoencephalogram and electromyogram at given tremor frequencies suggests that the tremor is imposed on the active muscle through descending pathways other than those originating in the primary motor cortex (Halliday et al., 2000; Figure 7.4). Coherence analysis of EMG signals in ET indicates that tremor in the right and left arm are related to independent

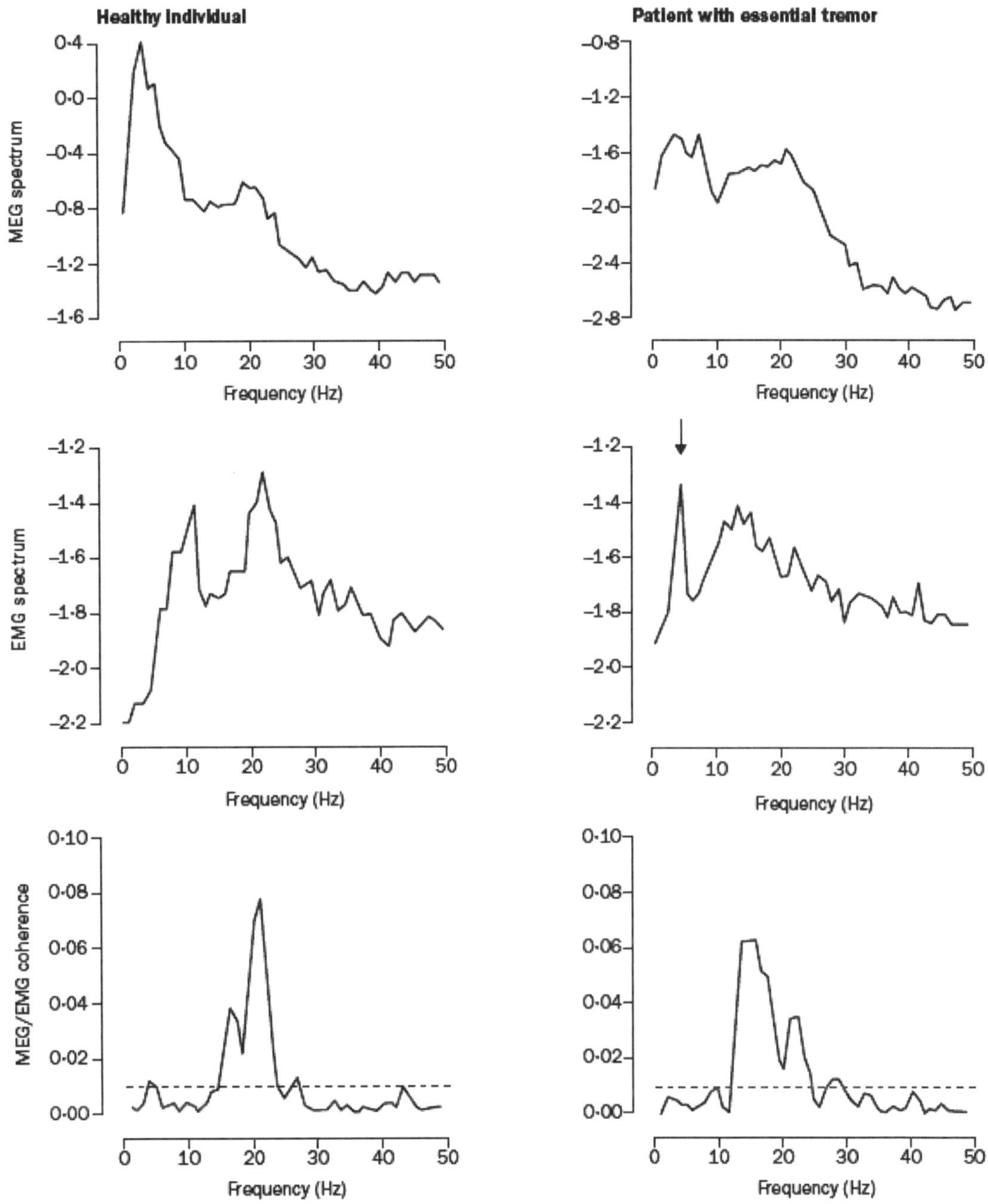

FIGURE 7.4: Spectral and coherence analysis of the magnetoencephalogram and contralateral electromyogram recorded from a healthy individual and a patient with essential tremor during voluntary maintained postural contraction of the first dorsal interosseous muscle. The horizontal dashed line indicates the level of the 95% CI for coherence. Values of coherence above this level are significant. Arrow indicates peak tremor frequency. From Halliday et al. (2000), with permission from Elsevier.

central oscillators. Hellwig and colleagues have proposed that central oscillators in the right and left brain are not entirely independent of each other and that they may dynamically synchronize, presumably by interhemispheric coupling via the corpus callosum (Hellwig et al., 2003; Figure 7.5).

Arguments for a primary involvement of cerebellum in ET genesis do exist. Proton magnetic resonance spectroscopic imaging (MRS) shows a decrease in cerebellar *N*-acetylaspartate/total creatine (NAA/tCr) ratio in ET, signifying cerebellar neuronal dysfunction or even predegeneration. A correlation between blood harmane concentration and brain NAA/tCR is found for the cerebellar cortex, but not in central cerebellar white matter, cerebellar vermis, thalamus, or basal ganglia (Louis et al., 2007). Although MRI voxel-based morphometry showed a lack of a consistent decrease in gray and white matter density in patients with ET (arguing against a progressive neurodegenerative

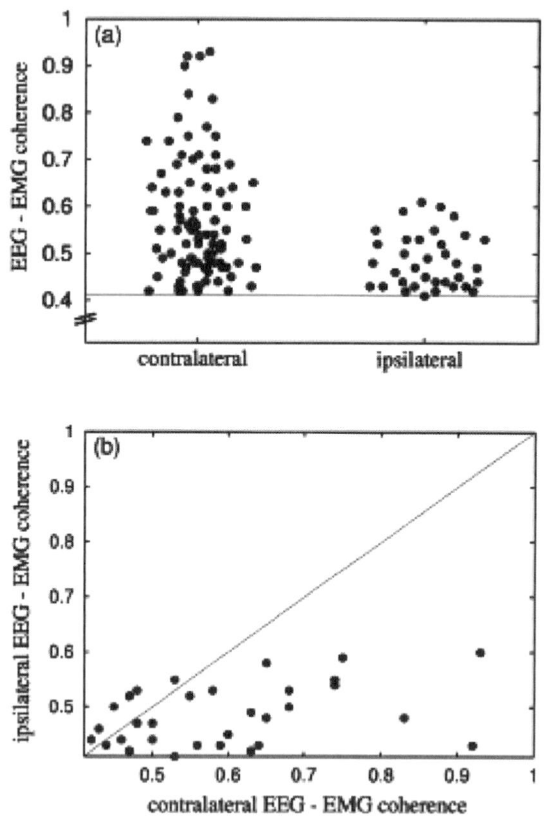

FIGURE 7.5: (a) All contralateral and all ipsilateral EEG–EMG coherences encountered in the study. The horizontal line at a coherence of about 0.4 indicates the level of significance. (b) For all cases of bilateral EEG–EMG coherence, the contralateral EEG–EMG coherences are plotted against the ipsilateral ones. From Hellwig et al. (2003), with permission from Elsevier.

process in ET; Daniels et al., 2006), heterogeneous and degenerative anatomopathological changes are reported in ET. Indeed, cerebellar changes and brainstem Lewy bodies distinguished 10 ET cases from 12 controls (Louis et al., 2006). A further study on 33 ET and 21 control brains confirmed the presence in ET brains of pathological changes in the cerebellum (75.8%)—probably as a consequence of an overactivity within cerebellar nuclei—with decreased numbers of Purkinje cells, presence of Lewy bodies in the brainstem (24.2%), degeneration of the dentate nucleus in two cases of ET without Lewy bodies, Purkinje cell heterotopias, and dendrite swellings (Louis et al., 2007). The reduction in Purkinje cell number in the brains of patients with ET has been demonstrated even in the absence of Lewy bodies, supporting the view that the cerebellum is anatomically, as well as functionally, abnormal in these cases of ET (Axelrad et al., 2008).

It is established that the cerebellum plays a role in the event-based timing of synchronized repetitive movements. Farkas et al. (2006) found a severe deficit of event-based rhythm generation on both sides, supporting the presumed bilateral cerebellar dysfunction in ET. Moreover, impairment of motor imagery (the process of mental representation of movements) was recently demonstrated. However, whether this is related to a genuine cerebellar or motor cortical dysfunction is still unclear (Lo et al., 2007).

In the subgroup of ET patients with predominant kinetic tremor, a relative expansion of gray matter areas involved in higher-order visuospatial processing is detectable by MRI voxel-based morphometry. Hypothetically, this is related to a long-term adaptive reorganization in the central nervous system, compensating the higher demands on the visuospatial control of skilled movements in case of trembling (Daniels et al., 2006).

7.7 REST TREMOR AND PARKINSON'S DISEASE

In a study on coherence in 22 subjects affected by Parkinson's disease (PD), no consistent pattern across patients was found, suggesting that tremor in PD is generated by multiple oscillatory circuits, which tend to operate on similar frequencies (Ben-Pazi et al., 2001; Raethjen et al., 2000). PD tremor is coupled within, but not between, limbs. The anatomy of basal ganglia loop (see Chapter 2) may explain the presence of several generators.

Three main neuronal mechanisms have been hypothesized:

- corticosubthalamopallidothalamic loop generating tremor
- pacemaker consisting of the external pallidum and the subthalamic nucleus
- an abnormal synchronization within the whole striatopallidothalamic pathway leading to a loss of segregation (Deuschl et al., 2000).

Typical PD resting tremor (4–6 Hz) is associated with strong coherence between the EMG of forearm muscles and activity in the contralateral primary motor cortex (M1) at tremor frequency but also at double tremor frequency. Tremor-related oscillatory activity within a cerebral network has been demonstrated. There is an abnormal coupling in a cerebellodiencephalic–cortical loop, cortical motor (primary motor cortex, cingulated/supplemental motor area, lateral premotor cortex), and sensory (secondary somatosensory cortex, posterior parietal cortex) areas contralateral to the tremor hand (Timmermann et al., 2003). The most striking differences between parkinsonian patients and healthy subjects imitating the resting tremor are a reduction of the coupling between primary sensorimotor cortex and a diencephalic structure—most likely the thalamus—and an enhancement of the coupling between premotor and primary sensorimotor cortex (Figure 7.6). These results indicate that the coupling of oscillatory activity within a cerebellodiencephalic–cortical loop constitutes a basic feature of physiological motor control, sustaining the hypothesis that parkinsonian resting tremor involves oscillatory cerebrocerebral coupling in a physiologically preexisting network (Pollok et al., 2004).

Animal studies suggest that neuronal oscillations are spontaneously generated within the basal ganglia system, especially from the GPe and the subthalamic nucleus (STN), but are mainly synchronized by cortical activity via the striatal inputs (see Cassim et al., 2002). Dopamine depletion causes a global enhancement of oscillations within the whole basal ganglia system, particularly in the GP–STN network. Interestingly, acute DA blockade (using high doses of haloperidol) generates synchronous oscillatory activity across the basal ganglia neuron populations, and prolonged DA blockade causes phase lag changes in pallidostriatal synchrony (Burkhardt et al., 2007).

The small amplitude postural tremor in PD arises from heterogeneous oscillator mechanisms. The associated increased corticomuscular coupling is related to the cortical involvement (Caviness et al., 2006).

Some authors have observed that low-frequency stimulation of the pedunculopontine region can induce tremor in the normal-behaving primate and that chronic high-frequency deep brain stimulation (DBS) of the zona incerta (which has reciprocal connections with several cortical areas, the upper brainstem, the cerebellum, and the thalamus) suppresses proximal limb tremor. This confirms the role of this area around the upper brainstem in the control of movement (Nandi et al., 2002).

Autopsy studies in PD and controls have shown that dopamine (DA) levels in the external globus pallidus (GPe) of normal brains are greater than in the internal pallidum (GPi). In PD, the mean loss of DA is marked (-82%) in GPe and moderate (-51%) in Gpi. However, DA levels are nearly normal in the ventral (rostral and caudal) GPi of PD cases with prominent tremor. There is marked loss of DA (-89%) in the caudate and severe loss (-98.4%) in the putamen in PD. The pattern of pallidal DA loss does not match the putaminal DA loss. The possible functional disequilibrium between GABAergic and DAergic influences the balance in favor of DA in the caudoventral

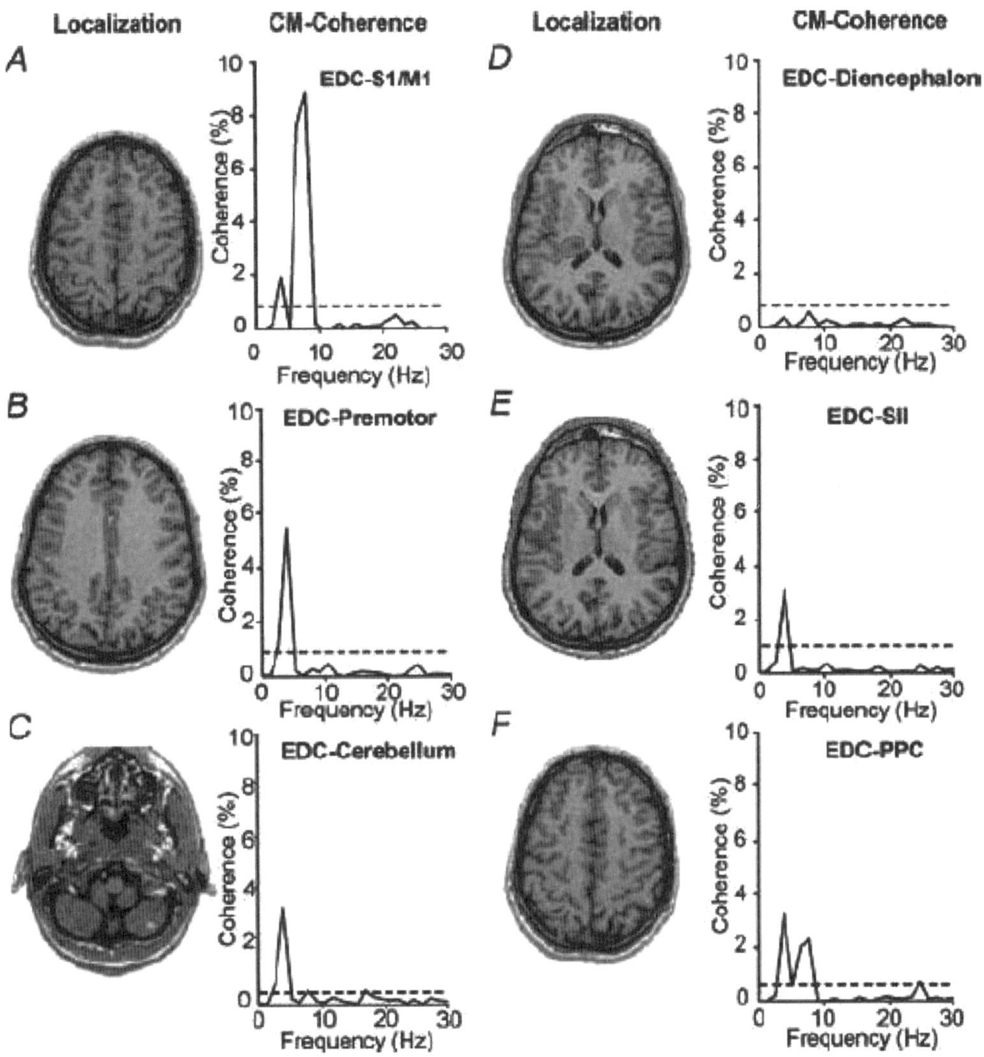

FIGURE 7.6: Localization of cerebral sources and coherence spectra to the right extensor digitorum communis (EDC) muscle in a representative subject imitating the PD resting tremor with the right hand. Note that only the S1/M1 source was localized with respect to the EDC. All other sources were detected with reference to the S1/M1 source. Coherent activity was found in the (a) contralateral primary sensorimotor cortex, (b) premotor cortex, (c) ipsilateral cerebellum, (d) diencephalon, (e) secondary somatosensory cortex, and (f) posterior parietal cortex. All sources were significantly coherent to the right EDC with the exception of the thalamic source where, in this subject, discernible spectral coherence peaks at tremor, and twice the tremor frequency failed to reach significance. Note that the strength of source localization is color coded. Red represents stronger coherent activation. The dotted lines indicate 95% confidence level of coherence. From Pollok et al. (2004), with permission from Blackwell Publishing.

parts of the Gpi, which may contribute to rest tremor in tremor-dominant and classic PD cases (Rajput et al., 2008).

The arguments against a pure peripheral mechanism generating rest tremor are the following (Llinas and Paré, 1995):

1. Rest tremor is not abolished by sectioning the dorsal roots, indicating that it does not reflect the sole action a spinal reflex loop.
2. It is very difficult to reset rest tremor by a mechanical perturbation, and the phase shift lasts for only a few cycles.
3. Recordings of Ia afferents (see Chapters 1 and 2) show patterns similar to the one found during a voluntary alternating movement.

7.8 KINETIC TREMOR

Reversible cerebellar lesions by cooling the dentate nuclei in monkeys cause ataxic movements with or without tremor (Flament and Hore, 1986). The tremor is attenuated or abolished by addition of mass and occurs in the absence of visual feedback, showing similar properties to cerebellar terminal tremor in humans. In reaching the target, movement without tremor is often hypermetric (overshoot), while movement with tremor shows an increased variability of the end position (Flament and Hore, 1986). Both discontinuities and tremor in movements during cerebellar dysfunction could result from impaired stretch reflexes and disorganized servo assistance mechanisms, both partly involving transcortical pathways (Hore and Falment, 1986). A detailed analysis of firings of neurons revealed that the population of neurons discharging strongly in relation to cerebellar tremor responded markedly and reciprocally to limb perturbation. No evidence was found that the 3- to 4-Hz cerebellar tremor is driven by a purely central oscillator (Hore and Flament, 1988). By comparing the properties of tremor following movements about the elbow with those following isometric contractions, Flament and Hore concluded that (1) a cerebellar tremor can occur in an isometric situation, and (2) movement of a given joint is required for the development of a rhythmic 3- to 5-Hz cerebellar intention tremor (Flament and Hore, 1988).

Concurrent cognitive tasks and visual feedback increase the kinetic tremor in humans. Trains of repetitive transcranial magnetic stimulation (at 10–30 Hz and intensities of 90–120% motor threshold) over the motor cortex while normal subjects make either voluntary wrist movements or during a postural task (maintenance of a fixed position) result in a tremor clinically very similar to cerebellar tremors (Topka et al., 1999). It seems that repetitive TMS may cause a terminal tremor by interfering with the adaptive cerebellar afferent inflow to the motor cortex.

Delayed second agonist EMG burst and delayed antagonist EMG burst associated with ballistic movements has been reported in advanced cases of ET during performance of ballistic movements, arguing again for a cerebellar dysfunction in case of kinetic tremor of ET (Koster et al., 2002).

7.9 ORTHOSTATIC TREMOR

Orthostatic tremor (OT) was originally classified as a variant of ET (FitzGerald and Jankovic, 1991). Several features such as a high degree of synchrony between different muscles, the lack or poor response to β-blocking agents (such as propranolol) and ethanol, a negative family history, and the higher frequencies of oscillations, suggest that OT and ET tremor are different disorders, although some overlap might exist (McManis et al., 1993; Boroojerdi et al., 1999).

A bilateral overactivity of cerebellar connections is considered as a common feature of several tremulous disorders, as demonstrated by the finding of abnormal bilateral cerebellar activation in orthostatic, essential, and writing tremors (Wills et al., 1996). OT secondary to pontine lesions may arise from dysfunction of the cerebellum or related pontine structures (Benito-León et al., 1997). The resetting by transcranial magnetic stimulation in OT associated with cerebellar cortical atrophy emphasizes the role of motor cortex in the genesis of this subtype of OT (Manto et al., 1999). Involvement of cranial muscles and failure of peripheral loading to modify tremor frequency argue strongly for a central origin (Köster et al., 1999). The existence of either a unique oscillator or a potent linking mechanism between key structures might explain the high intermuscular coherence. The fact that high-frequency discharges appear during isometric muscle contraction (irrespective of body posture) points toward a disorder of isometric force control rather than a disorder of stance regulation (Boroojerdi et al., 1999).

Sharott and colleagues showed that a 16-Hz activity may appear in healthy subjects, when unsteadiness is triggered by vestibular galvanic stimulation (Sharott et al., 2003). They suggested that the core abnormality in primary OT is an exaggerated sense of unsteadiness when standing still, which elicits activity from a 16-Hz oscillator normally engaged in postural responses. The importance of postural set is supported by the occurrence of EMG modification during the tilt test (patient tilted to different positions) but nor during the relief (patient gradually lifted) maneuver (Spiegel et al., 2006). Comparison of sway parameters in OT patients, controls, and patients with proprioceptive loss shows that one of the important components is a disruption of the normal generation or processing of proprioceptive signals (Bacsi et al., 2005). Piboolnurak and colleagues have proposed that the oscillator is most likely located in the brainstem and is influenced by the cerebral cortex, basal ganglia, and cerebellum (Piboolnurak et al., 2005).

7.10 OTHER TREMORS

7.10.1 Multiple Sclerosis

The predominance of action tremors in patients with multiple sclerosis (MS) points to a role for cerebellar structures or the cerebellothalamocortical pathway. Although the rarity of rest tremor in this demyelinating disease argues against an involvement of the basal ganglia, the tremors observed in MS might be generated by the basal ganglia. DBS of the nucleus ventralis oralis posterior (VOP) of the thalamus, which is the basal ganglia output nucleus of the thalamus, is effective (see also Chapter 8; Koch et al., 2007).

7.10.2 Psychogenic Tremor

Study of coherence between EMG signals from right and left arms (with the hands outstretched against gravity) shows that two different pathogenic mechanisms may play a role in psychogenic tremor: voluntary produced oscillations and an involuntary mechanism like clonus or enhanced physiological tremor (Raethjen et al., 2004). Evaluation of psychogenic tremor, while the subject carries out simultaneously two tasks, reveals the phenomenon of dual-task interference (Kumru et al., 2007; see also Chapter 6). The coexistence of psychogenic tremor with a tremor because of an anatomical lesion in the nervous system is not exceptional (comorbidity).

7.10.3 Dystonic Tremor

Reciprocal inhibition and other inhibitory reflex pathways at various levels (cerebral cortex, basal ganglia, spinal cord) seem to be reduced in dystonic patients (Deuschl et al., 2001; Deuschl, 2003). Indeed, excessive co-contraction and abnormalities in the time course of reciprocal inhibition between antagonist groups of muscles are considered as key features of some types of dystonia. Reduced speed of movement is often attributed to involuntary activation of antagonist muscles about a joint. Reciprocal inhibition studies include the Renshaw circuit, Ib inhibitory circuits, Ia inhibitory interneurons, and presynaptic inhibition.

Dystonic–postural tremor of the upper extremity may occur after a contralateral anterior thalamic infarct (Cho and Samkoff, 2000). The pathogenesis of thalamic tremor and dystonia is not precisely known. The main hypotheses are:

- interruption of afferents to the posterolateral thalamus from the brainstem and cerebellum (Miwa et al., 1996)
- destruction of thalamic neurons
- abnormal volleying of the circuitry between the thalamus and the cerebral cortex
- damage to the thalamic reticular nucleus (Cho and Samkoff, 2000).

7.10.4 Primary Writing Tremor

Studies using TMS showed shortened silent periods during maximal voluntary contraction (MVC) on both sides. This indicates that central inhibitory mechanisms are affected and strengthens the hypothesis that primary writing tremor is not a variant of focal task-specific dystonia but rather a separate nosological entity (Ljubisavljevic et al., 2006). In writer's cramp, there is also evidence of cortical motor reorganization (Byrnes et al., 2005).

Comparison between cortical and spinal excitability in patients with primary writing tremor (PWT) and patients with ET and writer's cramp suggests that the pathophysiology of PWT is distinct (Modugno et al., 2002). Nevertheless, PWT, dystonia, and ET are conditions which share pathophysiological hallmarks. Positron emission tomography reveals that both essential and writing tremors are associated with abnormal bilateral overactivity of cerebellar connections (Wills et al., 1995).

Studies on cerebral activation pattern during writing showed that cortex activation includes the contralateral premotor area (area 6), the ipsilateral prefrontal area (inferior frontal gyrus; areas 10, 44, and 47), the parietal lobule (area 40) bilaterally (>contralateral), and the cerebellum bilaterally (with a predominance ipsilaterally; Berg et al., 2000).

7.10.5 Complex Regional Pain Syndrome

Physiopathological mechanisms underlying the movement disorders associated with this condition are not yet clarified. Peripheral versus central origin as well as organic versus psychogenic origin have been considered. The following possibilities are investigated: vicious circle of reflexes involving anterior motor horn cells, direct action of substance P on motor neurons, abnormal sympathetic activity on muscle spindles, central reorganizaton mediated by retrograde axonal transport of trophic factors, diffuse activation of brain networks (Verdugo and Ochoa, 2000).

7.10.6 Holmes' Tremor

The involvement of the Guillain–Mollaret triangle (see Chapter 2) plays a key role in the genesis of midbrain tremor.

Recent neuroimaging–SPECT findings have demonstrated a dopaminergic denervation in a young case of Holmes' tremor (Guedj et al., 2007). In another case with dopamine sensitivity, brain SPECT has revealed impaired dopamine transporter activity in the basal ganglia and striatum, with diffusion tensor MRI (DTI) demonstrating missing ipsilateral tegmentofrontal connectivity and tractography showing reduced fiber connectivity of the superior and middle cerebellar peduncles on the lesioned side (Seidel et al., 2008). Taken together, these results suggest a reduced striatal dopaminergic input. Responsiveness to deep brain stimulation (DBS) of the Vim argues for a role of basal ganglia in the pathogenesis.

7.10.7 Cortical Tremor

Familial cortical myoclonic tremor with epilepsy (FCMTE) has been described in Chapter 6. The effects of low frequency repetitive transcranial magnetic stimulation (rTMS) have been recently studied in a case of cortical tremor caused by hyperexcitability of sensorimotor cortex. A decrease of acceleration total spectral power of the index finger postural tremor and a normalization of electrophysiologic parameters after 1 Hz rTMS over premotor cortex were found (Houdayer et al., 2007; Figure 7.7). Coherence analysis of EMG of forearm muscles and EEG of contralateral motor cortex showed strong cortico- and intermuscular coherence in the 8- to 30-Hz range, indicating a pathological cortical drive in FCMTE patients leading to tremulous movements (Van Rootselaar et al., 2006).

7.10.8 Palatal Tremor

Functional MR imaging (fMRI) reveals prominent bilateral neuronal activation in the putamen in essential palatal tremor (Haller et al., 2006; Figure 7.8). Inferior olivary hypertrophy may develop after pontine haemorrhage (lesion of the Guillain-Mollaret triangle, see Chapter 2) and may become a pacemaker for symptomatic palatal tremor (Yagura et al., 2007).

7.10.9 Tremor Following a Peripheral Nerve Injury

Pathophysiological mechanisms underlying this tremor are not fully established. Functional changes in afferent neuronal input to the spinal cord and secondary changes (central reorganization of the neuronal circuits) in the supraspinal centers are probably involved (Nobrega et al., 2002). EMG abnormalities in the absence of other findings (normal somatosensory evoked potentials, transcranial magnetic stimulation TMS, and MRI findings) and the persistence of tremor in REM and non-REM sleep in a patient with proximal right upper limb tremor (secondary to direct peripheral nerve lesion caused by thoracic surgery) suggest a peripheral generator (Costa et al., 2006).

7.10.10 Tremor in Peripheral Neuropathy

Nerve conduction velocities may be markedly reduced in paraproteinemic neuropathies. For instance, conduction velocities in the median nerves are reduced to 10–25 m/s (normal values >50 m/s). Although there is no correlation between slowing of nerve conduction and the presence of tremor, patients with the lowest velocities have usually the lowest frequencies (Smith, 1989). The fact that tremor peak frequency increases in patients recovering from severe polyneuropathy and in whom conduction velocities increase indicates that tremor frequency is linked to conduction velocities. The stretch reflex might be implicated in tremor genesis, the slowing in nerve conduction

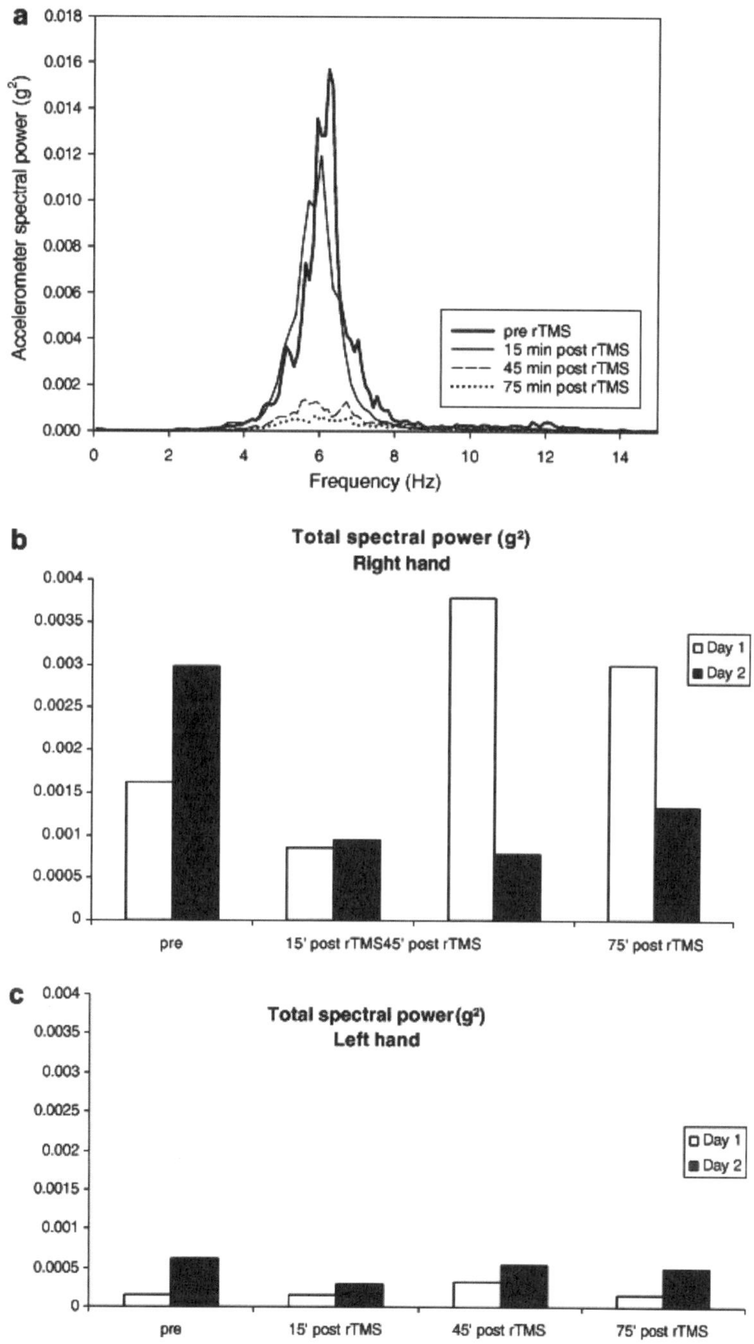

FIGURE 7.7: (a) Spectral analysis of postural tremor before and after rTMS applied over premotor cortex (PMC). (b) Accelerometer spectral power of the right hand before and three times after the first PMC stimulation. (c) Total spectral power of right (upper) and left (lower) accelerometers after the two consecutive days of PMC stimulation. From Houdayer et al. (2007), with permission from Elsevier.

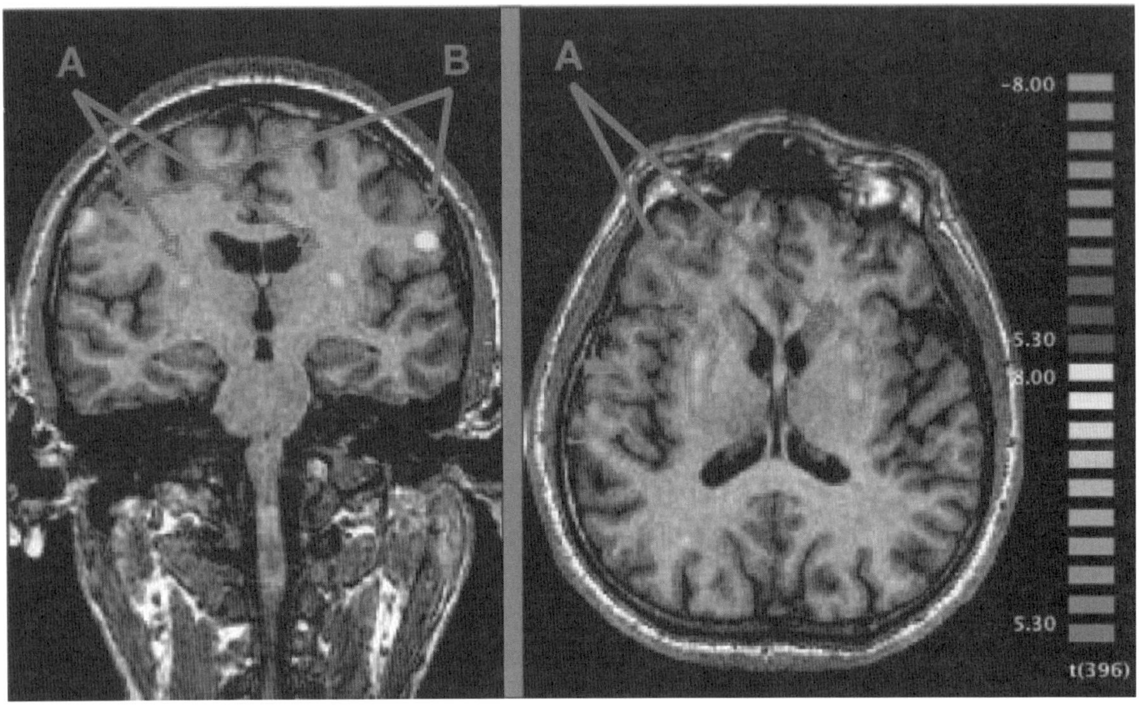

FIGURE 7.8: Neuronal activations associated with the initiation and maintenance of essential palatal tremor in a 41-year-old patient. (a) Peak activation assessed by fMRI is present in the putamen bilaterally. Additional activation is found in the (b) precentral gyrus bilaterally and right superior temporal and angular gyrus (not shown). No focal activation could be observed in the prefrontal motor cortex, the brainstem, or the cerebellum. The analysis is based on a fixed-effects general linear model with statistical threshold of $P < 0.01$ (corrected Bonferroni) and spatial threshold of 500 mm^3. From Haller et al. (2006), with permission from the American Society of Neuroradiology.

velocities causing an increase in the delay of the stretch reflex loop. There is no correlation between the conduction velocity and the amplitude of oscillations.

REFERENCES

Antonini A, Moresco RM, Gobbo C, De Notaris R, Panzacchi A, Barone P, Calzetti S, Negrotti A, Pezzoli G, Fazio F. The status of dopamine nerve terminals in Parkinson's disease and essential tremor: a PET study with the tracer [11-C]FE-CIT. *Neurol Sci.* 2001;22(1): 47–48.

Axelrad JE, Louis ED, Honig LS, Flores I, Ross GW, Pahwa R, Lyons KE, Faust PL, Vonsattel JP. Reduced Purkinje cell number in essential tremor: a postmortem study. *Arch Neurol.* 2008;65(1):101–107.

Bacsi AM, Fung VS, Colebatch JG. Sway patterns in orthostatic tremor: impairment of postural control mechanisms. *Mov Disord.* 2005;20(11):1469–1475.

Bain PG, Britton TC, Jenkins IH, Thompson PD, Rothwell JC, Thomas PK, Brooks DJ, Marsden CD. Tremor associated with benign IgM paraproteinaemic neuropathy. *Brain.* 1996;119(Pt 3):789–799.

Ben-Pazi H, Bergman H, Goldberg JA, Giladi N, Hansel D, Reches A, Simon ES. Synchrony of rest tremor in multiple limbs in Parkinson's disease: evidence for multiple oscillators. *J Neural Transm.* 2001;108(3):287–296.

Benito-León J, Rodríguez J, Ortí-Pareja M, Ayuso-Peralta L, Jiménez-Jiménez FJ, Molina JA. Symptomatic orthostatic tremor in pontine lesions. *Neurology.* 1997;49(5):1439–1441.

Benito-León J, Louis ED. Essential tremor: emerging views of a common disorder. *Nat Clin Pract Neurol.* 2006;2(12):666–678; quiz 2p following 691.

Berg D, Preibisch C, Hofmann E, Naumann M. Cerebral activation pattern in primary writing tremor. *J Neurol Neurosurg Psychiatry.* 2000;69(6):780–786.

Bilodeau M, Bisson E, Degrâce D, Després I, Johnson M. Muscle activation characteristics associated with differences in physiological tremor amplitude between the dominant and non-dominant hand. *J Electromyogr Kinesiol.* 2008 (in press).

Boroojerdi B, Ferbert A, Foltys H, Kosinski CM, Noth J, Schwarz M. Evidence for a non-orthostatic origin of orthostatic tremor. *J Neurol Neurosurg Psychiatry.* 1999;66(3):284–288.

Burkhardt JM, Constantinidis C, Anstrom KK, Roberts DC, Woodward DJ. Synchronous oscillations and phase reorganization in the basal ganglia during akinesia induced by high-dose haloperidol. *Eur J Neurosci.* 2007;26(7):1912–1924.

Byrnes ML, Mastaglia FL, Walters SE, Archer SA, Thickbroom GW. Primary writing tremor: motor cortex reorganisation and disinhibition. *J Clin Neurosci.* 2005;12(1):102–104.

Cantello R. Applications of transcranial magnetic stimulation in movement disorders. *J Clin Neurophysiol.* 2002;19(4):272–293.

Cassim F, Labyt E, Devos D, Defebvre L, Destée A, Derambure P. Relationship between oscillations in the basal ganglia and synchronization of cortical activity. *Epileptic Disord.* 2002;4(3):S31–S45.

Caviness JN, Shill HA, Sabbagh MN, Evidente VG, Hernandez JL, Adler CH. Corticomuscular coherence is increased in the small postural tremor of Parkinson's disease. *Mov Disord.* 2006;21:492–499.

Cho C, Samkoff LM. A lesion of the anterior thalamus producing dystonic tremor of the hand. *Arch Neurol.* 2000;Sep;57(9):1353–1355.

Costa J, Henriques R, Barroso C, Ferreira J, Atalaia A, de Carvalho M. Upper limb tremor induced by peripheral nerve injury. *Neurology.* 2006;67(10):1884–1886.

Daniels C, Peller M, Wolff S, Alfke K, Witt K, Gaser C, Jansen O, Siebner HR, Deuschl G. Voxel-based morphometry shows no decreases in cerebellar gray matter volume in essential tremor. *Neurology.* 2006;67(8):1452–1456.

Deuschl G, Raethjen J, Baron R, Lindemann M, Wilms H, Krack P. The pathophysiology of parkinsonian tremor: a review. *J Neurol.* 2000;247(Suppl 5):V33–V48.

Deuschl G, Raethjen J, Lindemann M, Krack P. The pathophysiology of tremor. *Muscle Nerve.* 2001;24(6):716–735.

Deuschl G. Dystonic tremor. *Rev Neurol (Paris).* 2003;159(10 Pt 1):900–905.

Duval C, Strafella AP, Sadikot AF. The impact of ventrolateral thalamotomy on high-frequency components of tremor. *Clin Neurophysiol.* 2005;116(6):1391–1399.

Farkas Z, Szirmai I, Kamondi A. Impaired rhythm generation in essential tremor. *Mov Disord.* 2006;21(8):1196–1199.

FitzGerald PM, Jankovic J. Orthostatic tremor: an association with essential tremor. *Mov Disord.* 1991;6(1):60–64.

Flament D, Hore J. Movement and electromyographic disorders associated with cerebellar dysmetria. *J Neurophysiol.* 1986;55(6):1221–1233.

Flament D, Hore J. Comparison of cerebellar intention tremor under isotonic and isometric conditions. *Brain Res.* 1988;439(1–2):179–186.

Fukuhara T, Gotoh M, Asari S, Ohmoto T. Magnetic resonance imaging of patients with intention tremor. *Comput Med Imaging Graph.* 1994;18(1):45–51.

Guedj E, Witjas T, Azulay JP, De Laforte C, Peragut JC, Mundler O. Neuroimaging findings in a case of Holmes tremor. *Clin Nucl Med.* 2007;32(2):139–140.

Haller S, Winkler DT, Gobbi C, Lyrer P, Wetzel SG, Steck AJ. Prominent activation of the putamen during essential palatal tremor: a functional MR imaging case study. *AJNR Am J Neuroradiol.* 2006;27(6):1272–1274.

Hallett M, Dubinsky RM. Glucose metabolism in the brain of patients with essential tremor. *J Neurol Sci.* 1993;114(1):45–48.

Halliday DM, Conway BA, Farmer SF, Shahani U, Russell AJ, Rosenberg JR. Coherence between low-frequency activation of the motor cortex and tremor in patients with essential tremor. *Lancet.* 2000;355(9210):1149–1153.

Hellwig B, Schelter B, Guschlbauer B, Timmer J, Lücking CH. Dynamic synchronisation of central oscillators in essential tremor. *Clin Neurophysiol.* 2003;114(8):1462–1467.

Herzog J, Hamel W, Wenzelburger R, Pötter M, Pinsker MO, Bartussek J, Morsnowski A, Steigerwald F, Deuschl G, Volkmann J. Kinematic analysis of thalamic versus subthalamic neurostimulation in postural and intention tremor. *Brain.* 2007;130(Pt 6):1608–1625.

Hore J, Flament D. Evidence that a disordered servo-like mechanism contributes to tremor in movements during cerebellar dysfunction. *J Neurophysiol.* 1986;56(1):123–136.

Hore J, Flament D. Changes in motor cortex neural discharge associated with the development of cerebellar limb ataxia. *J Neurophysiol.* 1988;60(4):1285–1302.

Houdayer E, Devanne H, Tyvaert L, Defebvre L, Derambure P, Cassim F. Low frequency repetitive transcranial magnetic stimulation over premotor cortex can improve cortical tremor. *Clin Neurophysiol.* 2007;118(7):1557–1562.

Kim JS, Park JW, Kim WJ, Kim HT, Kim YI, Lee KS. Disappearance of essential tremor after frontal cortical infarct. *Mov Disord.* 2006;21(8):1284–1285.

Koch M, Mostert J, Heersema D, De Keyser J. Tremor in multiple sclerosis. *J Neurol.* 2007;254(2):133–145.

Köster B, Lauk M, Timmer J, Poersch M, Guschlbauer B, Deuschl G, Lucking CH. Involvement of cranial muscles and high intermuscular coherence in orthostatic tremor. *Ann Neurol.* 1999;45(3):384–388.

Köster B, Deuschl G, Lauk M, Timmer J, Guschlbauer B, Lucking CH. Essential tremor and cerebellar dysfunction: abnormal ballistic movements. *J Neurol Neurosurg Psychiatry* 2002;73:400–405.

Kralic JE, Korpi ER, O'Buckley TK, Homanics GE, Morrow AL. Molecular and pharmacological characterization of GABA(A) receptor alpha1 subunit knockout mice. *J Pharmacol Exp Ther.* 2002;302(3):1037–1045.

Kralic JE, Criswell HE, Osterman JL, O'Buckley TK, Wilkie ME, Matthews DB, Hamre K, Breese GR, Homanics GE, Morrow AL. Genetic essential tremor in gamma-aminobutyric acidA receptor alpha1 subunit knockout mice. *J Clin Invest.* 2005;115(3):774–779.

Kumru H, Begeman M, Tolosa E, Valls-Sole J. Dual task interference in psychogenic tremor. *Mov Disord.* 2007;22(14):2077–2082.

Ljubisavljevic M, Kacar A, Milanovic S, Svetel M, Kostic VS. Changes in cortical inhibition during task-specific contractions in primary writing tremor patients. *Mov Disord.* 2006; 21(6):855–859.

Llinas R, Paré D. Role of intrinsic neuronal oscillations and network ensembles in the genesis of normal and pathological tremors. In: *Handbook of Tremor Disorders.* Findley LJ, Koller WC (eds), Marcel Dekker, New York. 1995.

Lo YL, Louis ED, Fook-Chong S, Tan EK. Impaired motor imagery in patients with essential tremor: a case control study. *Mov Disord.* 2007;22(4):504–508.

Louis ED, Vonsattel JP, Honig LS, Ross GW, Lyons KE, Pahwa R. Neuropathologic findings in essential tremor. *Neurology.* 2006;66(11):1756–1759.

Louis ED, Zheng W, Mao X, Shungu DC. Blood harmane is correlated with cerebellar metabolism in essential tremor: a pilot study. *Neurology.* 2007;69(6):515–520.

Louis ED, Faust PL, Vonsattel JP, Honig LS, Rajput A, Robinson CA, Rajput A, Pahwa R, Lyons KE, Ross GW, Borden S, Moskowitz CB, Lawton A, Hernandez N. Neuropathological changes in essential tremor: 33 cases compared with 21 controls. *Brain.* 2007;130(Pt 12):3297–3307.

Malfait N, Sanger TD. Does dystonia always include co-contraction? A study of unconstrained reaching in children with primary and secondary dystonia. *Exp Brain Res.* 2007;176(2):206–216.

Mally J, Baranyi M, Vizi ES. Change in the concentrations of amino acids in CSF and serum of patients with essential tremor. *J Neural Transm.* 1996;103(5):555–560.

Manto MU, Setta F, Legros B, Jacquy J, Godaux E. Resetting of orthostatic tremor associated with cerebellar cortical atrophy by transcranial magnetic stimulation. *Arch Neurol.* 1999;56(12):1497–1500.

McManis PG, Sharbrough FW. Orthostatic tremor: clinical and electrophysiologic characteristics. *Muscle Nerve.* 1993;16(11):1254-60.

Miwa H, Hatori K, Kondo T, Imai H, Mizuno Y. Thalamic tremor: case reports and implications of the tremor-generating mechanism. *Neurology.* 1996;46(1):75–79.

Miwa H. Rodent models of tremor. *Cerebellum.* 2007;6:66–72.

Miyagishima T, Takahashi A, Kikuchi S, Watanabe K, Hirato M, Saito N, Yoshimoto Y. Effect of ventralis intermedius thalamotomy on the area in the sensorimotor cortex activated by passive hand movements: fMR imaging study. *Stereotact Funct Neurosurg.* 2007;85(5): 225–234.

Modugno N, Nakamura Y, Bestmann S, Curra A, Berardelli A, Rothwell J. Neurophysiological investigations in patients with primary writing tremor. *Mov Disord.* 2002;17(6):1336–1340.

Morrison S, Newell KM. Inter- and intra-limb coordination in arm tremor. *Exp Brain Res.* 1996;110(3):455–464.

Morrison S, Newell KM. Postural and resting tremor in the upper limb. *Clin Neurophysiol.* 2000;111(4):651–663.

Nandi D, Aziz TZ, Liu X, Stein JF. Brainstem motor loops in the control of movement. *Mov Disord.* 2002;17(3):S22–S27.

Nobrega JC, Campos CR, Limongi JC, Teixeira MJ, Lin TY. Movement disorders induced by peripheral trauma. *Arq Neuropsiquiatr.* 2002;60(1):17–20.

O'Sullivan JD, Rothwell J, Lees AJ, Brown P. Bilaterally coherent tremor resembling enhanced physiological tremor: report of three cases. *Mov Disord.* 2002;17(2):387–3891.

Piboolnurak P, Yu QP, Pullman SL. Clinical and neurophysiologic spectrum of orthostatic tremor: case series of 26 subjects. *Mov Disord.* 2005;20(11):1455–1461.

Pinto AD, Lang AE, Chen R. The cerebellothalamocortical pathway in essential tremor. *Neurology.* 2003;60(12):1985–1987.

Pollok B, Gross J, Dirks M, Timmermann L, Schnitzler A. The cerebral oscillatory network of voluntary tremor. *J Physiol.* 2004;554(Pt 3):871–878.

Raethjen J, Lindemann M, Schmaljohann H, Wenzelburger R, Pfister G, Deuschl G. Multiple oscillators are causing parkinsonian and essential tremor. *Mov Disord.* 2000;15(1): 84–94.

Raethjen J, Pawlas F, Lindemann M, Wenzelburger R, Deuschl G. Determinants of physiologic tremor in a large normal population. *Clin Neurophysiol.* 2000;111(10):1825–1837.

Raethjen J, Lemke MR, Lindemann M, Wenzelburger R, Krack P, Deuschl G. Amitryptiline enhances the central component of physiological tremor. *J Neurol Neurosurg Psychiatry.* 2001;70(1):78–82.

Raethjen J, Lindemann M, Dümpelmann M, Wenzelburger R, Stolze H, Pfister G, Elger CE, Timmer J, Deuschl G. Corticomuscular coherence in the 6–15 Hz band: is the cortex involved in the generation of physiologic tremor? *Exp Brain Res.* 2002;142(1):32–40.

Raethjen J, Lindemann M, Morsnowski A, Dümpelmann M, Wenzelburger R, Stolze H, Fietzek U, Pfister G, Elger CE, Timmer J, Deuschl G. Is the rhythm of physiological tremor involved in cortico-cortical interactions? *Mov Disord.* 2004;19(4):458–465.

Raethjen J, Kopper F, Govindan RB, Volkmann J, Deuschl G. Two different pathogenetic mechanisms in psychogenic tremor. *Neurology.* 2004;63(5):812–815.

Raethjen J, Govindan RB, Kopper F, Muthuraman M, Deuschl G. Cortical involvement in the generation of essential tremor. *J Neurophysiol.* 2007;97(5):3219–3228.

Rajput AH, Sitte HH, Rajput AH, Fenton ME, Pifl C, Hornykiewicz O. Globus pallidus dopamine and Parkinson motor subtypes. Clinical and brain biochemical correlation. *Neurology.* 2008;15:1403–1401.

Rao SS, Hofmann LA, Shakil A. Parkinson's disease: diagnosis and treatment. *Am Fam Physician.* 2006;74(12):2046–2054.

Sadek AH, Rauch R, Schulz PE. Parkinsonism due to manganism in a welder. *Int J Toxicol.* 2003;22(5):393–401.

Schnitzler A, Gross J, Timmermann L. Synchronised oscillations of the human sensorimotor cortex. *Acta Neurobiol Exp (Wars).* 2000;60(2):271–287.

Seidel S, Kasprian G, Leutmezer F, Prayer D, Auff E. Disruption of nigrostriatal and cerebello-thalamic pathways in dopamine-responsive Holmes' tremor. *J Neurol Neurosurg Psychiatry.* 2008.

Shams'ili S, Grefkens J, de Leeuw B, van den Bent M, Hooijkaas H, van der Holt B, Vecht C, Sillevis Smitt P. Paraneoplastic cerebellar degeneration associated with antineuronal antibodies: analysis of 50 patients. *Brain.* 2003;126(Pt 6):1409–1418.

Sharott A, Marsden J, Brown P. Primary orthostatic tremor is an exaggeration of a physiological response to instability. *Mov Disord.* 2003;18(2):195–199.

Smith IS. Tremor associated with peripheral neuropathy. *Electroencephalogr Clin Neurophysiol.* 1989;72:41P.

Spiegel J, Krick C, Fuss G, Sood D, Becker G, Dillmann U. Orthostatic tremor during modification of standing. *Mov Disord.* 2006;21(2):173–178.

Stratton SE, Lorden JF. Effect of harmaline on cells of the inferior olive in the absence of tremor: differential response of genetically dystonic and harmaline-tolerant rats. *Neuroscience.* 1991;41(2–3):543–549.

Takanokura M, Sakamoto K. Neuromuscular control of physiological tremor during elastic load. *Med Sci Monit.* 2005;11(4):CR143–CR152.

Timmermann L, Gross J, Dirks M, Volkmann J, Freund HJ, Schnitzler A. The cerebral oscillatory network of parkinsonian resting tremor. *Brain.* 2003;126(Pt 1):199–212.

Topka H, Mescheriakov S, Boose A, Kuntz R, Hertrich I, Seydel L, Dichgans J, Rothwell J. A cerebellar-like terminal and postural tremor induced in normal man by transcranial magnetic stimulation. *Brain.* 1999;122(Pt 8):1551–1562.

van Rootselaar AF, Maurits NM, Koelman JH, van der Hoeven JH, Bour LJ, Leenders KL, Brown P, Tijssen MA. Coherence analysis differentiates between cortical myoclonic tremor and essential tremor. *Mov Disord.* 2006;21(2):215–222.

Verdugo RJ, Ochoa JL. Abnormal movements in complex regional pain syndrome: assessment of their nature. *Muscle Nerve.* 2000;23(2):198–205.

Vianello M, Morello F, Scaravilli T, Tavolato B, Giometto B. Tremor of the mouth floor and anti-glutamic acid decarboxylase autoantibodies. *Eur J Neurol.* 2003;10(5):513–514.

Wiener V, Honnorat J, Pandolfo M, Kentos A, Manto MU. Myorhythmia associated with Hodgkin's lymphoma. *J Neurol.* 2003;250:1382–1384.

Wills AJ, Jenkins IH, Thompson PD, Findley LJ, Brooks DJ. A positron emission tomography study of cerebral activation associated with essential and writing tremor. *Arch Neurol.* 1995;52(3):299–305.

Wills AJ, Thompson PD, Findley LJ, Brooks DJ. A positron emission tomography study of primary orthostatic tremor. *Neurology.* 1996;46(3):747–752.

Wilms H, Sievers J, Deuschl G. Animal models of tremor. *Mov Disord.* 1999 Jul;14(4):557–571.

Yagura H, Miyai I, Hatakenaka M, Yanagihara T. Inferior olivary hypertrophy is associated with a lower functional state after pontine hemorrhage. *Cerebrovasc Dis.* 2007;24(4):369–374.

Zappia M, Crescibene L, Bosco D, Arabia G, Nicoletti G, Bagalà A, Bastone L, Napoli ID, Caracciolo M, Bonavita S, Di Costanzo A, Gambardella A, Quattrone A. Anti-GM1 ganglioside antibodies in Parkinson's disease. *Acta Neurol Scand.* 2002;106(1):54–57.

• • • •

CHAPTER 8

Treatments

Given the disabling character of tremor, treatment is an important aspect of the management. In this chapter the current treatments applied for tremor disorders are discussed. Table 8.1 summarizes the therapies.

8.1 DRUGS

The first line for treatment of tremor is oral medication. β-Blockers such as propranolol, anticholinergic medications, and levodopa are effective drugs for rest tremor. Kinetic tremor may slightly respond to β-blockers, primidone, anticholinergic medication, and alcohol in case of ET (Habib-ur-Rehman, 2000). Some ataxic patients with cerebellar cortical atrophy may exhibit a benefit with gabapentin. Tremor associated with ataxia due to vitamin E deficiency (AVED) may respond to supplementation.

TABLE 8.1: Main treatments for tremors

Drugs	Primidone, β-blockers, benzodiazepines, antiepileptics, anticholinergics, levodopa and dopamine agonists, others
Lesioning surgery	Thalamotomy, pallidotomy, subthalamotomy
γ-Knife	Thalamotomy
Deep brain stimulation (DBS)	Main targets: thalamic ventralis intermedius nucleus (Vim), subthalamic nucleus (STN), pallidum
Transcranial magnetic stimulation	Motor cortical stimulation (under investigation)
Orthosis and prosthesis	
Rehabilitation	
Brain–machine interfaces	

TABLE 8.2: Treatment of ET		
GRADE OF SEVERITY	**DRUG SUGGESTED**	**GRADE OF EFFICACY/ SIDE EFFECTS**
Disabling only during periods of stress and anxiety	Propranolol and benzodiazepines	
Disabling	Either primidone or propranolol	
If primidone or propranolol do not provide adequate control	Primidone and propranolol in combination	Propranolol and primidone reduce limb tremor
If propranolol induces side effects	Consider other beta-adrenoceptor antagonists (such as atenolol or metoprolol)	
If combination of primidone and propranolol do not provide adequate control	The use of benzodiazepines provides some benefit	–Alprazolam is probably effective in reducing limb tremor –Clonazepam possibly reduces limb tremor
Other medications that may be helpful	–Atenolol, gabapentin (monotherapy), sotalol, and topiramate –Clozapine, nadolol, and nimodipine –Botulinum toxin in the hand muscles in case of hand tremor (risk of hand weakness)	
Others drugs	Carbonic anhydrase inhibitors (e.g. methazolamide), phenobarbital, clonidine, and mirtazapine	Questionable efficacy
Disabling head or voice tremor	Botulinum toxin injections	Breathiness, hoarseness, and swallowing difficulties may occur in attempting to treat voice tremor Injection should be performed in a specialized center
If disabling tremor persists after medication trials	Surgical options are considered	Chronic DBS and thalamotomy are usually effective

The currently available medications decrease tremor in approximately 50% of the patients. The combination of medical and surgical therapies provide benefit in up to 80% of cases. From Lyons et al. (2003), Zesiewicz et al. (2005), and Pahwa and Lyons (2003)

Table 8.2 and Figure 8.1 focus on therapeutic strategies for ET. Table 8.3 lists the medications, the doses and the potential side effects (Benito-León and Louis, 2006). Drugs are presented in detail in the text; further information about the studies quoted is given in Table 8.4.

8.1.1 Ethanol

Ethanol decreases postural essential tremor, but not parkinsonian rest tremor or the genuine cerebellar kinetic tremor (which is worsened by small doses of ethanol). Ethanol improves gait ataxia in patients with ET (Klebe et al., 2005). However, the improvement is temporary and followed by a rebound phenomenon when the alcohol effect wears off. Recently, a case of ethanol responsive tremor in a patient with MS has been reported (Hammond and Kerr, 2008).

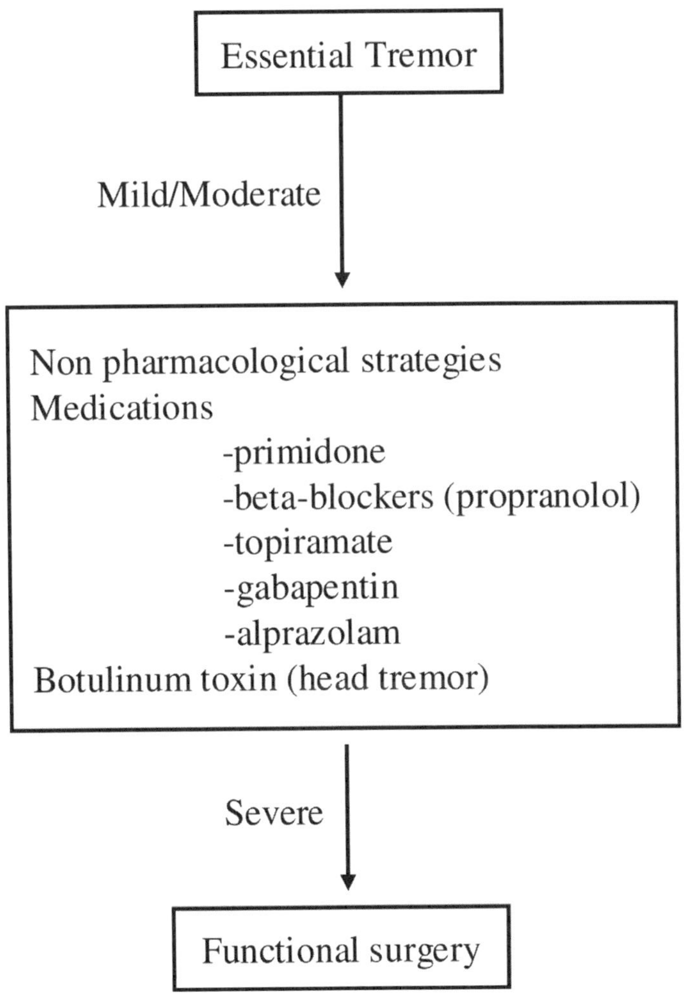

FIGURE 8.1: Therapeutic strategies of essential tremor.

TABLE 8.3: Pharmacological agents and nonpharmacological procedures for the treatment of ET

PHARMACOLOGICAL AGENTS	DOSAGE	SIDE EFFECTS/COMPLICATIONS
Alprazolam	0.125–3 mg/day	Sedation, fatigue, abuse
Atenolol	50–150 mg/day	Light-headedness, nausea, cough, dry mouth, sleepiness
Botulinum toxin A (hand tremor)	50–100 U	Hand or finger weakness, reduced grip strength, pain at injection site, hematoma
Botulinum toxin A (head tremor)	40–400 U	Neck weakness, postinjection pain
Clonazepam	0.5–6 mg/day	Drowsiness
Clozapine	6–75 mg/day	Sedation, agranulocytosis (risk of 0.8% at 1 year)
Gabapentin	1,200–1,800 mg/day	Lethargy, fatigue, decreased libido, dizziness, nervousness, shortness of breath
Levetiracetam	1,000 mg/day	Somnolence, asthenia, dizziness, headache
Nimodipine	120 mg/day	Headache, heartburn
Nadolol	120–240 mg/day	None
Olanzapine	20 mg/day	Drowsiness, sedation, weight gain, diabetes
1-Octanol	Up to 64 mg/kg	Unusual taste sensation
Primidone	Up to 750 mg/day	Sedation, drowsiness, fatigue, nausea, vomiting, ataxia, dizziness, unsteadiness, confusion, vertigo
Propranolol	60–800 mg/day	Reduced blood pressure, bradycardia, exertional dyspnea, confusion, drowsiness, headache, dizziness, impotency

PHARMACOLOGICAL AGENTS	DOSAGE	SIDE EFFECTS/COMPLICATIONS
Long-acting propranolol	80–320 mg/day	Skin eruption, dizziness
Sotalol	75–200 mg/day	Decreased alertness
Topiramate	Up to 400 mg/day	Appetite suppression, weight loss, paresthesias, attention deficit, kidney stones
Zonisamide	100–200 mg/day	Ataxia, dizziness, somnolence, agitation, anorexia
NONPHARMACOLOGICAL AGENTS	**INTENSITY OF RESPONSE**	**POTENTIAL SIDE EFFECTS**
Chronic thalamic stimulation DBS (hand)	60% to 90% improvement in CRS (clinical rating scale)	Dysarthria, dysequilibrium, paresthesias, weakness, headache, intracranial hemorrhage, subdural hemorrhage, ischemic changes, generalized motor seizures, lead dislodgement, decreased verbal fluency
Thalamotomy	55% to 90% improvement in CRS	Hemiparesis, transient speech and motor deficits, cognitive deficit, confusion, somnolence
Gamma knife surgery	70% to 85% improvement in CRS	Transient arm weakness, numbness in the contralateral arm, dysarthria, increased action tremor, dystonia of the contralateral upper and lower limbs, choreoathetosis, delayed side effects

From Benito-León and Louis (2006) and Zesiewicz et al. (2005).

Tolerance develops with time (Habib-ur-Rehman, 2000). For obvious reasons, regular alcohol intake cannot be recommended (see also Chapter 5). Therefore, ethanol is not used in the treatment of tremor, but as a clue for the diagnosis of ET. The improvement of tremor after ethanol ingestion is likely because of a direct effect on a central oscillator (Zeuner et al., 2003). Experimental studies in rodents suggest that ethanol antagonizes a dysregulation of glutamatergic pathways in the cerebellum. In addition, it enhances GABAergic transmission.

TABLE 8.4: Details of the studies reported

REFERENCE	DRUG	MODE OF ADMINISTRATION/ DOSE	TYPE OF TREMOR/ DISEASE	DESIGN OF STUDY	*N*° OF PATIENTS	RESULT
–Findley et al. 1985	Primidone	–Oral	–ET of the hands and head	–Double-blind, placebo-controlled trial	–22	–Significantly superior to placebo; efficacy comparable to propranolol
–Sasso et al. 1990	Primidone	–Oral 375–750 mg/day	–ET	–Placebo-controlled	–11	–Magnitude of tremor after 3, 6, and 12 months significantly reduced
–Serrano-Dueñas 2003	Primidone	–Oral 250 vs 750 mg/day	–ET	–Double-blind study with a 1-year follow-up	–113	–Low doses equally effective
–Calzetti et al. 1992	Propranolol	–Single oral dose (120 mg) and following 2 weeks of 120 or 240 mg/day	–ET of head	–Double-blind, placebo-controlled	–9	–Significant reduction in tremor magnitude after single oral dose but not on sustained administration

–Koller and Herbster 1987	–Long-acting propranolol, primidone and clonazepam	–Oral 160, 250, and 4 mg/day, respectively	–PD (resting, postural, kinetic)	Double-blind crossover design	–10	–Significantly decreased of resting and postural tremor by long-acting propranolol but not by primidone or clonazepam
–Henderson et al. 1997	–Propranolol	–Oral carbimazole plus propranolol (40 mg)	–Hyperthy-roidism	–Double-blind, cross-over and placebo-controlled	–7	–100%
–Metzer et al. 1993	–Long-acting propranolol vs placebo	–Oral	–Drug-induced parkinsonism	–Double-blind, placebo-controlled, crossover	–6	–No significant difference
Lee et al. 2003	Arotinolol vs propranolol	10 vs 40; 20 vs 80; 30 vs 160 mg/day	–ET	Crossover, multiple dose comparative trial	165	Arotinolol as effective as propranolol
–Gates 1993	–Clonazepam and chlordiazepoxid	–0.5–2.0; 30 mg/day	–OT		–9	–8 patients responded to clonazepam and 1 to chlordiazepoxid

continued on next page

TABLE 8.4: (*cont.*)

REFERENCE	DRUG	MODE OF ADMINISTRATION/ DOSE	TYPE OF TREMOR/ DISEASE	DESIGN OF STUDY	*N*° OF PATIENTS	RESULT
–Gunal et al. 2000	–Alprazolam, acetazolamide, primidone	–Alprazolam 0.75 mg/day	–ET	–Double-blind, cross-over, placebo-controlled	–22	–alprazolam superior to placebo and equal to primidone; acetazolamide vs placebo: no significant differences
–Elble et al. 2007	–Levetiracetam	–5-week titration from 500 to 3,000 mg/day; max tolerated dosage for 4 weeks	–ET	–Double-blind, placebo-controlled crossover	–15	–No efficacy
–Handforth et al. 2004	–Levetiracetam	–4-week titration to 3,000 mg/day; 2 weeks of stable dose	–ET	–Placebo-controlled, double-blind, randomized crossover	–15	–No significant difference
–Ondo et al. 2004	–Levetiracetam	–Evaluation in baseline, after 2 weeks 500 mg bid; at 4 weeks 1,500 mg bid	–ET	–Open label	–10	–No efficacy

–Bushara et al 2005	–Levetiracetam	–single dose of 1,000 mg	–ET	–Double-blind, placebo-controlled	–24	–Significant reduction of hand tremor for at least 2 h
–Striano et al 2006	–Levetiracetam	–500 mg/ bid the first week up to 50 mg/kg/ day by increments of 500 mg bid each week	–MS cerebellar tremor	–Open-label	–14	–11/14 completed the trial with significant improvement
–Ondo et al. 2006	–Topiramate vs placebo	–Mean final dose of 292 mg/day	–ET	–Multicenter, double-blind, placebo-controlled, parallel-design trial	–208	–Effective
–Gatto et al. 2007	–Topiramate	–50 mg/day during 3-12 months	–ET	–Report	–3	–Effective
–Connor et al. 2008	–Topiramate	–400 mg/day or maximum tolerated dose	–ET	–Randomized, double-blind, placebo-controlled, crossover trials	–62	–Effective
Raj et al. 2006	Oxcarbazepine	–450 mg bid	ET	Report	1	Improvement

continued on next page

TABLE 8.4: (*cont.*)

REFERENCE	DRUG	MODE OF ADMINISTRATION/ DOSE	TYPE OF TREMOR/ DISEASE	DESIGN OF STUDY	*N*° OF PATIENTS	RESULT
Rodrigues et al. 2006	Gabapentin	600-2,700 mg/d	OT	Double-blind placebo-controlled crossover	6	Tremor amplitude reduced from 100% to 79% ± 11%
Zesievicz et al. 2007	Pregabalin	50 mg/day up to 600 mg/day by 75 mg/day every 4 days	ET	Pilot, double-blind, placebo-controlled, randomized	22	Significant improvement
–Liou and Shih 2006	–Trihexy-phenidyl	–38 mg/day	–Rubral tremor	–Report	–1	–Significant benefit
–Wills et al. 1999	–Levodopa	–	–OT		–8	–Benefit in 5/8
–Yetimalar et al. 2005	–Olanzapine and propranolol	–20 and 120 mg/day	–ET	–Random-ized, crossover study	–28	–Olanzapine significantly improved tremor
–Ceravolo et al. 1999	–Clozapine	–Single dose of 12.5 mg; 39 ± 9 mg up to 50 mg in responders	–Drug-resistant ET	–Randomized, double-blind, crossover study	–15	–Single dose ef-fective in 13/15; significant im-provement with chronic treatment

–Rice et al. 1997	–Ondansetron	–Single dose intravenous	–Cerebellar tremor	–Placebo controlled, double blind, crossover study	–20	–Improvement in writing and 9-hole-peg test
–Gbadamosi et al. 2001	–Ondansetron	–Single dose intravenous	–Cerebellar MS tremor	–Open-label, prospective, and controlled study	–14	–No beneficial effects
–Bier et al. 2003	Ondansetron	–Oral 8 mg	–Cerebellar dysfunction	–Randomised, multicenter, double-blind trial	–45	–No beneficial effect
–Jiménez-Jiménez et al. 1994	Nicardipine	–Nicardipine 1 mg/kg/day and propranolol 160 mg/day; evaluation baseline and 1 month after each drug	–ET	–Crossover study	–14	–Improvement with a nonsignificantly higher efficacy of propranolol
–Muenter et al. 1991	–Methazolamide	–Max mean dose 203 mg/day (maintenance dose 129 mg/day)	–ET	–	–28	–21/28 mild to marked improvement
–Hallett et al 1991	–Isoniazid	up to 1,200 mg/day + 100 mg pyridoxine, for 4 weeks	–ET and postural action tremor	–Double-blind	–15 (11 and 4)	–Benefit in 2/11

8.1.2 Primidone

Primidone is significantly superior to placebo in reducing the magnitude of hand postural tremor. Its efficacy is similar to propranolol (Findley et al., 1985). Long-term efficacy of primidone (range of doses 375–750 mg/day) in ET has been evaluated. Tremorolytic effects last for up to 1 year or more (Sasso et al., 1990). In some studies, low doses of primidone (250 mg/day) were demonstrated to provide an equal or more effective effect than high doses (750 mg/day) in the control of ET with the subsequent advantage of fewer undesirable effects (Serrano-Dueñas, 2003). However, a progressive increase of the dose up to 750 mg/day is often required in daily practice. Low doses should be considered at the very beginning of the treatment. We start with doses in the range of 30–60 mg for 1 to 3 weeks, before increasing the dose.

8.1.3 β-Blockers

β-Adrenergic blocking drugs (mainly the nonselective blocker *propranolol* or a β2-selective blocker) have been the mainstays for the treatment of ET with limb tremor. They are probably less effective in the treatment of voice and head tremor (Habib-ur-Rehman, 2000). The efficacy of sustained propranolol on isolated or prominent essential head tremor is less predictable and satisfactory than expected on the basis of the single-dose response, compared with hand tremor (Calzetti et al., 1992).

A double-blind crossover study comparing the effects of long-acting propranolol hydrochloride (160 mg/d), primidone (250 mg at night), and clonazepam (4 mg/d) in parkinsonian patients, showed that long-acting propranolol is a useful adjuvant therapy for the parkinsonian tremor. In fact, long-acting propranolol reduced the mean amplitude of resting tremor by 70% and the mean amplitude of postural tremor by 50%, without the occurrence of noticeable side effects (Koller and Herbster, 1987). Long-acting propranolol is usually not effective to reduce drug-induced parkinsonian tremor (Metzer et al., 1993).

Interestingly, propranolol is a useful adjunct in the early treatment of both the tremor and tachycardia of hyperthyroidism (Henderson et al., 1997).

β-Blockers should be used very cautiously in the case of respiratory disease (asthma) or conduction block (heart disease), especially in the elderly.

8.1.4 Benzodiazepines

Alprazolam can be used in elderly patients with ET who do not tolerate primidone or propranolol (Gunal et al., 2000). *Clonazepam* alone or in combination with gabapentine or primidone improve orthostatic tremor (Gates, 1993) and may provide benefit in tremor associated with myoclonus.

8.1.5 Antiepileptic Drugs

Levetiracetam is not beneficial in the treatment of ET, as demonstrated by double-blind, placebo-controlled crossover trials (Elble et al., 2007; Handforth et al., 2004) and a 4-week, open label trial (Ondo et al., 2004). Levetiracetam has been proposed for the management of cerebellar tremor in multiple Sclerosis (Striano et al., 2006).

A case of ET responding to *oxcarbazepine* has been reported recently (Raj et al., 2006). *Topiramate* (up to a maximum of 400 mg/day) is effective in the treatment of moderate to severe ET. Tremor reduction is accompanied by functional improvements, such as writing or speaking (Ondo et al., 2006; Connor et al., 2008). Improvement of ET has been also reported with low doses of topiramate (50 mg/day; Gatto et al., 2007).

Gabapentin is the most effective treatment for OT. It is the first-line therapy, reducing both tremor and postural instability and improving quality of life (Rodrigues et al., 2006).

A recent study using *Pregabalin* in ET showed significant improvements in terms of accelerometry recordings and action tremor limb scores on the Fahn–Tolosa–Marin rating scale (see Annex 2). Larger studies are needed to confirm the beneficial effects (Zesievicz et al., 2007).

8.1.6 Levodopa, Dopamine Agonists, and Anticholinergic Agents

Dopamine agonists, such as *pramipexole* and *ropinirole*, are probably the most effective tremorolytic drugs among all dopaminergic treatments and should be considered in newly diagnosed Parkinson's disease (PD) patients exhibiting tremor in the absence of cognitive impairment. Improvement of tremor has also been reported with other dopamine agonists, including *pergolide* and *bromocriptine*. Dopamine agonists are also useful in advanced PD patients exhibiting a tremor refractory to levodopa and anticholinergics (Bhidayasiri, 2005). *Apomorphine (subcutaneous, puffs)* is used in advanced form of PD.

Dopaminergic and anticholinergic agents are equally effective in patients with parkinsonian tremor, but dopaminergic substances improve other parkinsonian signs also (Habib-ur-Rehman, 2000). Midbrain tremor in patients with symptomatic dystonia and mesencephalic lesions may significantly improve with levodopa (Vidailhet et al., 1999). *Levodopa* can also provide symptomatic relief in primary OT (Wills et al., 1999).

Anticholinergics, such as trihexylphenidyl, are effective, but side effects often limit their use, especially in the elderly population. Anticholinergics are not recommended in patients with cognitive decline, heart disease, or in elderly patients over 65 years of age (Bhidayasiri, 2005). In a study comparing the effects of *trihexiphenidyl*, *carbidopa-levodopa*, and *amantadine hydrochloride* in PD, tremor amplitude was reduced by 59% with trihexiphenidyl, 55% by carbidopa–levodopa, and 23% by amantadine (Koller, 1986). A case of successful monotherapy of midbrain tremor with high-dose trihexyphenidyl has been reported (Liou and Shih, 2006).

8.1.7 Other Drugs

Olanzapine might be effective for ET (Yetimalar et al., 2005). *Clozapine* could be considered in selected cases of resistant ET (Ceravolo et al., 1999). However, this drug requires a close follow-up.

The effects of *ondansetron* (a 5HT3-antagonist) on cerebellar dysfunction has been studied. The drug provides no benefit or minor reduction on cerebellar tremor (Rice et al., 1997; Gbadamosi et al., 2001; Bier et al., 2003).

The calcium-channel blocker *nicardipine* provides some minor benefits in ET (Jiménez-Jiménez et al.,1994).

Methazolamide might be effective in ET patients, particularly those with head and voice tremor (Muenter et al., 1991).

Jaw tremor can be successfully treated with *botulinum toxin* injections to the masseters (Gonzalez-Alegre et al., 2006).

Tremor associated with peripheral neuropathy in a context of chronic inflammatory demyelinating polyneuropathy (CIDP) may respond to steroid therapy, cytotoxic drugs, intravenous immunoglobulin therapy, and plasma exchanges (Cook et al., 1990; Dyck, 1990).

The currently available medications decrease tremor in approximately 50% of the patients. The combination of medical and surgical therapies provide benefit in up to 80% of cases. From Lyons et al. (2003), Zesiewicz et al. (2005), and Pahwa and Lyons (2003).

TABLE 8.5: Advantages of DBS compared to irreversible lesion therapy
No permanent brain lesion
Adaptable
Well tolerated in the elderly
May be performed even in patients who previously received other surgical treatments
Bilateral operations feasible

See text for references.

8.2 DEEP BRAIN STIMULATION AND SURGICAL TREATMENTS

DBS within the ventralis intermedius (Vim) nucleus is effective for tremor reduction, irrespective of the etiology. Nowadays, DBS has replaced lesioning techniques in most cases, making bilateral operations feasible, and allowing the postoperative selection of active contacts (for spatial adjustments after surgery). A great advantage is that DBS procedures replicate the effects of ablative interventions but do not require a destructive brain lesion (see Table 8.5 for the advantages of DBS; Espay et al., 2006).

A retrospective study in a total of 130 patients assessed that DBS surgery confirmed that DBS is an effective and safe method to treat PD (Seijo et al., 2007). Direct comparison of different stimulation sites in individual patients revealed that DBS in the subthalamic area is more effective in suppressing pharmacoresistant kinetic tremor than the ventrolateral thalamus proper.

The activation of A1 receptors has been proposed as a new pharmacological target to replace or improve the efficacy of DBS (Bekar et al., 2008). DBS is associated with a marked increase in the release of ATP, resulting in accumulation of its catabolic product, adenosine. Adenosine A1 receptor activation depresses excitatory transmission in the thalamus and reduces both tremor and DBS-induced side effects. Intrathalamic infusion of A1 receptor agonists directly reduce tremor.

A quality of life instrument specifically addressed to patients treated with DBS for movement disorders has been proposed (Kuehler et al., 2003).

8.2.1 Vim Stimulation

Anatomical structures possibly involved in tremor suppression include cerebellothalamic projections, the prelemniscal radiation, and the zona incerta (Hamel et al., 2007). Chronic stimulation of the thalamic ventralis intermedius (Vim) nucleus is reversible, adaptable, and well tolerated by elderly patients. Optimal electrode location for thalamic DBS in ET was found to correspond to the anterior margin of the Vim (Papavassiliou et al., 2004). Stimulator adjustments are frequently needed in ET patients, and it should be kept in mind that tremor may worsen despite a readjustment in the system (Lee and Kondziolka, 2005). Frequency of stimuli is usually between 110 and 150 Hz, with a pulse width around 60 μs.

8.2.2 STN Stimulation

STN is another effective target of DBS on tremor (Krack et al., 1998). Subthalamic stimulator implantation in a large consecutive series of patients with PD produced significant clinical improvement without mortality or major neurological morbidity (Goodman et al., 2006). STN stimulation in levodopa-intolerant patients has a great benefit in terms of reducing the level of required levodopa medication (Katayama et al., 2001).

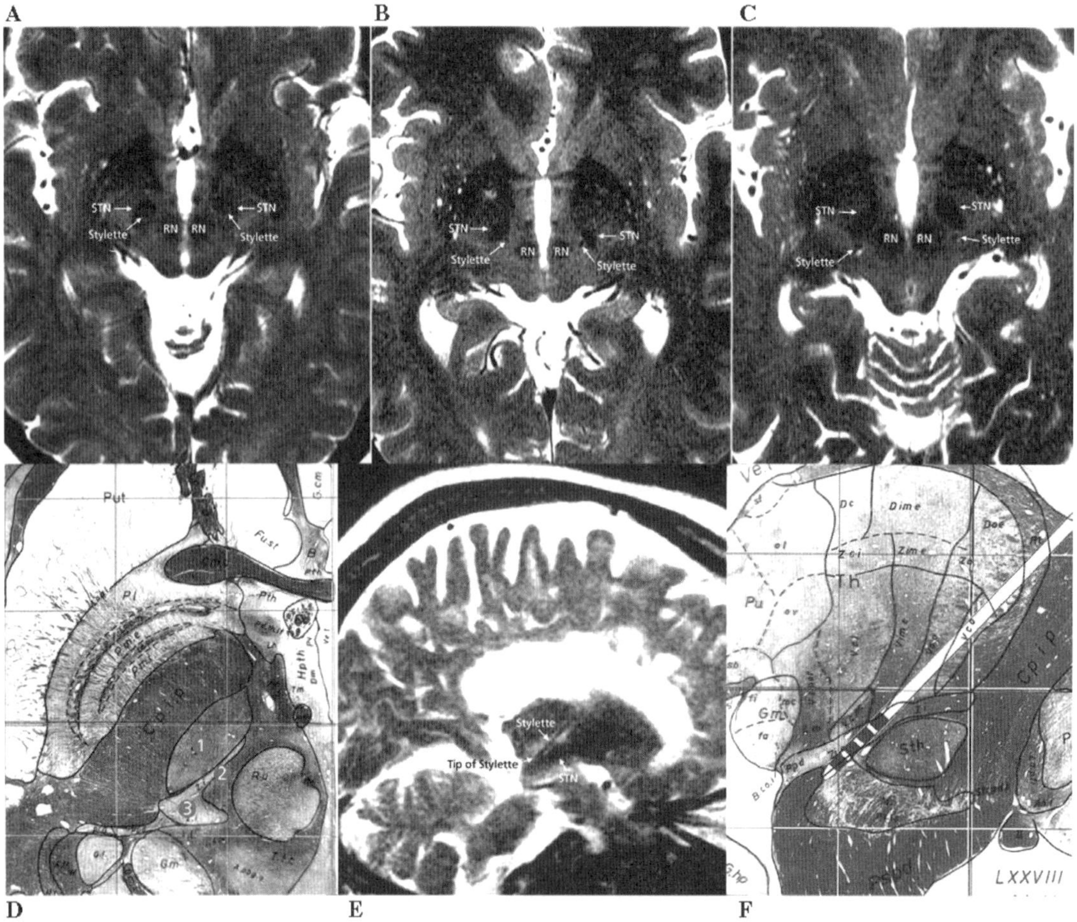

FIGURE 8.2: Intraoperative axial MRI scan showing bilateral stylettes in the **A.** STN, lateral to the anterior border of the red nucleus; **B.** medial/dorsomedial to the STN, and in the **C.** caudal zona incerta (RN: red nucleus, STN: subthalamic nucleus). **D.** Axial slice from the Schaltenbrand atlas showing the position of the three different subthalamic areas targeted. **E.** Intraoperative sagittal MRI scan showing the transfrontal trajectory of the stylette passing dorsal to the outlined STN and its tip posterior to it in the caudal zona incerta. **F.** Matching sagittal slice from the Schaltenbrand atlas with a DBS lead drawn to scale and the anatomical location of each contact on it. From Plaha et al. (2006), with permission from Oxford University Press.

The most effective contact lies dorsal/dorsomedial to the STN (region of the pallidofugal fibers and the rostral zona incerta) or at the junction between the dorsal border of the STN and the latter (Figures 8.2 and 8.3; Plaha et al., 2006). The prelemniscal radiation (RAPRL), a subthalamic bundle of fibers, has been also reported as an effective target—independent of the subthalamic nucleus—for the alleviation of tremor and rigidity in patients with PD by either lesioning or neuromodulation (Velasco et al., 2001).

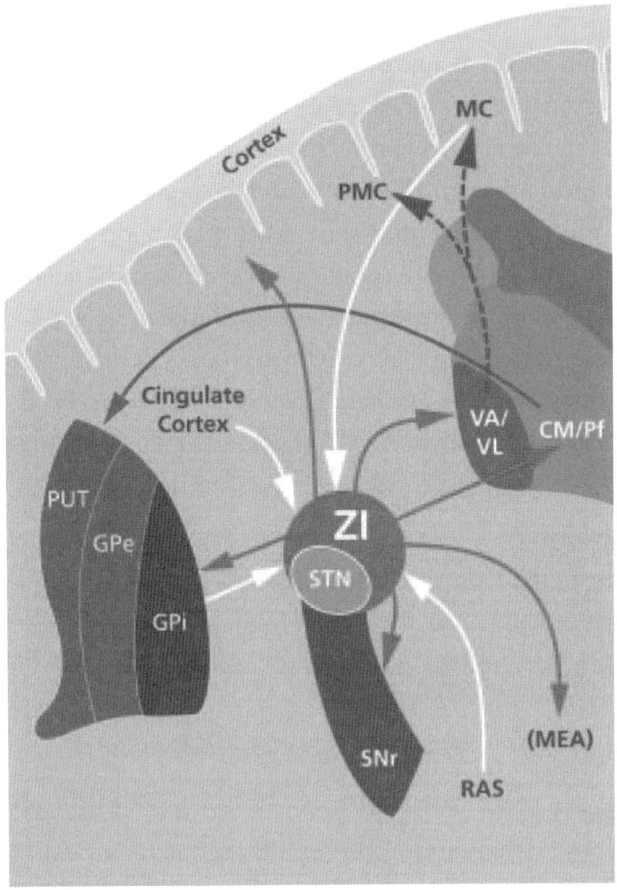

FIGURE 8.3: Schematic diagram showing the key afferent and efferent connections of the zona incerta. Afferent fibers are shown as white arrows, red arrows correspond to efferent fibers. Efferent fibers from CM/Pf to posterodorsal putamen are shown in green. PUT putamen, GPe globus pallidus externus, GPi globus pallidus internus, SNr substantia nigra reticulata, STN subthalamic nucleus, RAS reticular activating system, MEA midbrain extrapyramidal area, ZI zona incerta, CM/Pf centromedian and parafascicular nucleus of the thalamus, VA/VL ventral anterior and ventral lateral nucleus of the thalamus, MC motor cortex, PMC premotor cortex. From Plaha et al. (2006), with permission from Oxford University Press.

8.2.3 Complications of DBS

DBS is usually a safe procedure, even in patients who previously received other surgical treatments for PD, such as thalamotomy or cell grafting (Vergani et al., 2006; Fraix et al., 2005). However, serious complications leading to permanent neurologic deficit may develop after STN–DBS for advanced PD. Intracranial hemorrhage and infections are potential side effects given the surgical procedure. Intracranial hemorrhage (ICH) is considered as the most significant complication associated with the placement of stereotactic intracerebral electrodes (Sansur et al., 2007). Hypertension and the use of microelectrode recording are the main risk factors for cerebral hemorrhage. "Hardware complications," including lead migration or lead fracture are not exceptional (Lyons et al., 2004).

Reported rates of depression, cognitive impairment, mania, and behavior changes are, in principle, low when the DBS procedure has been carefully performed. However, according to some studies, there might be a higher rate of suicide in patients treated with DBS, particularly with thalamic and GPi stimulation. Therefore, a careful preoperative assessment and close postoperative psychiatric and behavioral follow-up are recommended (Burkhard et al., 2004; Appleby et al., 2007; Soulas et al., 2008).

A unique case of hyperhidrosis as a side effect of a misplaced DBS electrode near the Vim nucleus has been reported in a patient with ET (Diamond et al., 2007).

8.2.4 Lesion of the Vim

Radiofrequency lesioning in the Vim alleviates rest, postural, and action tremors affecting the distal and proximal parts of the left upper extremity in Holmes' (rubral) tremor (secondary to midbrain tumor; Kim et al., 2002).

8.2.5 Lesions of the STN or Pallidum

Unilateral radiofrequency STN lesions can be made safely and are an effective alternative to thalamotomy, pallidotomy, and unilateral STN–DBS for the treatment of asymmetrical tremor-dominant advanced PD. Combined lesioning of the dorsolateral STN and pallidofugal fibers and the zona incerta is particularly effective (Patel et al., 2003). For those PD patients whose tremor was not successfully controlled by pallidotomy, the combined pallidothalamotomy is an alternative (Llumiguano et al., 2006). Posterior globus pallidus internus pallidotomy combined with drug therapy is effective for parkinsonian leg tremor (Goto et al., 2002).

8.2.6 Complications of Lesions

PD Patients may develop aphasia (repetition, comprehension, fluency, and naming abnormalities) as a complication of thalamotomy (Bruce et al., 2004). Complications (intracranial hemorrhage, hemiparesis without hemorrhage, partial visual field deficit, speech disturbance) from stereotactic posteroventral pallidotomy (PVP) for patients with PD are relatively uncommon. However, residual symptoms may be serious (Higuchi and Iacono, 2003). Transient ataxia has been reported after nucleus ventralis intermedius thalamotomy for relief of severe parkinsonian tremor (Kimber et al., 2003). The risk of persistent morbidity following lesion therapy in patients after severe head trauma restricts the operation to selected cases with disabling tremor (Krauss et al., 1994).

8.2.7 Stimulation Versus Lesion?

Despite the usefulness and advantages of DBS (see Table 8.5) surgery in many cases of PD and ET, lesion therapy (subthalamotomy, pallidotomy, or thalamotomy) represents a viable alternative and has several advantages, including a reduced need for specialized care and close clinical follow-up, improved affordability, and a lower infection risk (Hooper et al., 2008). In ET, thalamotomy should be considered when deep brain stimulation does not succeed (Giller and Dewey, 2002). In their review of the comparison of the efficacy and safety of thalamotomy and DBS in the treatment of multiple sclerosis tremor, Yap and colleagues underline that both thalamotomy and thalamic DBS are comparable procedures for tremor suppression and that adverse effects (death, procedure-related hemorrhages, hemiparesis, dysarthria, swallowing difficulties, balance disorder) can occur with both procedures (Yap et al., 2007). In intractable MS tremor, thalamotomy has been pointed out as a more efficacious surgical treatment. However, the higher incidence of persistent neurological deficits in patients receiving lesional surgery brings support for the use of DBS as the preferred surgical strategy (Bittar et al., 2005).

8.2.8 Risk of Cognitive Deficits

A randomized trial where patients with severe drug-resistant tremor due to PD, ET, or multiple sclerosis underwent thalamotomy or thalamic stimulation showed that both thalamotomy and thalamic stimulation are associated with a minimal overall risk of cognitive deterioration and that verbal fluency decreased after left-sided thalamotomy and thalamic stimulation (Schuurman et al., 2002).

8.3 γ-KNIFE

Gamma knife (γK) radiosurgery represents a minimally invasive alternative to radiofrequency lesioning and DBS. Gamma knife thalamotomy is especially valuable for patients ineligible for radiofrequency thalamotomy or deep brain stimulation. γK radiosurgery is a viable alternative in the treatment of parkinsonian and ET in patients who are not good candidates for radiofrequency thalamotomy or thalamic DBS (Niranjan et al., 1999). Long-term follow-up of a group of 158 patients (diagnosis: PD, ET, tremor due to stroke, encephalitis, or cerebral trauma) who underwent MRI-guided radiosurgical nucleus ventralis intermedius (Vim) thalamotomy, indicates that relief of tremor is well maintained and free from radiation-induced complications (Young et al., 2000).

Gamma knife radiosurgical thalamotomy was found to be effective also for the palliative treatment of disabling multiple sclerosis intention tremor (Mathieu et al., 2007; Figure 8.4). A follow-up study showed that all patients experienced an improvement of tremor after a period of 2.5 months (median latency). One patient developed a transient hemiparesis which resolved after corticosteroid administration.

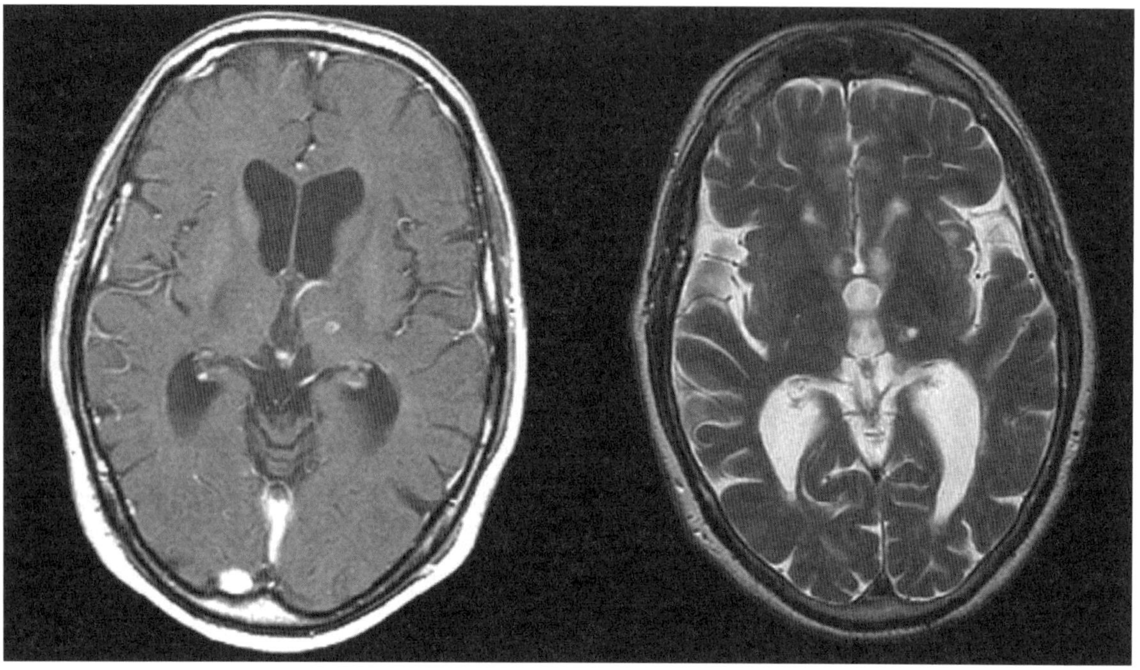

FIGURE 8.4: Brain MRI performed 2 years after γK thalamotomy. *Left*. Contrast-enhanced axial T1-weighted image demonstrating a ring-enhancing lesion in the left vim nucleus. *Right*. Axial T2-weighted image showing the lesion in the left vim nucleus with minimal changes in the surrounding brain. From Mathieu et al. (2007), with permission from Elsevier.

8.3.1 Complications

Potential complications may occur (Kondziolka et al., 2008). There is a low probability of a delayed neurological deficit. It is worthy to note that patients with ET are at risk for developing a complex, disabling movement disorder after gamma knife thalamotomy (Siderowf et al., 2001).

8.4 TRANSCRANIAL MAGNETIC STIMULATION

At this stage of development, the use of TMS has research implications. The technique is not entered yet in the field of therapeutics for tremor. Repetitive TMS (rTMS) modulates the cortico-spinal excitability and the effects last beyond the duration of the rTMS itself (Maeda et al., 2000). Motor cortical stimulation has been proposed as a possible therapeutic target for treatment of action tremor (Houdayer et al., 2007).

8.5 ORTHOSIS AND PROSTHESIS

In addition to medications, rehabilitation programs, and deep brain stimulation, biomechanical loading has appeared recently as a potential tremor-suppression alternative. Patients with severe

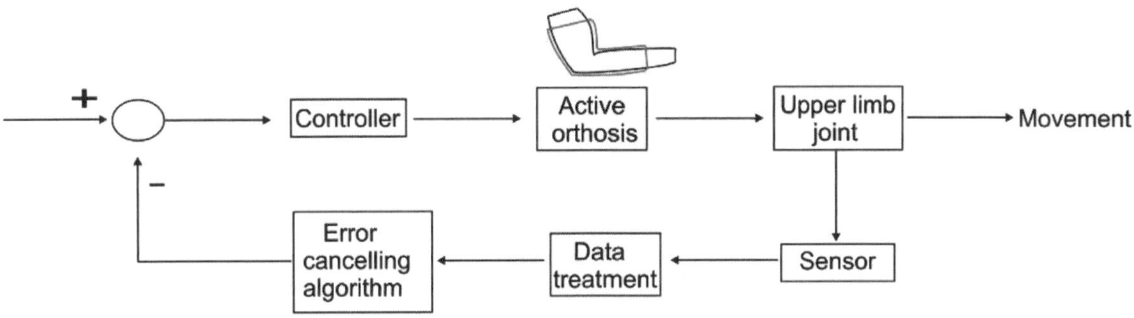

FIGURE 8.5: Tremor-suppression control system. Strategy used to control the upper limb using an active orthosis.

action tremor have uncontrollable, relatively rapid oscillatory motion superimposed on otherwise slower voluntary motor activity. The principle is that a mechanical damper produces an opposing force proportional to velocity; therefore applying damping loads to tremulous limbs should attenuate the tremulous component of movement. Aisen and colleagues have examined the effect of upper extremity damping in 10 patients with posttraumatic and multiple sclerosis cerebellar action tremor and found significant tremor reduction with the application of damping. The amounts of damping were applied by an energy-dissipating prosthesis which generated resistive viscous loads by means of computer-controlled magnetic particle brakes (Aisen et al., 1993).

A wearable orthosis applying effective dynamic force between consecutive segments of the upper limb and changing its biomechanical characteristics has been proposed as a new treatment for ET (Rocon et al., 2007). This exoskeleton acts in parallel to the affected limb and makes an online adaptative viscous control. The system makes a real-time estimation of tremor and consequently selects a cutoff frequency below the frequency of tremor (Figure 8.5). The wearable orthosis is able to detect position, rate, and acceleration of rotation of the joint by means of a chip gyroscope. The musculoskeletal system is modeled as a second-order biomechanical system exhibiting a low-pass filtering behavior.

Evaluation of this orthosis in six patients suffering from ET reveals this technique as a noninvasive alternative to medication and deep brain stimulation (Rocon et al., 2007). The robotic exoskeleton (Figure 8.6) provides a means of testing and validating nongrounded control strategies for orthotic tremor suppression (Rocon et al., 2007). Evaluation on a patient presenting a disabling ET resulted in a drastically decrease of tremor power (Figure 8.7).

Further developments are required to use a wearable orthosis in daily life. Indeed, wearable devices must present several characteristics in terms of aesthetics (appearance, size, and shape), cosmetics, and functional aspects. Functionality is related to the trade-off in terms of torque, velocity, and robustness. Design of the orthosis varies according to the shape, size, side, and functions of the limb. Current developments tend to use sensors and actuators embedded in comfortable textiles.

FIGURE 8.6: Overall view of Wotas (wearable orthosis for tremor assessment and suppression) activated by DC motors.

8.6 BRAIN–MACHINE INTERFACES

Brain–computer interfaces (BCIs) enable users to control devices using electroencephalographic (EEG) signals from the scalp or with single-neuron activity from inside the brain. BCIs aim to control limbs as smoothly as possible. Leuthardt and colleagues pointed out that electrocorticographic (ECoG) activity recorded from the surface of the brain allows controlling a one-dimensional computer cursor rapidly and accurately (Leuthardt et al., 2004). Therefore, they suggested that an ECoG-based BCI could be suggested for people with severe motor disabilities as a nonmuscular communication and control option that is more powerful than EEG-based BCIs. This BCI is potentially more stable and less traumatic than BCIs based upon electrodes penetrating the brain. The option of BCI to control tremor is under investigation.

8.7 ASSISTIVE THERAPY

A mouse computer is available for tremor sufferers, filtering the shaking component in movement. A digital smoothing filter is placed between the mouse motion sensors and the operating system software computing the coordinates of the cursor.

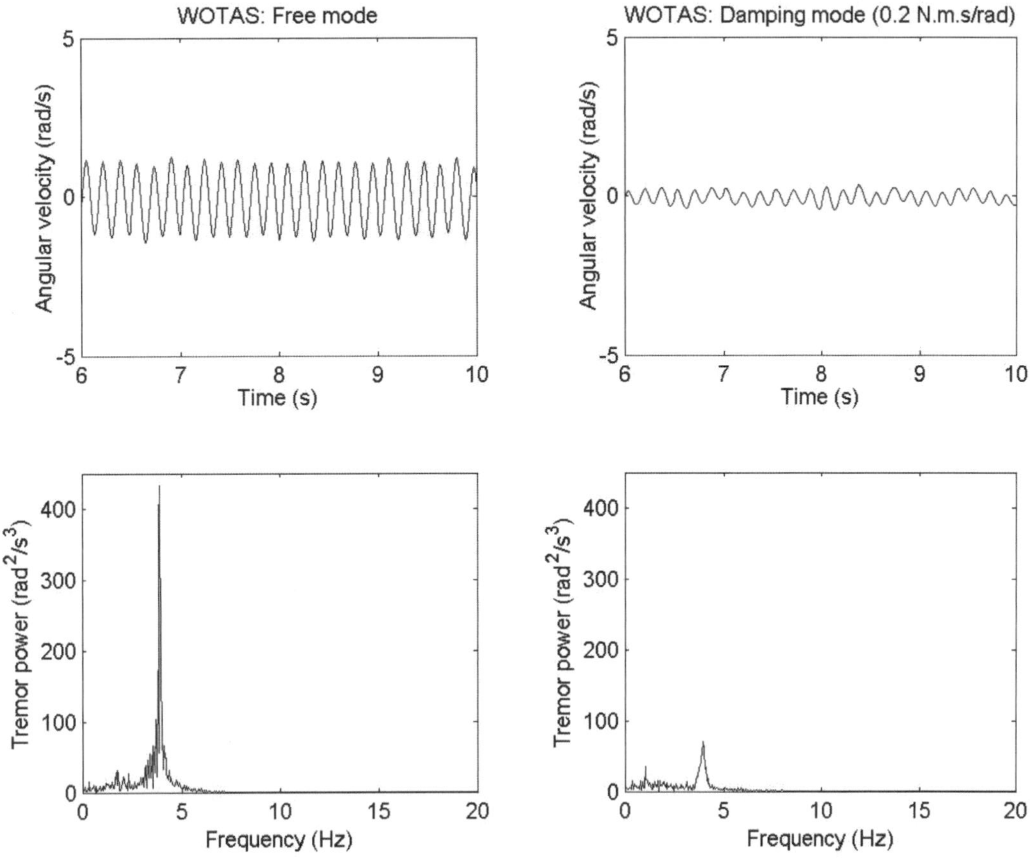

FIGURE 8.7: Oscillations of the elbow and the associated power spectral density (PSD), with the motor in a free mode (left panels) and providing viscosity of 0.2 N m s/rad (right panels) in one ET patient. Note the strong reduction in the PSD when viscosity is applied. Oscillations are expressed in rad/s and PSD is expressed in rad^2/s^3. From Manto et al. (2007), with permission from IPO Publishing Ltd.

Portable haptic devices allow to implement so-called gravity wells around selected regions of the screen.

8.8 FUZZY LOGICS

Fuzzy logic is an emerging field for tremor research and therapy (see also Chapter 6). A system for rehabilitation based on fuzzy assessment of fused features of the tremor movements has been proposed (Teodorescu et al., 2001). In this method, both classic and nonlinear indices are fused into a few, easy to represent, and grasp fuzzy measures. The purpose is the creation of a feedback to help patients to become aware of the characteristics of the tremor of their limbs and help them in controlling their tremor. One of the main requirements is the ease of use to understand and learn.

A representation of the tremor by images and sounds has been built up on the basis of the linguistic and fuzzy valuation of the main parameters of the tremor. Such use is justified by its simplicity (easy to understand by the patient). The rules used to globally characterize the tremor take into account tremor amplitude, main frequency, frequency power content, the correlation dimension, and the irregularity of the signal. The rules establish the relations between the pattern in the space of the tremor signal space and the classes of the tremor signal. The results of the inference are defuzzified and used in the feedback.

REFERENCES

Aisen ML, Arnold A, Baiges I, Maxwell S, Rosen M. The effect of mechanical damping loads on disabling action tremor. *Neurology.* 1993;43(7):1346–1350.

Alusi SH, Aziz TZ, Glickman S, Jahanshahi M, Stein JF, Bain PG. Stereotactic lesional surgery for the treatment of tremor in multiple sclerosis: a prospective case-controlled study. *Brain.* 2001;124(Pt 8):1576–1589.

Appleby BS, Duggan PS, Regenberg A, Rabins PV. Psychiatric and neuropsychiatric adverse events associated with deep brain stimulation: a meta-analysis of ten years' experience. *Mov Disord.* 2007;22(12):1722–1728.

Bekar L, Libionka W, Tian GF, Xu Q, Torres A, Wang X, Lovatt D, Williams E, Takano T, Schnermann J, Bakos R, Nedergaard M. Adenosine is crucial for deep brain stimulation-mediated attenuation of tremor. *Nat Med.* 2008;14(1):75–80.

Benito-León J, Louis ED. Essential tremor: emerging views of a common disorder. Nat *Clin Pract Neurol.* 2006;2(12):666–678.

Bhidayasiri R. Differential diagnosis of common tremor syndromes. *Postgrad Med J.* 2005;81:756–762.

Bier JC, Dethy S, Hildebrand J, Jacquy J, Manto M, Martin JJ, Seeldrayers P. Effects of the oral form of ondansetron on cerebellar dysfunction. A multi-center double-blind study. *J Neurol.* 2003;250(6):693–697.

Bittar RG, Hyam J, Nandi D, Wang S, Liu X, Joint C, Bain PG, Gregory R, Stein J, Aziz TZ. Thalamotomy versus thalamic stimulation for multiple sclerosis tremor. *J Clin Neurosci.* 2005;12(6):638–642.

Burkhard PR, Vingerhoets FJ, Berney A, Bogousslavsky J, Villemure JG, Ghika J. Suicide after successful deep brain stimulation for movement disorders. *Neurology.* 2004;63(11):2170–2172.

Bushara KO, Malik T, Exconde RE. The effect of levetiracetam on essential tremor. *Neurology.* 2005;64(6):1078–1080.

Bruce BB, Foote KD, Rosenbek J, Sapienza C, Romrell J, Crucian G, Okun MS. Aphasia and thalamotomy: important issues. *Stereotact Funct Neurosurg.* 2004;82(4):186–190.

Calzetti S, Sasso E, Negrotti A, Baratti M, Fava R. Effect of propranolol in head tremor: quantitative study following single-dose and sustained drug administration. *Clin Neuropharmacol.* 1992;15(6):470–476.

Ceravolo R, Salvetti S, Piccini P, Lucetti C, Gambaccini G, Bonuccelli U. Acute and chronic effects of clozapine in essential tremor. *Mov Disord.* 1999;14(3):468–472.

Connor GS, Edwards K, Tarsy D. Topiramate in essential tremor: findings from double-blind, placebo-controlled, crossover trials. *Clin Neuropharmacol.* 2008;31(2):97–103.

Cook D, Dalakas M, Galdi A, Biondi D, Porter H. High-dose intravenous immunoglobulin in the treatment of demyelinating neuropathy associated with monoclonal gammapathy. *Neurology.* 1990;40:212–214.

Diamond A, Kenney C, Almaguer M, Jankovic J. Hyperhidrosis due to deep brain stimulation in a patient with essential tremor. Case report. *J Neurosurg.* 2007;107(5):1036–1038.

Dyck PJ. Intravenous immunoglobulin in chronic inflammatory demyelinating polyradiculoneuropathy and in neuropathy associated with IgM monoclonal gammapathy of unknown significance. *Neurology.* 1990;40:327–328.

Elble RJ, Lyons KE, Pahwa R. Levetiracetam is not effective for essential tremor. *Clin Neuropharmacol.* 2007;30(6):350–356.

Espay AJ, Mandybur GT, Revilla FJ. Surgical treatment of movement disorders. *Clin Geriatr Med.* 2006;22(4):813–825, vi.

Findley LJ, Cleeves L, Calzetti S. Primidone in essential tremor of the hands and head: a double blind controlled clinical study. *J Neurol Neurosurg Psychiatry.* 1985;48(9):911–915.

Fraix V, Pollak P, Moro E, Chabardes S, Xie J, Ardouin C, Benabid AL. Subthalamic nucleus stimulation in tremor dominant parkinsonian patients with previous thalamic surgery. *J Neurol Neurosurg Psychiatry.* 2005;76(2):246–248.

Gates PC. Orthostatic tremor (shaky legs syndrome). *Clin Exp Neurol.* 1993;30:66–71.

Gatto EM, Roca MC, Raina G, Micheli F. Low doses of topiramate are effective in essential tremor: a report of three cases. *Clin Neuropharmacol.* 2003;26(6):294–296.

Gbadamosi J, Buhmann C, Moench A, Heesen C. Failure of ondansetron in treating cerebellar tremor in MS patients—an open-label pilot study. *Acta Neurol Scand.* 2001;104(5):308–311.

Giller CA, Dewey RB Jr. Ventralis intermedius thalamotomy can succeed when ventralis intermedius thalamic stimulation fails: report of 2 cases for tremor. *Stereotact Funct Neurosurg.* 2002;79(1):51–56.

Gonzalez-Alegre P, Kelkar P, Rodnitzky RL. Isolated high-frequency jaw tremor relieved by botulinum toxin injections. *Mov Disord.* 2006;21(7):1049–1050.

Goodman RR, Kim B, McClelland S 3rd, Senatus PB, Winfield LM, Pullman SL, Yu Q, Ford B, McKhann GM 2nd. Operative techniques and morbidity with subthalamic nucleus deep brain stimulation in 100 consecutive patients with advanced Parkinson's disease. *J Neurol Neurosurg Psychiatry.* 2006;77(1):12–17.

Goto S, Nishikawa S, Mita S, Ushio Y. Impact of posterior GPI pallidotomy on leg tremor in Parkinson's disease. *Stereotact Funct Neurosurg.* 2002;78(2):64–69.

Gunal DI, Af×ar N, Bekiroglu N, Aktan S. New alternative agents in essential tremor therapy: double-blind placebo-controlled study of alprazolam and acetazolamide. *Neurol Sci.* 2000;21(5):315–317.

Habib-ur-Rehman. Diagnosis and management of tremor. *Arch Intern Med.* 2000;160(16):2438–2444.

Hallett M, Ravits J, Dubinsky RM, Gillespie MM, Moinfar A. A double-blind trial of isoniazid for essential tremor and other action tremors. *Mov Disord.* 1991;6(3):253–256.

Hamel W, Herzog J, Kopper F, Pinsker M, Weinert D, Müller D, Krack P, Deuschl G, Mehdorn HM. Deep brain stimulation in the subthalamic area is more effective than nucleus ventralis intermedius stimulation for bilateral intention tremor. *Acta Neurochir (Wien).* 2007;149(8):749–758; discussion 758.

Hammond ER, Kerr DA. Ethanol responsive tremor in a patient with multiple sclerosis. *Arch Neurol.* 2008;65(1):142–143.

Handforth A, Martin FC. Pilot efficacy and tolerability: a randomized, placebo-controlled trial of levetiracetam for essential tremor. *Mov Disord.* 2004;19(10):1215–1221.

Henderson JM, Portmann L, Van Melle G, Haller E, Ghika JA. Propranolol as an adjunct therapy for hyperthyroid tremor. *Eur Neurol.* 1997;37(3):182–185.

Higuchi Y, Iacono RP. Surgical complications in patients with Parkinson's disease after posteroventral pallidotomy. *Neurosurgery.* 2003;52(3):558–571.

Hooper AK, Okun MS, Foote KD, Fernandez HH, Jacobson C, Zeilman P, Romrell J, Rodriguez RL. Clinical cases where lesion therapy was chosen over deep brain stimulation. *Stereotact Funct Neurosurg.* 2008;86(3):147–152

Houdayer E, Devanne H, Tyvaert L, Defebvre L, Derambure P, Cassim F. Low frequency repetitive transcranial magnetic stimulation over premotor cortex can improve cortical tremor. *Clin Neurophysiol.* 2007;118(7):1557–1562.

Jiménez-Jiménez FJ, Garcia-Ruiz PJ, Cabrera-Valdivia F. Nicardipine versus propranolol in essential tumor. *Acta Neurol (Napoli).* 1994;16(4):184–188.

Katayama Y, Kasai M, Oshima H, Fukaya C, Yamamoto T, Ogawa K, Mizutani T. Subthalamic nucleus stimulation for Parkinson disease: benefits observed in levodopa-intolerant patients. *J Neurosurg.* 2001;95(2):213–221.

Kim MC, Son BC, Miyagi Y, Kang JK. Vim thalamotomy for Holmes' tremor secondary to midbrain tumour. *J Neurol Neurosurg Psychiatry.* 2002;73(4):453–455.

Kimber TE, Brophy BP, Thompson PD. Ataxic arm movements after thalamotomy for Parkinsonian tremor. *J Neurol Neurosurg Psychiatry.* 2003;74(2):258–259.

Klebe S, Stolze H, Grensing K, Volkmann J, Wenzelburger R, Deuschl G. Influence of alcohol on gait in patients with essential tremor. *Neurology.* 2005;65(1):96–101.

Koller WC. Pharmacologic treatment of parkinsonian tremor. *Arch Neurol.* 1986;43(2):126–127.

Koller WC, Herbster G. Adjuvant therapy of parkinsonian tremor. *Arch Neurol.* 1987;44(9):921–923.

Kondziolka D, Ong JG, Lee JY, Moore RY, Flickinger JC, Lunsford LD. Gamma Knife thalamotomy for essential tremor. *J Neurosurg.* 2008;108(1):111–117.

Krack P, Benazzouz A, Pollak P, Limousin P, Piallat B, Hoffmann D, Xie J, Benabid AL. Treatment of tremor in Parkinson's disease by subthalamic nucleus stimulation. *Mov Disord.* 1998;13(6):907–914.

Krauss JK, Mohadjer M, Nobbe F, Mundinger F. The treatment of posttraumatic tremor by stereotactic surgery. Symptomatic and functional outcome in a series of 35 patients. *J Neurosurg.* 1994;80(5):810–819.

Kuehler A, Henrich G, Schroeder U, Conrad B, Herschbach P, Ceballos-Baumann A. A novel quality of life instrument for deep brain stimulation in movement disorders. *J Neurol Neurosurg Psychiatry.* 2003;74(8):1023–1030.

Lee JY, Kondziolka D. Thalamic deep brain stimulation for management of essential tremor. *J Neurosurg.* 2005;103(3):400–403.

Lee KS, Kim JS, Kim JW, Lee WY, Jeon BS, Kim D. A multicenter randomized crossover multiple-dose comparison study of arotinolol and propranolol in essential tremor. *Parkinson Relat Disord.* 2003;9(6):341–347.

Leuthardt EC, Schalk G, Wolpaw JR, Ojemann JG, Moran DW. A brain–computer interface using electrocorticographic signals in humans. *J Neural Eng.* 2004;1(2):63–71.

Liou LM, Shih PY. Successful treatment of rubral tremor by high-dose trihexyphenidyl: a case report. *Kaohsiung J Med Sci.* 2006;22(3):149–153.

Llumiguano C, Dóczi T, Baths I. Microeletrode guided stereotactic pallidotomy and pallidothalamotomy for treatment of Parkinson's disease. *Neurocirugia (Astur).* 2006;17(5):420–432.

Lyons KE, Pahwa R, Comella CL, Eisa MS, Elble RJ, Fahn S, Jankovic J, Juncos JL, Koller WC, Ondo WG, Sethi KD, Stern MB, Tanner CM, Tintner R, Watts RL. Benefits and risks of pharmacological treatments for essential tremor. *Drug Saf.* 2003;26(7):461–481.

Lyons KE, Wilkinson SB, Overman J, Pahwa R. Surgical and hardware complications of subthalamic stimulation: a series of 160 procedures. *Neurology.* 2004;63(4):612–616.

Maeda F, Keenan JP, Tormos JM, Topka H, Pascual-Leone A. Modulation of corticospinal excitability by repetitive transcranial magnetic stimulation. *Clin Neurophysiol.* 2000;111(5):800–805.

Manto M, Rocon E, Pons J, Belda JM, Camut S. Evaluation of a wearable orthosis and an associated algorithm for tremor suppression. *Physiol Meas.* 2007;28(4):415–425.

Mathieu D, Kondziolka D, Niranjan A, Flickinger J, Lunsford LD. Gamma knife thalamotomy for multiple sclerosis tremor. *Surg Neurol.* 2007;68(4):394–399.

Metzer WS, Paige SR, Newton JE. Inefficacy of propranolol in attenuation of drug-induced parkinsonian tremor. *Mov Disord.* 1993;8(1):43–46.

Muenter MD, Daube JR, Caviness JN, Miller PM. Treatment of essential tremor with methazolamide. *Mayo Clin Proc.* 1991;66(10):991–997.

Niranjan A, Jawahar A, Kondziolka D, Lunsford LD. A comparison of surgical approaches for the management of tremor: radiofrequency thalamotomy, gamma knife thalamotomy and thalamic stimulation. *Stereotact Funct Neurosurg.* 1999;72(2–4):178–184.

Ondo WG, Jimenez JE, Vuong KD, Jankovic J. An open-label pilot study of levetiracetam for essential tremor. *Clin Neuropharmacol.* 2004;27(6):274–277.

Ondo WG, Jankovic J, Connor GS, Pahwa R, Elble R, Stacy MA, Koller WC, Schwarzman L, Wu SC, Hulihan JF; Topiramate Essential Tremor Study Investigators. Topiramate in essential tremor: a double-blind, placebo-controlled trial. *Neurology.* 2006;66(5):672–677.

Pahwa R, Lyons KE. Essential tremor: differential diagnosis and current therapy. *Am J Med.* 2003;115(2):134–142.

Papavassiliou E, Rau G, Heath S, Abosch A, Barbaro NM, Larson PS, Lamborn K, Starr PA. Thalamic deep brain stimulation for essential tremor: relation of lead location to outcome. *Neurosurgery.* 2004;54(5):1120–1129.

Patel NK, Heywood P, O'Sullivan K, McCarter R, Love S, Gill SS. Unilateral subthalamotomy in the treatment of Parkinson's disease. *Brain.* 2003;126(Pt 5):1136–1145.

Plaha P, Ben-Shlomo Y, Patel NK, Gill SS. Stimulation of the caudal zona incerta is superior to stimulation of the subthalamic nucleus in improving contralateral parkinsonism. *Brain.* 2006 Jul;129(Pt 7):1732–1747.

Raj V, Landess JS, Martin PR. Oxcarbazepine use in essential tremor. *Ann Pharmacother.* 2006;40(10):1876–1879.

Rice GP, Lesaux J, Vandervoort P, Macewan L, Ebers GC. Ondansetron, a 5-HT3 antagonist, improves cerebellar tremor. *J Neurol Neurosurg Psychiatry.* 1997;62(3):282–284.

Rocon E, Belda-Lois JM, Ruiz AF, Manto M, Moreno JC, Pons JL. Design and validation of a rehabilitation robotic exoskeleton for tremor assessment and suppression. *IEEE Trans Neural Syst Rehabil Eng.* 2007;15(3):367–378.

Rocon E, Manto M, Pons J, Camut S, Belda JM. Mechanical suppression of essential tremor. *Cerebellum.* 2007;6(1):73–78.

Rodrigues JP, Edwards DJ, Walters SE, Byrnes ML, Thickbroom GW, Stell R, Mastaglia FL. Blinded placebo crossover study of gabapentin in primary orthostatic tremor. *Mov Disord.* 2006;21(7):900–905.

Sansur CA, Frysinger RC, Pouratian N, Fu KM, Bittl M, Oskouian RJ, Laws ER, Elias WJ. Incidence of symptomatic hemorrhage after stereotactic electrode placement. *J Neurosurg.* 2007;107(5):998–1003.

Sasso E, Perucca E, Fava R, Calzetti S. Primidone in the long-term treatment of essential tremor: a prospective study with computerized quantitative analysis. *Clin Neuropharmacol.* 1990;13(1):67–76.

Schuurman PR, Bruins J, Merkus MP, Bosch DA, Speelman JD. A comparison of neuropsychological effects of thalamotomy and thalamic stimulation. *Neurology.* 2002;59(8):1232–1239.

Seijo FJ, Alvarez-Vega MA, Gutierrez JC, Fdez-Glez F, Lozano B. Complications in subthalamic nucleus stimulation surgery for treatment of Parkinson's disease. Review of 272 procedures. *Acta Neurochir (Wien).* 2007;149(9):867–875.

Serrano-Dueñas M. Use of primidone in low doses (250 mg/day) versus high doses (750 mg/day) in the management of essential tremor. Double-blind comparative study with one-year follow-up. *Parkinsonism Relat Disord.* 2003;10(1):29–33.

Siderowf A, Gollump SM, Stern MB, Baltuch GH, Riina HA. Emergence of complex, involuntary movements after gamma knife radiosurgery for essential tremor. *Mov Disord.* 2001;16(5):965–967.

Soulas T, Gurruchaga JM, Palfi S, Cesaro P, Nguyen JP, Fénelon G. Attempted and completed suicides after subthalamic nucleus stimulation for Parkinson's disease. *J Neurol Neurosurg Psychiatry.* 2008.

Striano P, Coppola A, Vacca G, Zara F, Brescia Morra V, Orefice G, Striano S. Levetiracetam for cerebellar tremor in multiple sclerosis: an open-label pilot tolerability and efficacy study. *J Neurol.* 2006;253(6):762–766.

Teodorescu HL, Kandel A, Hall LO. Report of research activities in fuzzy AI and medicine at USF CSE. *Artif Intell Med.* 2001;21(1–3):177–183.

Velasco F, Jiménez F, Pérez ML, Carrillo-Ruiz JD, Velasco AL, Ceballos J, Velasco M. Electrical stimulation of the prelemniscal radiation in the treatment of Parkinson's disease: an old target revised with new techniques. *Neurosurgery.* 2001;49(2):293–306.

Vergani F, Landi A, Antonini A, Sganzerla EP. Bilateral subthalamic deep brain stimulation in a patient with Parkinson's disease who had previously undergone thalamotomy and autologous adrenal grafting in the caudate nucleus: case report. *Neurosurgery.* 2006;59(5):E1140; discussion E1140.

Vidailhet M, Dupel C, Lehéricy S, Remy P, Dormont D, Serdaru M, Jedynak P, Veber H, Samson Y, Marsault C, Agid Y. Dopaminergic dysfunction in midbrain dystonia: anatomoclinical study using 3-dimensional magnetic resonance imaging and fluorodopa F 18 positron emission tomography. *Arch Neurol.* 1999;56(8):982–989.

Wills AJ, Brusa L, Wang HC, Brown P, Marsden CD. Levodopa may improve orthostatic tremor: case report and trial of treatment. *J Neurol Neurosurg Psychiatry.* 1999;66(5):681–684.

Yap L, Kouyialis A, Varma TR. Stereotactic neurosurgery for disabling tremor in multiple sclerosis: thalamotomy or deep brain stimulation? *Br J Neurosurg.* 2007;21(4):349–354.

Yetimalar Y, Irtman G, Kurt T, Basoğlu M. Olanzapine versus propranolol in essential tremor. *Clin Neurol Neurosurg.* 2005;108(1):32–35.

Young RF, Jacques S, Mark R, Kopyov O, Copcutt B, Posewitz A, Li F. Gamma knife thalamotomy for treatment of tremor: long-term results. *J Neurosurg.* 2000;93(Suppl 3):128–135.

Zesiewicz TA, Elble R, Louis ED, Hauser RA, Sullivan KL, Dewey RB Jr, Ondo WG, Gronseth GS, Weiner WJ; Quality Standards Subcommittee of the American Academy of Neurology. Practice parameter: therapies for essential tremor: report of the Quality Standards Subcommittee of the American Academy of Neurology. *Neurology.* 2005;64(12):2008–2020.

Zesiewicz TA, Ward CL, Hauser RA, Salemi JL, Siraj S, Wilson MC, Sullivan KL. A pilot, double-blind, placebo-controlled trial of pregabalin (Lyrica) in the treatment of essential tremor. *Mov Disord.* 2007;22(11):1660–1663.

Zeuner KE, Molloy FM, Shoge RO, Goldstein SR, Wesley R, Hallett M. Effect of ethanol on the central oscillator in essential tremor. *Mov Disord.* 2003;18(11):1280–1285.

Annex 1

ANNEX 1

Extended ADL Scale according to Nourie and Lincoln (Nottingham Stroke Score)				
SCORING: **0= NO/WITH HELP** **1= ON MY OWN WITH DIFFICULTY/ ON MY OWN**	**NO**	**WITH HELP**	**ON MY OWN WITH DIFFICULTY**	**ON MY OWN**
Mobility				
–do you walk around outside?				
–do you climb stairs?				
–do you get in and out of the car?				
–do you walk over uneven ground?				
–do you cross roads?				
–do you travel on public transport?				
In the kitchen				
–do you manage to feed yourself?				
–do you manage to make yourself a hot drink?				
–do you take hot drinks from one room to another?				
–do you do the washing up?				

–do you make yourself a hot snack?				
Domestics tasks				
–do you manage your own money when you are out?				
–do you wash small items of clothing?				
–do you do your own housework?				
–do you do your own shopping?				
–do you do a full clothes wash?				
Leisure activities				
–do you read newspapers or books?				
–do you use the telephone?				
–do you write letters?				
–do you go out socially?				
–do you manage your own garden?				
–do you drive a car?				
Total score =				

From: Nouri and Lincoln (1987).

Schwab and England Activities of Daily Living scale

100% = Completely independent. Able to do all chores without slowness, difficulty or impairment. Essentially normal. Unaware of any difficulty.

90% = Completely independent. Able to do all chores with some degree of slowness, difficulty and impairment. Might take twice as long. Beginning to be aware of difficulty.

80% = Completely independent in most chores. Takes twice as long. Conscious of difficulty and slowness.

70% = Not completely independent. More difficulty with some chores. Three to four times as long in some. Must spend a large part of the day with chores.

60% = Some dependency. Can do most chores, but exceedingly slowly and with much effort. Errors; some impossible.

50% = More dependent. Help with half, slower, etc. Difficulty with everything.

40% = Very dependent. Can assist with all chores, but few alone.

30% = With effort, now and then does a few chores alone or begins alone. Much help needed.

20% = Nothing alone. Can be a slight help with some chores. Severe invalid.

10% = Totally dependent, helpless. Complete invalid.

0% = Vegetative functions such as swallowing, bladder, and bowel functions are not functioning. Bedridden.

From: Schwab and England Activities of Daily Living (Gillingham and Donaldson, 1969).

Annex 2

ANNEX 2

Clinical Tremor Rating Scale
1–10 Tremor Rate tremor at rest in items 1 and 2. Rate tremor with posture holding in items 3–10. Abbreviations: UE: upper extremities, LE: lower extremities. score: 0 = none; 1 = slight (amplitude < 0.5 cm), may be intermittent; 2 = moderate amplitude (0.5–1 cm), may be intermittent; 3 = marked amplitude (1–2 cm); 4 = severe amplitude (>2 cm).
1 Head
2 Trunk
3 UE: arms outstretched, wrist mildly extended, fingers spread apart
4 LE: legs flexed at hips and knees
5 Foot dorsiflexed
6 Tongue: when protruded
7 Head and trunk: when sitting or standing
8 Rate tremor with action and intention
9 UE: finger to nose and other actions
10 LE: toe to finger in flexed posture

11 Handwriting *Have patient write the standard sentence: "This is a sample of my best handwriting", sign his or her name and write the date.* score: 0 normal; 1= mildly abnormal, slight untidy, tremulous; 2= moderate abnormal, legible but with considerable tremor; 3 marked abnormal, illegible; 4 severely abnormal, unable to keep pencil or pen on paper without holding hand down with the other hand.
12–14 Drawings *Ask the patient to join both points of the various drawings without crossing the lines. Test each hand, beginning with the lesser involved, without leaning the hand or arm on the table* score: 0 = normal; 1 = slightly tremulous, may cross lines occasionally; 2 = moderately tremulous or crosses lines frequently; 3 = accomplishes the task with great difficulty, many errors; 4 = unable to complete drawing.
15 Pouring *Use firm plastic cups (8 cm tall), filled with water to 1 cm fro top. Ask the patient to pour water from one cup to another. Test each hand separately.*
16 Speaking *This includes spastic dysphonia if present* score: 0 = normal; 1 = mild voice tremulousness when "nervous" only; 2 = mild voice tremulousness, constant; 3 = moderate voice tremor; 4 = severe voice tremor, some word difficult to understand.
17 Feeding *Other than liquids* score: 0 = normal; 1 = mildly abnormal, can bring all solid to mouth, spilling only rarely; 2 = moderately abnormal, frequent spills of peas and similar foods, may bring head at least half way to meet food; 3 = markedly abnormal, unable to cut or uses two hands to feed; 4 = severely abnormal, needs help to feed.
18 Bringing liquids to mouth score: 0 = normal; 1 = mildly abnormal, can still use a spoon, is completely full; 2 = moderately abnormal, unable to use a spoon, uses cups or glasses; 3 = markedly abnormal, can drink from cup or glass, but needs help two hands; 4 = severely abnormal, must use a straw.

19 Hygiene score: 0 = normal; 1 = mildly abnormal, able to do everything but is more careful than the average person; 2 = moderately abnormal, able to do everything but with errors, uses electric razor because of tremor; 3 = markedly abnormal, unable to do most fine tasks, such as putting on lipstick or shaving (even with electric shaver), unless using two hands; 4 = severely abnormal, unable to do any fine movement tasks.
20 Dressing score: 0 = normal; 1 = mildly abnormal, able to do everything but is more careful than the average person; 2 = moderately abnormal, able to do everything but with errors; 3 = markedly abnormal, needs some assistance with buttoning or other activities, such as tying shoelaces; 4 = severely abnormal, requires assistance even for gross motor activities.
21 Writing score: 0 = normal; 1 = mildly abnormal, legible, continue to write letters; 2 = moderately abnormal, legible but no longer writes letters; 3 = markedly abnormal, illegible; 4 = severely abnormal, unable to sign checks or other documents requiring signature
22 Working score: 0 = tremor does not interfere with job; 1 = able to work, but needs to be more careful than the average person; 2 = able to do everything but with errors, poorer than usual performance because of tremor; 3 = unable to do regular job, may have changed to a different job because of tremor, tremor limits housework, such as ironing; 4 = unable to do any outside job, housework very limited.
Total score

From: Fahn et al. (1988).

Annex 3

ANNEX 3
Unified Parkinson's Disease Rating Scale (UDPRS)

I. MENTATION, BEHAVIOR, AND MOOD

1. Intellectual impairment

0 = None.

1 = Mild. Consistent forgetfulness with partial recollection of events and no other difficulties.

2 = Moderate memory loss, with disorientation and moderate difficulty handling complex problems. Mild but definite impairment of function at home with need of occasional prompting.

3 = Severe memory loss with disorientation for time and often to place. Severe impairment in handling problems.

4 = Severe memory loss with orientation preserved to person only. Unable to make judgements or solve problems. Requires much help with personal care. Cannot be left alone at all.

2. Thought disorder (due to dementia or drug intoxication)

0 = None.

1 = Vivid dreaming.

2 = "Benign" hallucinations with insight retained.

3 = Occasional to frequent hallucinations or delusions; without insight; could interfere with daily activities.

4 = Persistent hallucinations, delusions, or florrid psychosis. Not able to care for self.

3. Depression

1 = Periods of sadness or guilt greater than normal, never sustained for days or weeks.

2 = Sustained depression (1 week or more).

3 = Sustained depression with vegetative symptoms (insomnia, anorexia, weight loss, loss of interest).

4 = Sustained depression with vegetative symptoms and suicidal thoughts or intent.

4. Motivation/initiative
0 = Normal.
1 = Less assertive than usual; more passive.
2 = Loss of initiative or disinterest in elective (nonroutine) activities.
3 = Loss of initiative or disinterest in day to day (routine) activities.
4 = Withdrawn, complete loss of motivation.

II. ACTIVITIES OF DAILY LIVING (for both "on" and "off")

5. Speech
0 = Normal.
1 = Mildly affected. No difficulty being understood.
2 = Moderately affected. Sometimes asked to repeat statements.
3 = Severely affected. Frequently asked to repeat statements.
4 = Unintelligible most of the time.

6. Salivation
0 = Normal.
1 = Slight but definite excess of saliva in mouth; may have nighttime drooling.
2 = Moderately excessive saliva; may have minimal drooling.
3 = Marked excess of saliva with some drooling.
4 = Marked drooling, requires constant tissue or handkerchief.

7. Swallowing
0 = Normal.
1 = Rare choking.
2 = Occasional choking.
3 = Requires soft food.
4 = Requires NG tube or gastrotomy feeding.

8. Handwriting
0 = Normal.
1 = Slightly slow or small.
2 = Moderately slow or small; all words are legible.

3 = Severely affected; not all words are legible.
4 = The majority of words are not legible.

9. Cutting food and handling utensils
0 = Normal.
1 = Somewhat slow and clumsy, but no help needed.
2 = Can cut most foods, although clumsy and slow; some help needed.
3 = Food must be cut by someone, but can still feed slowly.
4 = Needs to be fed.

10. Dressing
0 = Normal.
1 = Somewhat slow, but no help needed.
2 = Occasional assistance with buttoning, getting arms in sleeves.
3 = Considerable help required, but can do some things alone.
4 = Helpless.

11. Hygiene
0 = Normal.
1 = Somewhat slow, but no help needed.
2 = Needs help to shower or bathe; or very slow in hygienic care.
3 = Requires assistance for washing, brushing teeth, combing hair, going to bathroom.
4 = Foley catheter or other mechanical aids.

12. Turning in bed and adjusting bed clothes
0 = Normal.
1 = Somewhat slow and clumsy, but no help needed.
2 = Can turn alone or adjust sheets, but with great difficulty.
3 = Can initiate, but not turn or adjust sheets alone.
4 = Helpless.

13. Falling (unrelated to freezing)
0 = None.
1 = Rare falling.
2 = Occasionally falls, less than once per day.

3 = Falls an average of once daily.
4 = Falls more than once daily.

14. Freezing when walking
0 = None.
1 = Rare freezing when walking; may have starthesitation.
2 = Occasional freezing when walking.
3 = Frequent freezing. Occasionally falls from freezing.
4 = Frequent falls from freezing.

15. Walking
0 = Normal.
1 = Mild difficulty. May not swing arms or may tend to drag leg.
2 = Moderate difficulty, but requires little or no assistance.
3 = Severe disturbance of walking, requiring assistance.
4 = Cannot walk at all, even with assistance.

16. Tremor (symptomatic complaint of tremor in any part of body)
0 = Absent.
1 = Slight and infrequently present.
2 = Moderate; bothersome to patient.
3 = Severe; interferes with many activities.
4 = Marked; interferes with most activities.

17. Sensory complaints related to parkinsonism
0 = None.
1 = Occasionally has numbness, tingling, or mild aching.
2 = Frequently has numbness, tingling, or aching; not distressing.
3 = Frequent painful sensations.
4 = Excruciating pain.

III. MOTOR EXAMINATION

18. Speech
0 = Normal.

1 = Slight loss of expression, diction and/or volume.
2 = Monotone, slurred but understandable; moderately impaired.
3 = Marked impairment, difficult to understand.
4 = Unintelligible.

19. Facial expression
0 = Normal.
1 = Minimal hypomimia, could be normal "Poker Face."
2 = Slight but definitely abnormal diminution of facial expression
3 = Moderate hypomimia; lips parted some of the time.
4 = Masked or fixed facies with severe or complete loss of facial expression; lips parted 1/4 inch or more.

20. Tremor at rest (head, upper and lower extremities)
0 = Absent.
1 = Slight and infrequently present.
2 = Mild in amplitude and persistent. Or moderate in amplitude, but only intermittently present.
3 = Moderate in amplitude and present most of the time.
4 = Marked in amplitude and present most of the time.

21. Action or postural tremor of hands
0 = Absent.
1 = Slight; present with action.
2 = Moderate in amplitude, present with action.
3 = Moderate in amplitude with posture holding as well as action.
4 = Marked in amplitude; interferes with feeding.

22. Rigidity (judged on passive movement of major joints with patient relaxed in sitting position; cog-wheeling to be ignored.)
0 = Absent.
1 = Slight or detectable only when activated by mirror or other movements.
2 = Mild to moderate.
3 = Marked, but full range of motion easily achieved.
4 = Severe, range of motion achieved with difficulty.

23. Finger taps (patient taps thumb with index finger in rapid succession)
0 = Normal.

1 = Mild slowing and/or reduction in amplitude.
2 = Moderately impaired. Definite and early fatiguing. May have occasional arrests in movement.
3 = Severely impaired. Frequent hesitation in initiating movements or arrests in ongoing movement.
4 = Can barely perform the task.

24. Hand movements (patient opens and closes hands in rapid succession)
0 = Normal.
1 = Mild slowing and/or reduction in amplitude.
2 = Moderately impaired. Definite and early fatiguing. May have occasional arrests in movement.
3 = Severely impaired. Frequent hesitation in initiating movements or arrests in ongoing movement.
4 = Can barely perform the task.

25. Rapid alternating movements of hands (pronation–supination movements of hands, vertically, and horizontally, with as large an amplitude as possible, both hands simultaneously)
0 = Normal.
1 = Mild slowing and/or reduction in amplitude.
2 = Moderately impaired. Definite and early fatiguing. May have occasional arrests in movement.
3 = Severely impaired. Frequent hesitation in initiating movements or arrests in ongoing movement.
4 = Can barely perform the task.

26. Leg agility (Patient taps heel on the ground in rapid succession picking up entire leg. Amplitude should be at least 3 in.)
0 = Normal.
1 = Mild slowing and/or reduction in amplitude.
2 = Moderately impaired. Definite and early fatiguing. May have occasional arrests in movement.
3 = Severely impaired. Frequent hesitation in initiating movements or arrests in ongoing movement.
4 = Can barely perform the task.

27. Arising from chair (patient attempts to rise from a straightbacked chair, with arms folded across chest)
0 = Normal.
1 = Slow; or may need more than one attempt.
2 = Pushes self up from arms of seat.

3 = Tends to fall back and may have to try more than one time, but can get up without help.
4 = Unable to arise without help.

28. Posture
0 = Normal erect.
1 = Not quite erect, slightly stooped posture; could be normal for older person.
2 = Moderately stooped posture, definitely abnormal; can be slightly leaning to one side.
3 = Severely stooped posture with kyphosis; can be moderately leaning to one side.
4 = Marked flexion with extreme abnormality of posture.

29. Gait
0 = Normal.
1 = Walks slowly, may shuffle with short steps, but no festination (hastening steps) or propulsion.
2 = Walks with difficulty, but requires little or no assistance; may have some festination, short steps, or propulsion.
3 = Severe disturbance of gait, requiring assistance.
4 = Cannot walk at all, even with assistance.

30. Postural stability (response to sudden, strong posterior displacement produced by pull on shoulders while patient erect with eyes open and feet slightly apart; patient is prepared)
0 = Normal.
1 = Retropulsion, but recovers unaided.
2 = Absence of postural response; would fall if not caught by examiner.
3 = Very unstable, tends to lose balance spontaneously.
4 = Unable to stand without assistance.

31. Body bradykinesia and hypokinesia (combining slowness, hesitancy, decreased arm swing, small amplitude, and poverty of movement in general)
0 = None.
1 = Minimal slowness, giving movement a deliberate character; could be normal for some persons. Possibly reduced amplitude.
2 = Mild degree of slowness and poverty of movement which is definitely abnormal. Alternatively, some reduced amplitude.
3 = Moderate slowness, poverty or small amplitude of movement.
4 = Marked slowness, poverty or small amplitude of movement.

IV. COMPLICATIONS OF THERAPY (in the past week)

A. DYSKINESIAS

32. Duration: What proportion of the waking day are dyskinesias present? (Historical information)

0 = None
1 = 1–25% of day.
2 = 26–50% of day.
3 = 51–75% of day.
4 = 76–100% of day.

33. Disability: How disabling are the dyskinesias? (Historical information; may be modified by office examination)

0 = Not disabling.
1 = Mildly disabling.
2 = Moderately disabling.
3 = Severely disabling.
4 = Completely disabled.

34. Painful dyskinesias: How painful are the dyskinesias?

0 = No painful dyskinesias.
1 = Slight.
2 = Moderate.
3 = Severe.
4 = Marked.

35. Presence of early morning dystonia (historical information)

0 = No
1 = Yes

B. CLINICAL FLUCTUATIONS

36. Are "off" periods predictable?

0 = No
1 = Yes

37. Are "off" periods unpredictable?
0 = No
1 = Yes

38. Do "off" periods come on suddenly, within a few seconds?
0 = No
1 = Yes

39. What proportion of the waking day is the patient "off" on average?
0 = None
1 = 1–25% of day.
2 = 26–50% of day.
3 = 51–75% of day.
4 = 76–100% of day.

C. OTHER COMPLICATIONS

40. Does the patient have anorexia, nausea, or vomiting?
0 = No
1 = Yes

41. Any sleep disturbances, such as insomnia or hypersomnolence?
0 = No
1 = Yes

42. Does the patient have symptomatic orthostasis?
(Record the patient's blood pressure, height, and weight on the scoring form)
0 = No
1 = Yes

V. MODIFIED HOEHN AND YAHR STAGING

STAGE 0 = No signs of disease.
STAGE 1 = Unilateral disease.
STAGE 1.5 = Unilateral plus axial involvement.

STAGE 2 = Bilateral disease, without impairment of balance.
STAGE 2.5 = Mild bilateral disease, with recovery on pull test.
STAGE 3 = Mild to moderate bilateral disease; some postural instability; physically independent.
STAGE 4 = Severe disability; still able to walk or stand unassisted.
STAGE 5 = Wheelchair bound or bedridden unless aided.

From: Fahn et al. (1987).

List of Abbreviations

5-FU: 5-fluorouracil

5HT: 5-hydroxytryptamine

6-OHDA: 6-hydroxydopamine

9HPT: Nine Hole Peg Test

A: adenosine

Ach: acetylcholine

ADL: Activity of Daily Living

Ago: agonist

AIN: anterior interpositus nucleus

AMPA: α-amino-3-hydroxy-5-methylisoazol-4-propionate

Antago: antagonist

ATP: adenosine triphosphate

ATPase: adenosine triphosphatase

BCA: beta-carboline alkaloids

BCIs: brain–computer interfaces

BG: basal ganglia

BHFSD: bilateral high-frequency synchronous discharges

BSN: body sensor networks

C7–C8: cervical roots 7–8

Ca: calcium

CBD: corticobasal ganglionic degeneration

CC: cerebellar cortex

CD: cervical dystonia

cf: climbing fibers

CI: confidence interval

CIDP: chronic inflammatory demyelinating polyneuropathy

Cl: chloride

CM/Pf: centromedian/parafascicular nucleus of the thalamus

CM: centromedian nucleus

CN: cerebellar nuclei

CNS: central nervous system

CO: carbon monoxide

CRPS: complex regional pain syndrome

CRS: clincal rating scale

CS: cross-spectrum

CSF: cerebrospinal fluid

CT: computerized tomography

CWT: continuous wavelet transform

DA: dopamine

DAT: dopamine transporter

DBS: deep brain stimulation

DR: dorsal root ganglia

DSCT: dorsal spinocerebellar tract

DTI: diffusion tensor MRI

DWT: discrete wavelet transform

E: electrical potential across the membrane

ECoG: electrocorticography

ECR: extenson carpi radialis

EDC: extensor digitorum communis

EEG: electroencephalography

EMD: empirical mode decomposition

EMG: electromyography

EPSP: excitatory postsynaptic potentials

EPT: enhanced physiological tremor

ET: essential tremor

ETV: essential tremor of voice

F: Faraday unit

FCMTE: familial cortical myoclonic tremor with epilepsy

FCR: flexor carpi radialis

FFT: fast Fourier transform

FMR1: fragile X mental retardation 1 gene

fMRI: functional magnetic resonance imaging

FT: Fourier transform_

FXTAS: fragile-X-associated tremor/ataxia syndrome

GABA: amino-butyric acid

GC: granule cells

Gly: glycine

GP: globus pallidus

GPe: globus pallidus—external segment

GPi: globus pallidus—internal segment

GTP: guanosine triphosphate

HIV: human immunodeficiency virus

HS: Hilbert spectrum

Hz: Hertz

Ia: Ia sensory afferent fibers

IC: internal capsula

ICH: intracranial hemorrhage

III: third ventricle

IMFs: intrinsic mode functions

IO: inferior olive

ION: inferior olivary nucleus

IPSP: inhibitory postsynaptic potential

K: potassium

LB: Lewy body

LBD: Lewy body disease

LLR: long-latency reflexes

LMN: lower motoneuron

LN: Lewy neuritis

LTP: long-term potentiation

M1: primary motor cortex

MEA: midbrain extrapyramidal area

MEG: magnetoencephalography

mf: mossy fibers

MGUS: monoclonal gammopathy of undetermined significance

MHS: marginal Hilbert spectrum

mM: millimole

MN α: alpha motoneuron

MN γ: gamma motoneuron

MPTP: 1-methyl-4-phenyl-1,2,3,6-tetrahydropyridine

MRI: magnetic resonance imaging

ms: millisecond

MS: multiple sclerosis

MSA: multiple system atrophy

MU: motor unit

MUAP: motor unit action potentials

mV: millivolt

MVC: maximal voluntary contraction

Na: sodium

NAA/tCr: cerebellar *N*-acetylaspartate/total creatine

NIP: neural input processor

nm: nanometer

NMDA: *N*-methyl-D-aspartate

NRTP: nucleus reticularis tegmenti pontis

OPCA: olivopontocerebellar atrophy

OT: orthostatic tremor

PD: Parkinson's disease

PE: parallel viscous element

PET: positron emission tomography

PF: parallel fibers

PMC: premotor cortex

PN: peripheral neuropathy

PN: Purkinje neurons

POT: primary orthostatic tremor

PPC: posterior parietal cortex

PSD: power spectral density

PSP: progressive supranuclear palsy

PT: physiological tremor

Put: putamen

PVP: posteroventral pallidotomy

PWT: primary writing tremor

R: resistance

RAPRL: prelemniscal radiation

RAS: reticular activating system

RC: Renshaw cells

REM: rapid eye movements

RetST: reticulospinal tract

RFL: radiofrequency lesioning

RN: red nucleus

rTMS: repetitive transcranial magnetic stimulation

RuST: rubrospinal tract

SE: series elastic component

SEMG: surface EMG electrodes

SEPs: somatosensory evoked potentials

SN: substantia nigra

SND: striatonigral degeneration

SNr: substantia nigra pars reticulata

SOT: secondary orthostatic tremor

SPECT: single-photon emission computed tomography

STN: subthalamic nucleus

STT: spino-thalamic tract

TCA: Tremor Coherence Analyzer

Thal: thalamus

TMS: transcranial magnetic stimulation

TRS: Tremor Rating Scale

UMN: upper motoneuron

UPDRS: Unified Parkinson Disease Rating Scale

V: vermis

VA: ventral anterior

Vce: ventralis caudalis externa

Vci: ventralis caudalis interna

Vim: ventralis intermedius nucleus

VLN: ventral lateral nucleus

Vm: membrane potential

VOA: ventralis oralis anterior nucleus

VOP: ventralis oralis posterior nucleus

VP: ventral posterior

VSCT: ventral spinocerebellar tract

VST: vestibulospinal tract

VTSS: Vocal Tremor Scoring System

WOTAS: wearable orthosis for tremor assessment and suppression

Xi: concentration of the ion inside the membrane

Xo: concentration of the ion outside the membrane

ZI: zona incerta

Author Biography

Giuliana Grimaldi received her MD degree from the University of Palermo, Italy. She is working on the applications of haptic technology in neuroengineering and rehabilitation, with an emphasis on myoelectric applications. She is a member of the Society for Cerebellar Research (SRC). She is affiliated with the Clinical Neuroscience Department of the University of Palermo, Italy. She is a visiting scientist of the Free University of Brussels, Belgium.

Mario Manto received his MD degree from the Free University of Brussels, Belgium. He has a PhD degree in neurophysiology. He has published more than 90 full papers in refereed journals. He is the founding editor of the international journal *The Cerebellum* and is the associate editor of the *Journal of NeuroEngineering and Rehabilitation* (*JNER*). He is the founder of the Society for Cerebellar Research (SRC). He has served as reviewer for more than 10 international journals. He teaches human physiology at the Faculty of Applied Sciences, Free University of Brussels. He is currently a researcher at the Belgian National Research Foundation (FNRS). He is active in international (European and transcontinental) networks on tremor. He has patented several devices in the field of tremor.

Index